SpringerBriefs on Pioneers in Science and Practice

Nobel Laureates

Volume 50

Series editor

Hans Günter Brauch, Mosbach, Germany

More information about this series at http://www.springer.com/series/13510
http://www.afes-press-books.de/html/SpringerBriefs_PSP.htm
http://afes-press-books.de/html/SpringerBriefs_PSP_Crutzen.htm

Paul J. Crutzen · Hans Günter Brauch
Editors

Paul J. Crutzen: A Pioneer on Atmospheric Chemistry and Climate Change in the Anthropocene

 Springer

Editors
Paul J. Crutzen
Max-Planck-Institut für Chemie
  (Otto-Hahn-Institut)
Mainz
Germany

Hans Günter Brauch
Studies (AFES-PRESS)
Peace Research and European Security
  Studies
Mosbach
Germany

A book website with additional information on Prof. Dr. Paul J. Crutzen, including videos is at: http://afes-press-books.de/html/SpringerBriefs_PSP_Crutzen.htm.

ISSN 2194-3125   ISSN 2194-3133   (electronic)
SpringerBriefs on Pioneers in Science and Practice
Nobel Laureates
ISBN 978-3-319-27459-1   ISBN 978-3-319-27460-7   (eBook)
DOI 10.1007/978-3-319-27460-7

Library of Congress Control Number: 2015957129

Cover photo: Stockholm, 10 December 1995—Paul Crutzen receives the Nobel Prize for Chemistry from Carl XVI Gustaf, King of Sweden. *Photo: Reportagebild* represented by dana press photo a/s, Industrivej 1, DK-6000 Kolding, Denmark which granted the permission to use it here. The photos that were taken during the symposium on the occasion of Paul Crutzen's 80th birthday on 2 December 2013 in Mainz and a more recent photo from July 2015 were all taken by Carsten Costard who granted permission to use them here. The personal photos are from the author's personal photo collection.

Copyediting: PD Dr. Hans Günter Brauch, AFES-PRESS e.V., Mosbach, Germany

Printed on acid-free paper

This Springer imprint is published by SpringerNature
The registered company is Springer International Publishing AG Switzerland

To Terttu

Ms. Terttu Crutzen; Prof. Dr. Paul J. Crutzen and Minister Doris Ahnen, Rhineland-Palatinate Minister for Education, Research, Further Education and Culture during the symposium "The Anthropocene" on 2 December 2013 in Mainz on the occasion of his 80th birthday. Photo by Carsten Costard

Prof. Dr. Paul J. Crutzen received the Order of Merit of the land Rhineland-Palatinate from Doris Ahnen, Rhineland-Palatinate Minister for Education, Research, Further Education and Culture, during the symposium "The Anthropocene" on 2 December 2013 in Mainz. Photo by Carsten Costard

Foreword

Paul Crutzen grew up near the center of Amsterdam, the Netherlands, during a period of economic depression and World War II. Clearly, he was not born with a silver spoon in his mouth. However, his family held up and made the best of it, using simple means to overcome the difficult period. This included playing soccer in the streets and having arithmetic and language contests between father and son. Actually young Paul excelled in soccer, languages as well as mathematics, but could not develop these talents in full, at least not immediately. In such an environment practical solutions are needed, and he took up an education in civil engineering to design bridges in his water-rich country. This was a good career start because later Paul showed that bridges can also be built between scientific disciplines and communities.

With the love of his wife from Finland, Terttu Soininen, Sweden became their country of choice to reside and raise two wonderful daughters. It was a lucky decision, because the University of Stockholm offered Paul the opportunity to advance one of his talents, mathematics, which he pursued within a practical context: meteorology. Another fortunate coincidence was that he became involved in computer modeling early on. In those days, he was debugging the code for predicting weather and storms by listening to the hums of the mechanical processors. The development of these skills was instrumental for an unparalleled vocation in atmospheric and environmental science.

During his graduate studies, Paul addressed the middle and upper atmosphere, which initially seemed to be of academic interest only. However, he became aware of the great importance of stratospheric ozone and in particular the risks of ozone depletion. Although not trained as a chemist, he found that some of the reactions used to model the stratospheric fate of nitrogen oxides and their impact on ozone were severely misrepresented as they could not explain the observations of ozone.

By adopting more realistic reaction rate constants he discovered that nitrogen oxides from human-made sources could actually damage the ozone layer, which was absolutely unexpected. It was published in Paul's third first-author paper "The influence of nitrogen oxides on the atmospheric ozone content," for which he would be awarded the Nobel Prize 25 years later.

By combining models of meteorology and ozone, Paul pioneered the field of atmospheric chemistry, and showed how local emissions can have a global effect, even though the substances in question occur in minute, i.e., trace amounts. With his work, that has had an impact well beyond his own field, he followed in the footsteps of pioneers in chemistry in the past centuries such as Scheele, Priestley, Lavoisier, and Laplace. Like Paul, they were also intrigued by the chemical composition of air, what controls it, and tried to unravel its importance for life on Earth. The central role of nitrogen oxides in stratospheric ozone chemistry was the first of Paul's impressive series of discoveries.

Prof. Dr. Jos Lelieveld opening the Symposium on 2 December 2013 in Mainz on the occasion of Prof. Paul J. Crutzen's 80th birthday devoted to The Anthropocene. The photo was taken by Carsten Costard

After his theoretical studies in Stockholm, Paul spent some years at Oxford University to work with experimental data and enhance his proficiency in chemistry. During this period an international discussion sparked about the environmental consequences of high-flying supersonic aircraft, such as the Concorde, of which entire fleets were planned. Since these aircraft release their nitrogen oxide containing exhausts directly into the ozone layer, Paul and other scientists engaged in a debate about the risks. Suddenly, the formerly academic research landed in the center of attention. This became the story of Paul's life, as many of his discoveries inspired both scientific and public discussion. He does not shy away from expressing his opinion, also with an unwelcome message. Fortunately, the fleet of high-flying aircraft was limited to a few Concordes—of which the last were decommissioned in 2003.

Paul continued to make major contributions to stratospheric chemistry. For example, he explained how nitric acid clouds cause the Antarctic ozone hole. At the same time, he also turned his attention to the troposphere, which is the air layer that connects with the biosphere and where weather and climate take place. The troposphere is also prone to air pollution, while it is cleaned by oxidation reactions. The self-cleaning capacity relies on the presence of reactive hydroxyl radicals that convert pollutant gases into more soluble compounds that are removed by rain. The primary formation of hydroxyl radicals in turn is from ozone. While most ozone is located in the stratosphere, protecting life on Earth against harmful ultraviolet radiation from the Sun, a small amount is needed in the troposphere to support the self-cleaning capacity. While previous theories had assumed that tropospheric ozone originates in the stratosphere, Paul discovered that much of it is actually chemically formed within the troposphere. The formation mechanism is similar to the creation of ozone pollution in photochemical "smog".

Prof. Dr. Jos Lelieveld opening the Symposium on 2 December 2013 in Mainz on the occasion of Prof. Paul J. Crutzen's 80th birthday devoted to "The Anthropocene". The photo was taken by Carsten Costard

In the mid-1970s, Paul and his family moved to the USA at the invitation of renowned research institutions in Boulder, Colorado. He helped establish strato-spheric research programs, though mostly pursued tropospheric chemistry, and also investigated large-scale sources of air pollution. One such source is biomass burning, which had previously gone unnoticed because it mostly takes place in scarcely populated regions. Paul found that it is actually a major source worldwide, especially in the tropics. This was confirmed by field measurement campaigns, and later by satellite measurements. The effects are particularly large in the southern hemisphere where other human-related pollution sources are smaller than in the strongly industrialized northern hemisphere.

The next major step in the life of the Crutzen's was the move to Mainz, Germany, when Paul accepted a directorship at the Max Planck Institute for Chemistry where he succeeded another pioneer in atmospheric chemistry, Christian Junge. In Mainz, Paul became involved in scientific as well as political discussions about the impacts of using nuclear weapons through his seminal paper "The atmosphere after a nuclear war: twilight at noon." It built on knowledge gained from the biomass burning studies. A main effect that was expected from nuclear conflict is that giant smoke plumes from large-scale fires submerge the world into darkness, similar to some of the super volcano eruptions or asteroid impacts that changed the fate of the Earth in prehistoric times.

In Mainz he also developed the next generation of comprehensive meteorology–chemistry models to simulate the biogeochemical cycles of reactive nitrogen and carbon compounds, for example methane, and their control of tropospheric ozone and the self-cleaning capacity. His group created the first numerical tool of this kind, ultimately leading to the development of atmospheric chemistry-climate and Earth system models. In this period, I first met Paul, which was a milestone in my career and personal life. He bestowed me the honor of writing a preface in my book "Air pollution in the troposphere" (in Dutch), and I am delighted to now continue the tradition. Subsequently I started a Ph.D. project under Paul's guidance, to study the role of clouds in tropospheric chemistry. In the past three decades we developed a close personal friendship, which includes our wives Terttu and Tineke.

In my pre-Mainz life, I was involved in aircraft measurements, and Paul had turned me into a computer modeler, or perhaps something in-between. Actually we learned that this is a useful combination and a good basis for collaboration, also after my career continued elsewhere. One example is the "Indian Ocean Experiment," planned with the climate change pioneer V. Ramanathan (Ram). We performed measurements with several instrumented aircraft, ships, and satellites and combined the results with computer modeling. This provided indisputable evidence of the environmental impacts of an atmospheric brown cloud with the dimension of several million square kilometers. This huge pollution haze was shown to not only affect air quality on a large scale, but also influence the monsoon and climate in South Asia. Later, several more brown clouds were identified, for example in East Asia.

Paul's engagement in climate change studies and his persuasive communication of the results made an important contribution to policy making. For example, he can convincingly articulate the lessons learned from stratospheric ozone depletion. The ozone hole was a total surprise. No one could have anticipated the catastrophic ozone loss during Antarctic spring. And it cannot be excluded that climate change also holds tipping points where the Earth system swaps from one state to another, perhaps a very undesirable one. These have been central topics since the 1980s, when Paul participated in the German parliamentary commission on "Preventive measures to protect the Earth's atmosphere." He helped publish an influential report with compelling arguments that shaped national and international policies on the atmospheric environment and climate change.

Because of his concern about climate change, Paul advocated studies on "geo-engineering", for example solar radiation management, to investigate if the increase in atmospheric reflectivity could help cool the planet and moderate climate change—just in case impacts might become calamitous. One method is to release sulfur dioxide at high altitudes, much like a volcano eruption, after which the gas is converted into sulfate particles that reflect sunlight and linger in the stratosphere for a few years. The proposal has given rise to controversial discussion. Opponents argue that developing geoengineering options might distract from the real problem, namely reducing greenhouse gas emissions. It should be mentioned that we are still far from practicable geoengineering solutions, and I doubt if one will ever be found, but it cannot hurt thinking about it.

Paul's comprehensive work on many global change issues has almost inevitably led to the next level of reasoning, as he defined a name for the geological epoch in which it all takes place: the "Anthropocene". While geologists have traditionally named the most recent 12,000 years the Holocene, Paul argues that in the past centuries the impact of humanity on the Earth's surface is so large, and unique, that a renaming of the geological timescale is justified. It sends the message that humans have so strongly transformed the planetary environment that it will leave an ineradicable imprint in the rock strata, which will be detectable by future geologists, even if the world population could instantly stop global change. While the proposal on the Anthropocene is still debated among geologists, I conclude that Paul has left an ineradicable imprint in science that will be detectable for a very long time.

Mainz   Prof. Dr. Jos Lelieveld
September 2015   Director, Max Planck Institute for Chemistry
                                                                            Otto Hahn Institute

Acknowledgments

I gratefully acknowledge stimulating discussions and exchange with numerous colleagues and friends from the scientific community, and I wish to express my gratitude to Sarah Alznauer, Susanne Benner, Hans Günter Brauch, Anna-Christina Hans, Astrid Kaltenbach, Jos Lelieveld, and Ulrich Pöschl for their help with the compilation and completion of this book.

Paul J. Crutzen

Prof. Dr. Paul J. Crutzen during the symposium "The Anthropocene" on 2 December 2013 in Mainz on the occasion of his 80th birthday. Photo by Carsten Costard

Prof. Dr. Paul J. Crutzen thanking the speakers during the symposium "The Anthropocene" on 2 December 2013 on the occasion of his 80th birthday in Mainz. Photo by Carsten Costard

Contents

Chapter 1
The Background of an Ozone Researcher: A Brief Autobiography

Paul J. Crutzen

1.1 How I Became a Scientist: A Personal History

I was born in Amsterdam on December 3, 1933, the son of Anna Gurk and Jozef Crutzen.[1] I had one sister. My mother's parents moved to the industrial Ruhr region in Germany from East Prussia towards the end of the 19th century. They were of mixed German and Polish origin. In 1929 at the age of 17, my mother moved to Amsterdam to work as a housekeeper. There she met my father. He came from Vaals, a little town in the south-eastern corner of the Netherlands, bordering Belgium and Germany and very close to the historical city of Aachen. He had relatives in the Netherlands, Germany, and Belgium. Thus, from both parents I inherited a cosmopolitan view of the world. Despite having worked in several countries outside the Netherlands since 1958, I have remained a Dutch citizen.

In May 1940 the Netherlands were overrun by the German army. In September of the same year I entered elementary school, "de grote School" (the big school), as it was popularly called. My six years of elementary school largely overlapped with the Second World War. Our school class had to move several times to different premises in Amsterdam after the German army had confiscated our original school building. The last months of the war, between the fall of 1944 and Liberation Day

---

[1]This text is based on: P.J. Crutzen: "My Life with O_3, NO_x, and Other YZO_x Compounds (Nobel Lecture)", in: *Angew. Chem. Int. Ed Engl.*, **35**, 1758–1777, 1996 and was updated by the author in 2015 Permission to reproduce the original text was granted by the Nobel Foundation in Stockholm in September 2015.

© The Author(s) 2016
P.J. Crutzen and H.G. Brauch (eds.), *Paul J. Crutzen: A Pioneer on Atmospheric Chemistry and Climate Change in the Anthropocene*,
Nobel Laureates 50, DOI 10.1007/978-3-319-27460-7_1

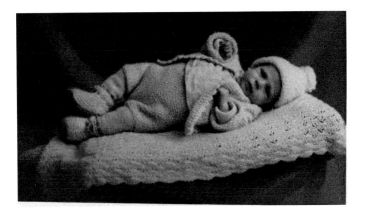

Paul J. Crutzen as a baby in 1933 in Amsterdam

With his grandmother Elisabeth Gurk in Gelsenkirchen-Buer, 1933

on May 5, 1945, were horrible. During the cold "hongerwinter" (winter of famine) of 1944–1945 there was a severe lack of food and heating fuels. Also water for drinking, cooking, and washing was available only in limited quantities for a few hours per day, which caused poor hygienic conditions. Many died of hunger and disease, including several of my schoolmates. Some relief came at the beginning of 1945 when the Swedish Red Cross dropped food supplies by parachute from airplanes. To welcome them we waved our red, white, and blue Dutch flags in the streets.

Paul J. Crutzen as a pupil in the elementary school in Amsterdam in 1940. *Source* Personal photo collection

In 1945 after a successful exam, I entered the "Hogere Burgerschool" (HBS), Higher Citizen School, a five-year long middle school which prepared students for University. I finished this school in June 1951, with natural sciences as my focal subjects. However, besides Dutch, we all also had to become proficient in 3 foreign languages: French, English, and German. I was given considerable help in learning languages from my parents: German from my mother, French from my father.

Paul J. Crutzen as a boy playing soccer in Amsterdam in 1942. *Source* Personal photo collection

During those years, chemistry definitely was not one of my favorite subjects. They were mathematics and physics, but I also did very well in the three foreign languages. During my school years I spent considerable time at sport: football, cycling, and my greatest passion, skating on the Dutch canals and lakes. I also played chess. I read widely about travels in distant lands, about astronomy, as well as about bridge and tunnel building. Unfortunately, because of a heavy fever, my grades in the final exam of the HBS were not good enough to qualify for a university study stipend, which was very hard to obtain at that time, only six years after the end of the Second World War and a few years after the end of the colonial war in Indonesia, which was a large drain on Dutch resources. As I did not want to be a further financial burden on my parents for another four years or more (my father, a waiter, was often unemployed; my mother worked in the kitchen of a hospital), I chose to attend the "Middelbare Technische School" (MTS), now called the Higher Technical School (HTS), to train as a civil engineer. Although the MTS took three years, the second year was a practical year during which I earned a modest salary, enough to live on for about two years.

From the summer of 1954 until February 1958, with a 21-month interruption for compulsory military service, I worked at the Bridge Construction Bureau of the City of Amsterdam. In the meanwhile, on a vacation in Switzerland I met a sweet girl, Terttu Soininen, a student of Finnish history and literature at the University of Helsinki. A few years later I was able to entice her to marry me. What a great choice I made! She has been the center of a happy family; without her support, I would never have been able to devote so much of my time to studies and science. After our marriage in February 1958, we settled in Gävle, a little town about 200 km north of Stockholm, where I had found a job in a building construction bureau. In December at that same year our daughter Ilona was born. In March 1964, she got a little sister, Sylvia. Ilona is a registered nurse, now living in Boulder, Colorado. Sylvia is a marketing assistant in Munich, Germany. Terttu and I have now three grandchildren.

All this time I had longed for an academic career. One day at the beginning of 1958, I saw an advertisement in a Swedish newspaper by the Department of Meteorology of Stockholm Högskola (from 1961, Stockholm University) announcing an opening for a computer programmer. Although I had not the slightest experience in such work, I applied for the job and had the great luck to be chosen among many candidates. On July 1, 1959, we moved to Stockholm and I started with my second profession. At that time the Meteorology Institute of Stockholm University (MISU) and the associated International Meteorological Institute (IMI) were at the forefront of meteorological research, and many top researchers worked in Stockholm for extended periods. Only about a year earlier the founder of the institutes, Prof. Gustav Rossby, one of the greatest meteorologists ever, had died suddenly and was succeeded by Dr. Bert Bolin, another famous meteorologist, starting director of the Intergovernmental Panel on Climate Change (IPCC). At that time Stockholm University housed the fastest computers in the world (BESK and its successor FACIT).

With the exception of participation in a field campaign in northern Sweden, led by Georg Witt to measure the properties of noctilucent clouds that appear during summer at about 85 km altitude in the coldest parts of atmosphere, and some programming work related to this, I was mainly involved in various meteorological projects until about 1966, especially helping to build and run some of the first numerical (barotropic) weather prediction models.

The great advantage of being at a university department was that I got the opportunity to follow some of the courses that were offered. By 1963 I could thus fulfil the requirement for the "filosofie kandidat" (corresponding to a Master of Science degree), combining the subjects mathematics, mathematical statistics, and meteorology. Unfortunately, I could include neither physics nor chemistry in my formal education, because this would have required my participation in time consuming laboratory courses.

However, around 1965 I was given the task of helping a scientist from the United States to develop a numerical model of the oxygen allotrope distribution in the stratosphere, mesosphere, and lower thermosphere. This project got me highly interested in the photochemistry of atmospheric ozone, and I started an intensive study of the scientific literature. This convinced me of the limited status of scientific knowledge on stratospheric chemistry, thus setting the "initial conditions" for my scientific career. Instead of the initially proposed Ph.D. research project, I preferred research on stratospheric chemistry, which was generously accepted. At that time the main topics of research at the MISU were dynamics, cloud physics, the carbon cycle, studies of the chemical composition of rainwater, and especially the "acid rain" problem which was largely "discovered" at MISU through the work of Svante Odén and Erik Eriksson. Several researchers at MISU, among them Prof. Bolin and my good friend and fellow student Henning Rodhe, now emeritus Professor in Chemical Meteorology at MISU, became heavily involved in the issue, which drew considerable political interest at the first United Nations Conference on the Environment in Stockholm in 1972 [1]. However, I wanted to do pure science related to natural processes and therefore I picked stratospheric ozone as my subject, without the slightest anticipation of what lay ahead.

In 1973 Paul J. Crutzen obtained his D.Sc. (Filosofie Doctor) in Stockholm, Sweden. This photo shows him at his exam with Prof. Richard P. Wayne and Sir John Theodore Houghton. Courtesy of R.P. Wayne

Box 1.1: Curriculum Vitae of Paul Josef Crutzen, born 3 December 1933, in Amsterdam, Holland, former Director at the Max Planck Institute for Chemistry in Mainz, Germany. He is married, with two children.

Education

High School: 1946–1951, Amsterdam, Holland.

Civil Engineering, 1951–1954, Amsterdam, Holland.

Academic Studies and Research Activities 1959–1973 at the University of Stockholm

M.Sc. (Filosofie Kandidat), 1963.

Ph.D. (Filosofie Licentiat), Meteorology, 1968,

Title: *"Determination of parameters appearing in the 'dry' and the 'wet' photochemical theories for ozone in the stratosphere"*, Examiner: Prof. Dr. Bert Bolin, Stockholm.

D.Sc. (Filosofie Doctor), 1973, Stockholm, Sweden,

Title: *"On the photochemistry of ozone in the stratosphere and troposphere and pollution of the stratosphere by high-flying aircraft"*, Promoters: Prof. Dr. John Houghton, FRS, Oxford, and Dr. R.P. Wayne, Oxford.

(Ph.D. and D.Sc. degrees were given with the highest possible distinctions).

Employment

1954–1958	Bridge Construction Bureau of the City of Amsterdam, The Netherlands
1956–1958	Military Service, The Netherlands
1958–1959	House Construction Bureau (HKB), Gävle, Sweden
1959–1974	Various computer consulting, teaching and research positions at the department of Meteorology of the University of Stockholm, Sweden, Latest positions: Research Associate and Research Professor
1969–1971	Post-doctoral fellow of the European Space Research Organization at the Clarendon Laboratory of the University of Oxford, England
1974–1977	(1) Research Scientist in the Upper Atmosphere Project, National Center for Atmospheric Research (NCAR), Boulder, Colorado, USA. (2) Consultant at the Aeronomy Laboratory, Environmental Research Laboratories, National Oceanic and Atmospheric Administisitration (NOAA), Boulder, Colorado, USA
1977–July 1980	Senior Scientist and Director of the Air Quality Division, National Center for Atmospheric Research (NCAR), Boulder, Colorado, USA
1976–1981	Adjunct professor at the Atmospheric Sciences Department, Colorado State University, Fort Collins, Colorado
1980–2000	Member of the Max Planck Society for the Advancement of Science and Director of the Atmospheric Chemistry Department, Max Planck Institute for Chemistry, Mainz, Germany

(continued)

(continued)

1987–1991	Professor (part-time) at the Department of Geophysical Sciences, University of Chicago, USA
1992–2008	Professor (part-time), Scripps Institution of Oceanography, University of California, San Diego, LaJolla, USA
1997–2000	Professor (part-time), Utrecht University, Institute for Marine and Atmospheric Sciences, The Netherlands
Since Nov. 2000	Emeritus, Max Planck Institute for Chemistry, Mainz
Since 2004	Institute Scholar, International Institute for Applied Systems Analysis (IIASA), Laxenburg, Austria
Since May 2008	Emeritus Professor, Scripps Institution of Oceanography, University of California, San Diego, La Jolla, USA
Since March 2009	Visiting Professor, Seoul National University, Korea (1–2 months/year)

Other Functions

- Member of the International Ozone Commission and of the International Commission of the Upper Atmosphere of IAMAP (International Association for Meteorology and Atmospheric Physics) (1974–1984).
- Member of NASA Stratospheric Research Advisory Committee (1975–1977).
- Member, Atmospheric Sciences Advisory Committee, National Foundation (1977–1979).
- Member, Committee of Atmospheric Sciences (CAS), National Academy of Sciences, U.S.A. (1978–1980).
- Member, Advisory Committee, High Altitude Pollution Program, Federal Aviation Authority (FAA), U.S.A. (1978–1982).
- Member, Commission of Air Chemistry and Global Air Pollution (CACGP) of the International Association of Meteorology and Atmospheric Physics (IAMAP) (1979–1990).
- Member of Special Inter-Ministerial Advisory Commission on Forest Damage in the Federal Republic of Germany (1983–1987).
- Member (present and past) of various research advisory committees of the German National Science Foundation (DFG) and Ministry of Research and Technology (BMFT) of the Federal Republic of Germany.
- Member of Steering Committees of the SCOPE/ICSU effort to estimate the environmental consequences of a nuclear war (SC OPE/ENUWAR) (1984–1988).

- Member of the Kuratorium (Board of Trustees) of the Max-Planck-Institut für Meteorologie, Hamburg (1984–2000).
- Member, Editorial Board "Climate Dynamics" (1985–2000).
- Member, Executive Board of SCOPE (Scientific Committee on Problems of the Environment) of the International Council of Scientific Unions and Chairman of the National SCOPE Committee of the FRG (1986–1989).

- Member of Special Committee and Executive Committee, and Chairman of Coordinating Panel I of the International Geosphere-Biosphere Programme (IGBP) (1986–1990).
- Member, Commission of the Parliament of the F.R.G. for the "Protection of the Earth's Atmosphere" (Enquete-Kommission zum Schutz der Erdatmosphäre) (1987–1990).
- Chairman of the Steering Committee of the International Global Atmospheric Chemistry (IGAC) Programme, a Core Project of the IGBP (1987–1990); Vice-Chairman (1990–1996).
- Chairman of the European IGAC Project Office (1992–1998).
- Member of the Kuratorium (Board of Trustees) of the Fraunhofer-Institut für atmosphärische Umweltforschung, Garmisch-Partenkirchen (1987–1994); Chairman (1992–1994).
- Member, Editorial Board "Tellus".
- Member Editorial Advisory Board "Issues in Environmental Science and Technology", Royal Society of Chemistry, Britain.
- Member of the Advisory Board of the Institute for Marine and Atmospheric Research, University Utrecht, Netherlands (1993–1997).
- Reviewing Editor "Science" (1993–1999).
- Member of STAP (Scientific and Technical Advisory Panel) Roster of Experts of the United Nations Environment Programme (1993–1998).
- Member of the European Environmental Research Organisation (EERO) (1993–present).
- Member of the Advisory Council of the Volvo Environment Prize (1993–present).
- Member of the *Prix Lemaitre Committee* (1994–present).
- Member of the Steering Committee on Global Environmental Change of International Institute for Applied Systems Analysis (IIASA), Laxenburg, Austria (1994–1997).

- Member of the *SPINOZA* Prize Committee of Nederlandse Organisatie voor Wetenschappelijk Onderzoek (Dutch Organization for Scientific Research) (1994–1995).
- Member of the Scientific Advisory Group of the School of Environmental Sciences, University of East Anglia, Norwich, Britain (1995–present)
- Member, Executive Board of Gesellschaft Deutscher Naturforscher und Ärzte (GDNÄ, German Society of Natural Scientists and Physicians) (1995–2000).
- Member of the General Advisory Board of "Encyclopedia of Life Support Systems (EOLSS)" (1996–2002).
- Member of Vereinigung Deutscher Wissenschaftler—VDW— (Association of German Scientists) (1997–present).
- Member of the Advisory Board of "International Journal of Environmental Studies" (1997–present).
- Member of the International Advisory Board on the "Encyclopedia of Global Change Project" (1997–present).
- Member of the Board of Consulting Editors of "European Review" (Interdisciplinary Journal of the Academia Europaeae) (1997–present).
- Member Editorial Board of "Earth and Planetary Science Letters" (1997–present).
- Member Editorial Advisory Board of "Current Topics in Meteorology" (1997–present).
- Past Editor "Journal of Atmospheric Chemistry"; Member of Editorial Board.
- Member of the Editorial Board of "Mitigation and Adaptation Strategies for Global Change".
- Member of ESTA (European Science and Technology Assembly) of the European Union, Brussels (1997–1999).
- Vice-Chairman of the Scientific Committee for the International Geosphere-Biosphere Project (SC-IGBP) (1998–2003).
- Member Editorial Board of AMBIO (1998–present).
- Jury-member of Deutscher Studienpreis (Körber Foundation, Germany) (1998–1999).
- Member of the Editorial Board of "Encyclopedia of Physical Science and Technology, 3rd edition" (1998–present).
- Member of the "Global Change" Committee of the German Research Council and the Federal Ministry of Research and Technology, Germany (1998–1999).
- Member of the Fifth Framework Programme External Advisory Group on "Global change, climate and biodiversity", European Commission, Brussels (1998–2000).
- Member of the Board of Governors, Weizmann Institute, Israel (1999–2012).

- Member of the Scientific and Academic Advisory Committee, Weizmann Institute, Israel (1999–2010).
- Member of the Honorary Committee of EUROSCIENCE (European Association for the Promotion of Science and Technology) (1999–present).
- Co-Chief Scientist of the Indian Ocean Experiment (INDOEX) (with Prof. V. Ramanathan, Scripps Institution of Oceanography) (1999).
- Member of the International Steering Committe of INDOEX (1999–).
- Member of the Advisory Committee (Beirat), Jahrbuch Ökologie, Berlin (1999–present).
- Member of the Editorial Advisory Board of "ChemPhysChem" (2000–present).
- Member of the Advisory Body on Science and Technology in Europe (2000–2002).
- Member of the Working Group on Establishing an Independent Advisory Body on European Research (2000–2002).
- Member of the steering committee of the "Atmospheric Brown Clouds" programme, in collaboration with the United Nations Environmental Program (UNEP), La Jolla/Nairobi (2001–).
- Member of the Council of Chancellors of the Global Foundation of the Consejo Cultural Mundial, Mexico (2001–).
- Ambassador for the Environment of the European Commission for the Environment (2001–).
- Member of the Steering Committee of the Center for Atmospheric Sciences, University of California, Berkeley (2000–present).
- Member of the Framework Programme Expert Advisory Group (EAG) on "Global change, climate and biodiversity", European Commission, Brussels (2001–2003).
- Member of the advisory committee of the Institute: Urbanization, Emissions, and the Global Carbon Cycle, START, Washington DC (2002–present).
- Member of the ABC (Atmospheric Brown Clouds) Steering Committee (2002–).
- Co-Chief scientist of the ABC (Atmospheric Brown Clouds) Science Team (2002–2005).
- Member of the "Council on the Future", UNESCO, Paris (2003–).
- Member of the International Polar Foundation, Brussels (2003–).
- Member of the Founders' Assembly, Foundation Lindau Nobel Prize Winners Meetings at Lake Constance, Lindau (2003–)
- Member of the Honorary Board of the International Raoul Wallenberg Foundation and the Angelo Roncalli International Committee, Jerusalem (2003–).
- Member of the Scientific Council of the International Centre for Theoretical Physics (Abdus Salam), Trieste (2004–2008).

- Fellow of the Literary & Historical Society, University College of Dublin (2004–).
- Fellow of the American Association for the Advancement of Science (AAAS), Washington (2004).
- Associate Fellow of TWAS (the Third World Academy of Sciences), Trieste (2004–).
- Member of the Board of Directors of the Mariolopoulos-Kanaguinis Foundation for Environmental Sciences, Athens (2004–).
- Member of the ABC science team (2005–).
- Member of the Advisory Board of the National Society of High School Scholars, Atlanta (2005–).
- Advisor for the "Human World" radio show, Earth & Sky Radio Series, NSF, Austin (2005–).

Awards and honours

1969–1971	Visiting Fellow of St. Cross College, Oxford, England
1976	Outstanding Publication Award, Environmental Research Laboratories, National Oceanic and Atmospheric Administration (NOAA), Boulder, Colorado, U.S.A
1977	Special Achievement Award, Environmental Research Laboratories, NOAA, Boulder, Colorado, U.S.A
1984	Rolex-Discover Scientist of the Year
1985	Recipient of the Leo Szilard Award for "Physics in the Public Interest" of the American Physical Society
1986	Elected to Fellow of the American Geophysical Union
	Honorary Doctoral Degree, York University, Canada
	Foreign Honorary Member of the American Academy of Arts and Sciences, Cambridge, U.S.A
1987	Lindsay Memorial Lecturer, Goddard Space Flight Center, National Aeronautics and Space Administration
1988	Founding Member of Academia Europaea
1989	Recipient of the Tyler Prize for the Environment
1990	Tracy and Ruth Scorer Lecturer at the University of California, Davis, U.S.A
	Corresponding Member of The Royal Netherlands Academy of Science
1991	Recipient of the Volvo Environmental Prize
1992	Honorary Doctoral Degree, Universite Catholique de Louvain, Belgium
	Member of the Royal Swedish Academy of Sciences
	Member of the Royal Swedish Academy of Engineering
	Member Leopoldina, Halle

(continued)

(continued)

1993	Ida Beam Visiting Professor, The University of Iowa, U.S.A
1994	Honorary Professor at the Johannes Gutenberg-University of Mainz
	Raymond and Beverly Sackler Distinguished Lecturer in Geophysics and Planetary Sciences, Tel Aviv University, Israel
	Recipient of the Deutscher Umweltpreis of the Umweltstiftung (German Environmental Prize of the Federal Foundation for the Environment)
	Foreign Associate of the U.S. National Academy of Sciences
	Honorary Doctoral Degree, School of Environmental Sciences, University of East Anglia, Norwich, U.K
	Recipient of the Max-Planck-Forschungspreis (with Dr. M. Molina, U.S.A.)
1995	Recipient of the Nobel Prize in Chemistry (with Dr. M. Molina and Dr. F.S. Rowland, U.S.A.)
	Recipient of United Nations Environment Ozone Awards for Outstanding Contribution for the Protection of the Ozone Layer
1996	Member of the Pontifical Academy of Sciences
	"Commandeur in de Orde van de Nederlandse Leeuw" (knighted by the Queen of The Netherlands)
	Recipient of the Bundesverdienstorden (Order of Merit of the Federal Republic of Germany)
	Recipient of The Louis J. Battan Author's Award (with Dr. T.E. Graedel, U.S.A.) by the American Meteorological Society for "their book entitled *Atmosphere, Climate and Change*, an authoritative and beautifully illustrated introduction to the role of the atmosphere in global change"
	Honorary Member of the International Ozone Commission
	Recipient of the Minnie Rosen Award for "High Achievement in Service to Mankind" of Ross University, New York
	Recipient of the Médail d'Or de la Ville de Grenoble, France
	Election to the Global 500 Roll of Honour of the United Nations Environment Programme (UNEP)
	Symons Memorial Lecture of the Royal Meteorological Society of England, Imperial College, London
	Aristoteles Lecture at the Aristotle University of Thessaloniki, Greece
	Member of the World Institute of Science, Brussels, Belgium
	Titular Member of the European Academy of Arts, Sciences and Humanities, Paris
	Holder of the 'Ehrenring' (ring of honour) of the City of Mainz
	Group Achievement Award (to HALOE Science Data Validation, Data Processing, and Flight Operations Team) for outstanding contributions to the success of the HALOE/UARS satellite experiment by NASA Langley Research Center

(continued)

(continued)

1997	Honorary Member of the American Meteorological Society
	Honorary Member of the European Geophysical Society (EGS)
	Corresponding Member of the "Société Royale des Sciences de Liège"
	Honorary Fellow of Physical Research Laboratory (PRL), Ahmedabad, India (1997)
	Ceremonial Lecturer (Festvortrag) at the Annual Assembly of the Deutsche Physikalische Gesellschaft (German Physical Society), Regensburg
	Ceremonial Lecturer (Festvortrag) at the Annual General Assembly of the Max-Planck-Society, Bremen
	Erasmus Lecturer and Medalist of the Academia Europaeae
	Foreign Member of Accademia Nazionale dei Lincei (Italian Academy of Sciences, Roma, Italy
	Shipley Distinguished Lecturer, Clarkson University, Postdam, U.S.A.
1998	Honorary Fellow of St. Cross College, Oxford, England
	"First Lecturer in the Thompson Lecture Series", Advanced Study Program, National Center for Atmospheric Research, Boulder, Colorado, USA
	Public lecturer at Michigan Technological University, Houghton, U.S.A.
	G.N. Lewis Lecturer at the University of California, Berkeley, U.S.A.
	Honorary Member of the Commission on Atmospheric Chemistry and Global Pollution (CACGP)
1999	Foreign Member of the Russian Academy of Sciences
	H. Julian Allan Award 1998. In recognition of the outstanding scientific paper (co-authored) for 1998 at NASA Ames
2000	Named "Hero of the Planet" by Time Magazine, Special Edition, Earth Day, (April–May 2000)
	Honorary Member of the Swedish Meteorological Society
	"Beatty Memorial" Lecturer at McGill University, Montreal, Canada (8 March, 2000)
	The Third "Rosenblith" Lecturer at the Massachusetts Institute of Technology: Atmospheric Chemistry in the 21st Century (14 March, 2000)
2001	Worldwide most cited author in the Geosciences with 2911 citations from 110 publications during the decade 1991–2001, ISI (Institute for Scientific Information, Philadelphia, USA), issue November/December 2001
	Honored by the Karamanlis Institute for Democracy Athens, Greece, for outstanding contributions to Science and Society
	Member of the Council of the Pontifical Academy of Sciences
	Honorary Chairman "Climate Conference 2001", 20–24 August, Utrecht, The Netherlands

(continued)

(continued)

2003	Paul Crutzen Prize awarded to the best paper for participants in the International Young Scientists' Global Change START Conference, Trieste, Italy, November 16–19, 2003
	Golden Medal (highest destination) given by the Academy of Athens, October 22, 2003
2004	Honorary Member of the European Geosciences Union (EGU)
2005	Distinguished Lecturer in Science, The Hongkong University of Science and Technology, School of Science
	Recipient of the UNEP/WMO Vienna Convention Award
2006	Recipient of the Jawaharlal Nehru Birth Centenary Medal 2006, Indian National Science Academy, New Delhi, India
	Foreign Member of the British Royal Society (allowed to use the title Paul Josef Crutzen, ForMemRS), London, UK
	Fellow of the World Academy of Art and Science, San Francisco, USA
2007	Election to International Member of the American Philosophical Society, Class 1, USA
2008	Recipient of the Capo d'Orlando Award, Discepolo Foundation, Vico Equense, Italy
	Honorary Member of European Academy of Sciences and Arts, Salzburg, Austria
	Honorary Fellow of the Institute of Green Professionals, Weston, Florida, USA. Entitled to use the identifiere "Hon. FIGP"
	Einstein Lecturer, Free University Berlin, Germany, June 6, 2008
2011	Honorary Member of the „Naturforschende Gesellschaft zu Emden von 1814"
2013	Recipient of the Landesverdienstorden of Rheinland-Pfalz (Order of Merit of the land Rhineland-Palatinate)
2014	Honorary Member of the „Nationale Akademie der Wissenschaften Leopoldina", Halle, Germany

1.2   Stratospheric Ozone Chemistry

As early as 1930 the famous British scientist Sydney Chapman [2] had proposed that the formation of "odd oxygen" O_x (=$O + O_3$) is due to photolysis of O_2 by solar radiation at wavelengths shorter than 240 nm [Eq. (1)].

$$O_2 + h\nu \rightarrow 2\,O\,(\lambda < 240\,\text{nm}),\ \text{followed by} \tag{1}$$

$$O + O_2 + M \rightarrow O_3 + M \tag{2}$$

Fast reactions (2), where M is a mediator, and (3) next lead to the rapid establishment of a steady state for the concentrations of O and O_3 without affecting the concentration of odd oxygen. Destruction of odd oxygen, counteracting its production by reaction (1), occurs by reaction (4).

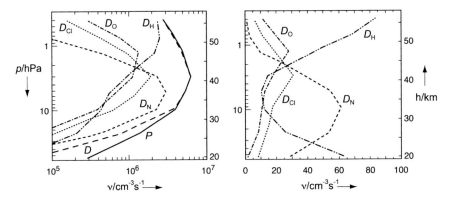

Fig. 1.1 Ozone Production and destruction rates, including absolute and relative contributions by the Chapman reaction R4 (D_O), NO_x catalysis R11 + R12 (D_N), HO_x catalysis R5 + R6 (D_H) and ClO_x catalysis R21 + R22 (DCl)$_x$. The calculations neglect the heterogeneous halogen activation which become very important below 25 km under cold conditions

$$O_3 + h\nu \rightarrow O + O_2 (\lambda < 1180 \, nm) \tag{3}$$

$$O + O_3 \rightarrow 2O_2 \tag{4}$$

Until about the middle of the 1960s it was generally believed that reactions (1)–(4) sufficed to explain the ozone concentration distribution in the stratosphere. However, by the mid 1960s, especially following a study by Benson and Axworthy [3], it became clear that reaction (4) is much too slow to balance the production of odd oxygen by reaction (1) (Fig. 1.1). In 1950 David Bates and Marcel Nicolet [4], together with Sydney Chapman, the great pioneers of upper atmospheric photochemistry research, proposed that catalytic reactions involving OH and HO_2 radicals could counterbalance the production of odd oxygen in the mesosphere and thermosphere. Building on their work and on laboratory studies conducted by one of the 1967 Nobel Prize Laureates in Chemistry, Prof. R. Norrish of Cambridge University and his co-workers [5, 6], the ozone destruction reaction pair (5) and (6) involving OH and HO_2 radicals as catalysts were postulated by Hampson [7] and incorporated into an atmospheric chemical model by Hunt [8].

$$OH + O_3 \rightarrow HO_2 + O_2 \tag{5}$$

$$HO_2 + O_3 \rightarrow OH + 2O_2 \tag{6}$$

$$2O_3 \rightarrow 3O_2 \tag{5} + (6)$$

He proposed that the primary source for the OH radicals was photolysis of O_3 by solar ultraviolet radiation of wavelengths shorter than about 320 nm (Eq. 7), which leads to electronically excited $O(^1D)$ atoms, a small fraction of which reacts with water vapour (Eq. 8). Most $O(^1D)$ reacts with O_2 and N_2 to reproduce ozone via reactions (7), (9) and (2).

$$O_3 + h\nu \rightarrow O(^1D) + O_2 (\lambda < 320\,\text{nm}) \tag{7}$$

$$O(^1D) + H_2O \rightarrow 2OH \tag{8}$$

$$O(^1D) + M \rightarrow O + M \tag{9}$$

In the absence of laboratory measurements for the rate constants of reactions (5) and (6), and in order for these reactions to counterbalance the production of odd oxygen by reaction (1), Hunt adopted the rate constants $k_5 = 5 \times 10^{-13}$ $\text{cm}^3\,\text{molec}^{-1}\,\text{s}^{-1}$ and $k_6 = 10^{-14}\,\text{cm}^3\,\text{molec}^{-1}\,\text{s}^{-1}$, respectively.

In my filosofie licenciat thesis of 1968 I analyzed the proposal by Hampson and Hunt and concluded that the rate constants for reactions (5) and (6) that they had chosen could not explain the vertical distribution of ozone in the lower stratosphere. Furthermore, I pointed out [9] that the above choice of rate constants would also lead to unrealistically rapid loss of ozone (on a timescale of only a few days) in the troposphere. Anticipating a possible role of OH in tropospheric chemistry, I also briefly mentioned the potential importance of a reaction between OH with CH_4. We now know that reactions (5) and (6) proceed about 25 and 10 times slower, respectively, than postulated by Hunt and Hampson and that the CH_4 oxidation cycle plays a very large role in tropospheric chemistry, a topic to which I will return.

With respect to stratospheric ozone chemistry, I discarded the theory of Hampson and Hunt and concluded "… at least part of the solution of the problem of the ozone distribution might be the introduction of photochemical processes other than those treated here. The influence of nitrogen compounds on the photochemistry of the ozone layer should be investigated."

Unfortunately, no measurements of stratospheric NO_x (NO and NO_2) were available to confirm my thoughts about their potential role in stratospheric chemistry. By the summer of 1969 I had joined the Department of Atmospheric Physics at the Clarendon Laboratory of Oxford University as a postdoctoral fellow of the European Space Research Organization and stayed there for a two year period. The head of research group, Dr. (now Sir) John Houghton, hearing of my idea on the potential role of NO_x, handed me a solar spectrum, taken on board a balloon by Dr. David Murcray and co-workers of the University of Denver, and indicated to me that it might reveal the presence of HNO_3 [10] After some analysis I could derive the approximate amounts of stratospheric HNO_3, including a rough idea of its vertical distributions. I did not get the opportunity to write up the result, before Rhine et al. [11] published a paper, showing a vertical HNO_3 column density of 2.8×10^{-4} atm cm ($\approx 7.6 \times 10^{15}$ molecules per cm^2) above 18.8 km. With this information I knew that NO_x should also be present in the stratosphere as a result of reactions (10a) and (10b).

$$OH + NO_2(+M) \rightarrow HNO_3(+M) \tag{10a}$$

$$HNO_3 + h\nu \rightarrow OH + NO_2 (\lambda < 320\,nm) \tag{10b}$$

This gave me enough confidence to submit my paper [12] on catalytic ozone destruction by NO and NO$_2$, based on the set of reactions [Eqs. (11) + (12)].

$$NO + O_3 \rightarrow NO_2 + O_2 \tag{11}$$

$$NO_2 + O \rightarrow NO + O_2 \tag{12}$$

$$O + O_3 \rightarrow 2O_2 \tag{11}+(12)$$

The net result of reactions (11) and (12) is equivalent to the direct reaction (4). However, the rate of the net reaction can be greatly enhanced by relatively small quantities of NO$_x$ on the order of a few nanomole per mole. I also included a calculation of the vertical distribution of stratospheric HNO$_3$. As the source of stratospheric NOx, I initially accepted the proposal by Bates and Hays [13] that about 20 % of the photolysis of N$_2$O would yield N and NO. Subsequent work showed that this reaction does not take place. However, it was soon shown that NO could also be formed to a lesser extent, but still in significant quantities, by the oxidation of nitrous oxide (N$_2$O) by O(^{1}D) [14–16].

$$N_2O + O(^1D) \rightarrow 2NO \tag{13}$$

It was further shown by Davis et al. [17]. that reaction (12) proceeds about 3.5 times faster than I had originally assumed based on earlier laboratory work. A few years later it was also shown that earlier estimates of O$_3$ production by reaction (1) and (2) had been too large due to overestimations of both the absorption cross-sections of molecular oxygen [18] and solar intensities in the ozone-producing 200–240 nm wavelength region [19, 20]. As a result of these developments it became clear that enough NO is produced in reaction (13) to make reactions (11) and (12) the most important ozone loss reactions in the stratosphere in the altitude region between 25 and 45 km (see Fig. 1.1).

N$_2$O is a natural product of microbiological processes in soils and waters. A number of anthropogenic activities, such as the application of nitrogen fertilizers in agriculture, also lead to significant N$_2$O emissions. The rate of increase in atmospheric N$_2$O concentrations for the past decades has been about 0.2–0.3 % per year [21]. That, however, was not known in 1971. The discovery of the indirect role of a primarily biospheric product on the chemistry of the ozone layer has greatly stimulated interest in bringing biologists and atmospheric scientists together.

1.3 Man's Impact on Stratospheric Ozone

In the fall of 1970, still in Oxford, I obtained a preprint of a study sponsored by the Massachusetts Institute of Technology (MIT), the Study on Critical Environmental Problems (SCEP), which was held in July of that year [22]. This report also considered the potential impact of the introduction of large stratospheric fleets of supersonic aircraft (US: Boeing, Britain, France: Concorde, Soviet Union: Topolev; in the following, supersonic stratospheric transport is abbreviated SST) and gave me the first quantitative information on the stratospheric inputs of NO_x that would result from these operations. By comparing these with the production of NO_x by reaction (13), I realized immediately that we could be faced with a severe global environmental problem. Although the paper in which I proposed the important catalytic role of NO_x on ozone destruction had already been published in April 1970, the participants in the study conference had clearly not taken any note of it, since they concluded "The direct role of CO, CO_2, NO, NO_2, SO_2, and hydrocarbons in altering the heat budget is small. It is also unlikely that their involvement in ozone photochemistry is as significant as water vapour." I was quite upset by the statement. Somewhere in the margin of this text I wrote "Idiots".

After it became quite clear to me that I had stumbled on a hot topic, I decided to extend my 1970 study by treating in much more detail the chemistry of the oxides of nitrogen (NO, NO_2, NO_3, N_2O_4, N_2O_5), hydrogen (OH, HO_2), and HNO_3, partially building on a literature review by Nicolet [23]. I soon got into big difficulties. In the first place, adopting Nicolet's reaction scheme I calculated high concentrations of N_2O_4, a problem that I could soon resolve when I realized that this compound is thermally unstable, a fact not considered by Nicolet. An even greater headache was caused by gas phase reactions (14) and (15)

$$N_2O_5 + H_2O \rightarrow 2\,HNO_3 \tag{14}$$

$$O + HNO_3 \rightarrow OH + NO_3 \tag{15}$$

for which the only laboratory studies available at that time had yielded rather high rate coefficients: $k_{14} = 1.7 \times 10^{-18}$ and $k_{15} = 1.7 \times 10^{-11} - 17 \times 10^{-11}$ cm^3 $molec^{-1}$ s^{-1} at room temperatures. A combination of reactions (14) and (15) with these rate constants would provide a large source of OH radicals, much larger than supplied by reaction (8), and would lead to prohibitively rapid catalytic ozone loss. This was a terribly nervous period for me. At that time no critical reviews and recommendations of rate coefficients were available. With no formal background in chemistry, I basically had to compile and comprehend much of the needed chemistry by myself from the available publications, although I profited greatly from discussions at the University of Oxford, especially Dr. Richard Wayne of the Physical Chemistry Laboratory, a former student of the Nobel laureate, Prof. R. Norrish. I discussed all these difficulties and produced extensive model calculations on the vertical distributions of trace gases in the $O_x/NO_x/HO_x/HNO_x$ system in a paper which was submitted by the end of 1970 to the Journal of Geophysical Research (received there on January 13, 1971) and which, after revision, was finally published in the October 29 issue of 1971 [15]. The publication of this paper was much delayed because of an extended mail strike in Britain. Because of the major problems I had encountered, I did not make any calculations of ozone depletions, but instead drew attention to the potential seriousness of the problem by stating:

> An artificial increase of the mixing ratio of the oxides of nitrogen in the stratosphere by about 1×10^{-8} may lead to observable changes in the atmospheric ozone level … It is estimated that global nitrogen oxide mixing ratios may increase by almost 10^{-8} from a fleet of 500 SSTs in the stratosphere. Larger increases, up to 7×10^{-8}, are possible in regions of high traffic densities … Clearly, serious decreases in the total atmospheric ozone level and changes in the vertical distributions of ozone, at least in certain regions, can result from such an activity …

1.4   The Supersonic Transport Controversy in the USA

Unknown to me, a debate on the potential environmental impact of supersonic stratospheric transport had erupted in the USA. Initially the concern was mainly enhanced catalytic ozone destruction by OH and HO_2 radicals resulting from the release of H_2O in the engine exhausts [24]. By mid-March 1971, a workshop was organized in Boulder, Colorado, by an advisory board of the Department of Commerce, to which Prof. Harold Johnston of the University of California, Berkely, was invited. As an expert in laboratory kinetics and reaction mechanisms of NO_x compounds [25–27], he immediately realized that the role of NO_x in reducing stratospheric ozone had been grossly underestimated. Very quickly (submission 14 April, revision 14 June) on August 6, 1971, his paper appeared in Science [27] with the title "Reduction of Stratospheric Ozone by Nitrogen Oxide Catalysts from Supersonic Transport Exhaust". In the abstract of this paper Johnston stated "… oxides of nitrogen from SST exhaust pose a much greater threat to the ozone layer than does the increase of water. The projected increase in stratospheric oxides of nitrogen could reduce the ozone shield by about a factor of 2, thus permitting the harsh radiation below 300 nm to permeate the lower atmosphere." During the summer of 1971, I received a preprint of Johnston's study via a representative of British Aerospace, one of the Concorde manufacturers. This was the first time I had heard of Harold Johnston, for whom I quickly developed a great respect both as a scientist and a human being. Although I had expressed myself rather modestly about the potential impact of stratospheric NO_x emissions from SSTs, for the reasons given above, I fully agreed with Prof. Johnston on the potential severe consequences for stratospheric ozone, and I was really happy to have support for my own ideas from such an eminent scientist. For a thorough resumé of the controversies between scientists and industry, and between meteorologists and chemists (recurring themes also in later years) I refer to Johnston's article "Atmospheric Ozone" [28]. It should also be mentioned here that

Prof. Johnston's publications in the early 1970s removed several of the major reaction kinetic problems that I had encountered in my 1971 study. It was shown, for instance, that neither reaction (14) nor (15) occur to a significant degree in the gas phase, and that the earlier laboratory studies had been significantly influenced by reactions on the walls of the reaction vessels [29], advice that was earlier also given to me in a private communication by Prof. Sidney Benson of the University of Southern California.

In July 1971, I returned to the University of Stockholm and devoted myself mainly to studies concerning the impact of NO_x releases from SSTs on stratospheric ozone. In May 1973, I submitted my inaugural dissertation "On the Photochemistry of Ozone in the Stratosphere and Troposphere and Pollution of the Stratosphere by High-Flying Aircraft" to the Faculty of Natural Sciences and was awarded the degree of Doctor of Philosophy with the highest possible distinction, the third time this had ever happened during the history of Stockholm University (and earlier Stockholm "Högskola"). This was one of the last occasions in which the classical and rather solemn "Filosofie Doktor, similar to the Habilitation in Germany and France, was awarded. I had to dress up. First and second "opponents" were Dr. John Houghton and Dr. Richard Wayne of the University of Oxford, who wore their college gowns for the occasion. Dr. Wayne also served as a non-obligatory third opponent, whose task it was to make fun of the candidate. The classical doctoral degree has been abolished (I was one of the last to go through the procedure). The modern Swedish Filosofie Doktor degree corresponds more closely to the former Filosofie Licentiat degree.

In large part as a result of the proposal by Johnston [27] that NOx emissions from SSTs could severely harm the ozone layer, major research programs were started: the Climate Impact Assessment Program (CIAP), organized by the US Department of Transportation [30], and the COVOS/COMESA [31, 32] program, jointly sponsored by France and Great Britain (the producers of the Concorde aircraft). The aim of these programs was to study the chemical and meteorological processes that determine the abundance and distribution of ozone in the stratosphere, about which so little was known then that the stratosphere was sometimes dubbed the "ignorosphere". The outcome of the CIAP study was summarized in a publication by the US National Academy of Sciences in 1975 [33]. "We recommend that national and international regulatory authorities be alerted to the existence of potentially serious problems arising from growth of future fleets of stratospheric airlines, both subsonic and supersonic. The most clearly established problem is a potential reduction of ozone in the stratosphere, leading to an increase in biologically harmful ultraviolet light at ground level".

The proposed large fleets of SSTs never materialized, largely for economic reasons. The CIAP and COVOS/COMESA research program, however, greatly enhanced knowledge about stratospheric chemistry. They confirmed the catalytic role of NO_x in stratospheric ozone chemistry. A convincing example of this was provided by a major solar proton event which occurred in August 1972 and during which, within a few hours, large quantities of NO, comparable to the normal NO_x content, were produced at high geomagnetic latitudes ($>65°$), as shown in Fig. 1.2.

24 P.J. Crutzen

Fig. 1.2 Production of NO at high geomagnetic latitudes during the solar proton event of 1972 for two assumptions about the electronic states of the N atoms formed ($P_N = 0$, or 1). Also shown are the average NO_x concentrations for these locations.

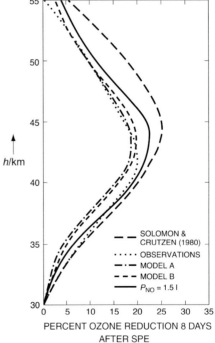

Fig. 1.3 Observed and calculated percentage ozone depletions resulting from the 1972 solar proton event. The various calculated curves correspond to assumed values of parameters that were not well known

With such a large input of NO, a clear depletion of stratospheric ozone was expected [34], a hypothesis which was confirmed by analysis of satellite observations [35]. Figure 1.3 shows results of the calculated and observed ozone depletions, the former obtained with a model that also considered chlorine chemistry [36].

Although I had started my scientific career with the ambition to do basic research related to natural processes, the experiences of the early 1970s had made it utterly clear to me that human activities had grown so much that they could compete and interfere with natural processes. Since then this has been an important factor in my research efforts. Already by the end of 1971 I wrote in an article published in the "the Future of Science Year Book" of the USSR in 1972:

> ... the upper atmosphere is an important part of our environment. Let us finish by expressing a sincere hope that in the future environmental dangers of new technological development will be recognizable at an early stage. The proposed supersonic air transport is an example of a potential threat to the environment by future human activities. Other serious problems will certainly arise in the increasingly complicated world of tomorrow.

1.5 Tropospheric Ozone

My first thoughts on tropospheric photochemistry go back to about 1968 [9]. However, in the following three years, my research was largely devoted to stratospheric ozone chemistry. Then in 1971 a very important paper with the title "Normal Atmosphere: Large Radical and Formadehyde Concentrations Predicted" was published by Hiram Levy III, then of the Smithsonian Astrophysical Observatory in Cambridge, Massachusetts [37]. Levy proposed that OH radicals could also be produced in the troposphere by the action of solar ultraviolet radiation on ozone [Eqs. (7) and (8)], and that they are responsible for the oxidation of CH_4 and CO, an idea that was also quickly adopted by Jack McConnell, Michael McElroy, and Steve Wofsy [38] of Harvard University. The recognition of the important role of OH was a major step forward in our understanding of atmospheric chemistry. Despite very low atmospheric concentrations, currently estimated at about 10^6 molecules per cm^3, corresponding to a mean tropospheric volume mixing ratio of 4×10^{-14} [39], it is this ultraminor constituent—and not 10^{13} times more abundant O_2—that is responsible for the oxidation of almost all compounds emitted into the atmosphere by natural processes and anthropogenic activities. The lifetimes of most atmospheric gases are, therefore, largely determined by the concentrations of OH and the corresponding reaction coefficients [40] (Table 1.1). Those gases that do not react with OH have very long atmospheric residence times and are largely destroyed in the stratosphere. Examples of the latter class of compounds are N_2O and several fully halogenated, industrial organic compounds, such as $CFCl_3$, CF_2Cl_2, and CCl_4. These play a major role in stratospheric ozone chemistry, an issue to which we will return.

Following Levy's paper my attention turned to tropospheric chemistry. Starting with a presentation at the 1972 International Ozone Symposium in Davos, Switzerland, I proposed that in situ chemical processes could produce or destroy ozone in quantities larger than the estimated downward flux of ozone from the stratosphere to the troposphere [41, 42]. Destruction of ozone occurs by reactions

Table 1.1 Schematic representations of importance of OH radicals in atmospheric chemistry

PRIMARY PRODUCTION OF OH RADICALS

$O_3 + h\nu \ (\leq 320 \text{ nm}) \quad \rightarrow \quad O\,(^1D) + O_2$

$O\,(^1D) + H_2O \quad \rightarrow \quad 2\,OH$

GLOBAL, 24 HOUR, AVERAGE (OH) $\approx 10^6$ MOLECULES/CM^3

MOLAR MIXING RATIO IN TROPOSPHERE $\approx 4 \times 10^{-14}$

REACTION WITH OH DETERMINES THE LIFETIME OF MOST GASES IN ATMOSPHERE

EXAMPLES:

CH_4:	8 YEARS
C_2H_6:	2 MONTHS
C_3H_8:	10 DAYS
C_5H_8:	HOURS
$(CH_3)_2S$:	2–3 DAYS
CH_3Cl:	≈ 1 YEAR
CH_3CCl_3:	≈ 5 YEARS
NO_2:	≈ 1 DAY

$CFCl_3$, CF_2Cl_2, N_2O do not react with OH. They are broken down in the stratosphere and have a large influence on ozone chemistry.

(5) + (6) and (7) + (8). Ozone production takes place in environments containing sufficient NO_x by reactions (16), (17), and (2), where R = H, CH_3, or other organoperoxy radicals.

$$RO_2 + NO \rightarrow RO + NO_2 \tag{16}$$

$$NO_2 + h\nu \rightarrow NO + O \ (\lambda \leq 405\,\text{nm}) \tag{17}$$

$$O + O_2 + M \rightarrow O_3 + M \tag{2}$$

$$\text{Net}: \quad RO_2 + O_2 \rightarrow RO + O_3 \tag{(16)+(17)+(2)}$$

The catalytic role of NO in atmospheric chemistry is, therefore, twofold. At altitudes above about 25 km, where O atom concentrations are high, ozone destruction by reactions (11) + (12) dominates over ozone production by reactions (16) + (17) + (2). The latter chain of reactions is at the base of all photochemical ozone formation in the troposphere, including that taking place during photochemical smog episodes, originally discovered in southern California, as discussed by Johnston [28]. Such reactions can, however, also take place in background air with ubiquitous CO and CH_4 serving as fuels: for instance in the case of CO oxidation

$$CO + OH \rightarrow CO_2 + H \quad (18)$$

$$H + O_2 + M \rightarrow HO_2 + M \quad (19)$$

$$HO_2 + NO \rightarrow OH + NO_2 \quad (16)$$

$$NO_2 + h\nu \rightarrow NO + O \quad (17)$$

$$O + O_2 + M \rightarrow O_3 + M \quad (2)$$

$$\text{Net}: \quad CO + 2O_2 \rightarrow CO_2 + O_3 \quad (18) + (19) + (16) + (17) + (2)$$

This reaction chain requires the presence of sufficient NO. At low NO volume mixing ratios, below about 10 pmol mol^{-1}, oxidation of CO may lead to ozone destruction, since the HO_2 radical then reacts mostly with O_3 [see Eq (6)]. The result of the participating reactions [(18) + (19) + (6)] is: $CO + O_3 \rightarrow CO_2 + O_2$. In a similar way, the oxidation of CH4 in the presence of sufficient NO_x will lead to tropospheric ozone production.

Besides reacting with NO or O_3, HO_2 can also react with itself [Eq. (20)] to produce H_2O_2 which serves as a strong oxidizer of S^{IV} compounds in cloud and rain water.

$$HO_2 + HO_2 \rightarrow H_2O_2 + O_2 \quad (20)$$

My talk at the International Ozone Symposium was not well received by some members of the scientific establishment of the time. However, in the following years the idea gradually received increased support. In particular, Bill Chameides and Jim Walker [43], then of Yale University, took it up and went as far as proposing that even the diurnal variation of lower tropospheric ozone could be explained largely by in situ photochemical processes. Although I did not agree with their hypothesis (CH$_4$ and CO oxidation rates are just not rapid enough), it was good to note that my idea had been taken seriously. (I should also immediately add that especially Bill Chameides in subsequent years added much to our knowledge of tropospheric ozone.) A couple of years later, together with two of my finest students, Jack Fishman and Susan Solomon, we presented observational evidence for a strong in situ tropospheric ozone chemistry [44, 45]. Laboratory measurements by Howard and Evenson [46] next showed that reaction (16) proceeded about 40 times faster than determined earlier, strongly promoting ozone production and increased OH concentrations with major consequences for tropospheric and stratospheric chemistry [47]. A consequence of the faster rate of this reaction is a reduction in the estimated ozone depletions by stratospheric aircraft as the ozone production reactions (16) + (17) + (2) are favored over the destruction reaction (6). Furthermore, a faster reaction (16) leads to enhanced OH concentrations and thus a faster conversion of reactive NO_x to far less reactive HNO_3. Table 1.2 summarizes an ozone budget calculated with an early version of a three-dimensional chemical

Table 1.2 Tropospheric ozone budgets, globally and for the northern (NH) and southern (SH) hemisphere in 10^{13} mol/year

	Global	NH	SH
Sources			
$HO_2 + NO$	6.5	4.1	2.4
$CH_3O_2 + NO$	1.7	1.0	0.7
Transport from stratosphere	1.0	0.7	0.3
Sinks			
$O(^1D) + H_2O$	3.8	2.2	1.6
$HO_2 + O_3$ and $OH + O_3$	2.8	1.8	1.0
Deposition on surface	2.7	1.8	0.9
Net chemical source	1.6	1.1	0.5

Only CH_4 and CO oxidation cycles were considered. Calculations were made with an early version of the global, three-dimensional MOGUNTIA model [48]

transport model of the troposphere. The results clearly show the dominance of in situ tropospheric ozone production and destruction. With the same model, estimates were also made of the present and pre-industrial meridional and seasonal average ozone concentration distributions (Figs. 1.4 and 1.5). The calculations, derived with the first three-dimensional MOGUNTIA model of Peter Zimmermann and the author, indicate a clear increase in tropospheric ozone concentrations over the past centuries [48].

With the same model we have also calculated the OH concentration distributions for pre-industrial and present conditions. Since pre-industrial times, the CH_4 volume mixing ratio in the atmosphere has increased [49] from about 0.7 to 1.7 ppmv (v indicates a volume/volume comparison). Because reaction with CH_4 is one of the main sinks for OH, an increase in CH_4 should have led to a decrease in OH concentrations. On the other hand, increased ozone concentrations, which enhance OH production by reactions (7) + (8), and reactions (6) and (16'), both stimulated by strongly enhanced anthropogenic NO production, should exert the opposite effect. Figures 1.5, 1.6 and 1.7 show the zonally averaged, meridional distributions of the diurnally averaged OH concentrations, both for the pre-industrial and industrial periods. They indicate:

(a) strong maxima of OH concentrations in the tropics, largely due to high intensities of ultraviolet radiation as a consequence of a minimum in the vertical ozone column. Consequently the atmospheric oxidation efficiency is strongly determined by tropical processes. For instance, most CH_4 and CO is removed from the atmosphere by reaction with OH in the tropics.

(b) the possibility of a different OH concentrations between pre-industrial to industrial conditions, a question, which has not been answered.

The results presented in Figs. 1.7 and 1.8 are of great importance, as they allow estimations of the sink of atmospheric CH_4 by reaction with OH. Prior to the discovery of the fundamental role of the OH radical [37], estimates of the sources

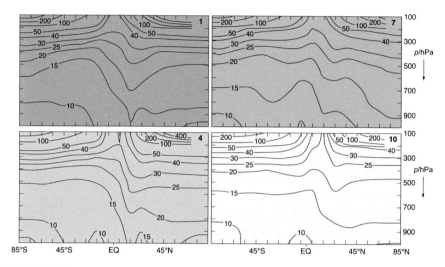

Fig. 1.4 Calculated zonal average ozone volume mixing ratgios in units of nanomole/mole (or ppbv) for the pre-industrial era (nano = 10^{-9}) for different months

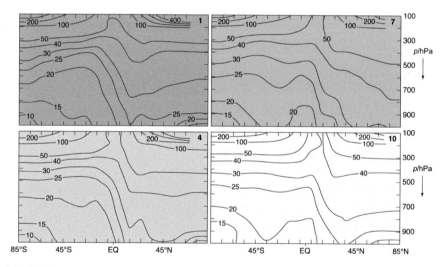

Fig. 1.5 Same as Fig. 1.4, but for mid-1980s

and sinks of trace gases were largely based on guess work without a sound scientific basis. As shown in Table 1.3, this recognition has led to very large changes in the budget estimates of CH_4 and CO. "Authoritative" estimates of the CH_4 budget of 1968 (no reference will be given) gave much higher values for CH_4 releases from natural wetlands. With such a dominance of natural sources, it would have been impossible to explain the annual increase in atmospheric CH_4 concentrations by almost 1 % per year. Early estimates of CO sources, on the other hand, were much too low.

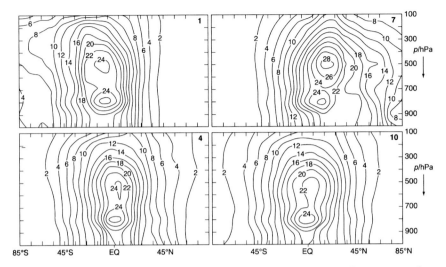

Fig. 1.6 Calculated zonal and 24-h average OH concentrations in units of 10^5 molecules/cm^3 for the pre-industrial period for January, April, July and October

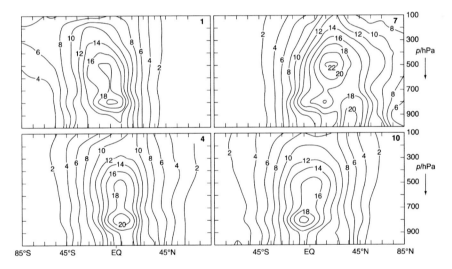

Fig. 1.7 Same as Fig. 1.6 for 1985

The dominance of OH concentrations and the high photochemical activity in the tropics clearly points at the great importance of the tropics and subtropics in atmospheric chemistry.

Contrary to what was commonly believed prior to the early 1980s, the chemical composition of the tropical and *subtropical atmosphere is substantially affected by human activities, in particular biomass burning, which takes* place during the dry season. The high temporal and spatial variability of ozone in the tropics is shown in

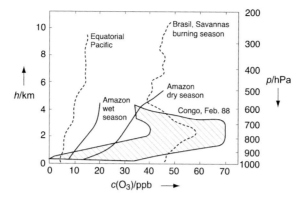

Fig. 1.8 Variability of ozone profiles in the tropics, including contrast between dry and wet season, and continents versus marine soundings

Table 1.3 Estimated budgets of important atmospheric trace gases made in 1968 and at present

	1968	1995
CH$_4$ BUDGET (Tg/year)		
Natural wetlands	1180	275
Anthropogenic	270	265
	1450	540
CO BUDGET (Tg/year)		
Natural	75	860
Anthropogenic	274	1640
	350	2500
S BUDGET (Tg S/year)		
Pollutants	76	78
Oceanic emissions	30 (H$_2$S)	25 (DMS)
Land emissions	70 (H$_2$S)	few (various compounds)
	176	105
NO$_x$ BUDGET (Tg N/year)		
Biological	150	10
Pollution	15	24
Lightning	–	2–10
	165	36–44
N$_2$O BUDGET (Tg N/year)		
Biological	340	15
Anthropogenic	–	3.5
	340	18.5

DMS denotes dimethylsulfide

Fig. 1.8. Highest ozone concentrations are observed over the polluted regions of the continents during the dry season, lowest values in the clean air over the Pacific. I will return to the topic of tropical tropospheric ozone, but will first review the stormy developments in stratospheric ozone depletion by halogen compounds that started in 1974.

1.6   Pollution of the Stratosphere by ClO$_x$

Towards the end of the CIAP programme some researchers had turned their interest
to the potential input of reactive chlorine radicals on stratospheric ozone. In the
most thorough of these studies, Stolarski and Cicerone [50] calculated substantial
ozone depletions if inorganic chlorine were present in the stratosphere at a volume
of mixing ratio of 1 nmol/mol of air. Odd oxygen destruction would take place by
the catalytic reaction cycle (21) + (22). This reaction sequence is very similar to the

$$Cl + O_3 \rightarrow ClO + O_2 \qquad\qquad (21)$$

$$O + ClO \rightarrow Cl + O_2 \qquad\qquad (22)$$

$$O + O_3 \rightarrow 2O_2 \qquad\qquad (21) + (22)$$

catalytic NO$_x$ cycle (11) + (12) introduced before. The study by Stolarski and
Cicerone, first presented at a conference in Kyoto, Japan, in the fall of 1973, mainly
considered volcanic injections as a potential source of ClX compounds (their initial
interest in chlorine chemistry was, however, concerned with the impact of the
exhaust of solid rocket fuels of the space shuttle). Two other conference papers [51,
52] also dealt with ClO$_x$ chemistry. All three papers struggled, however, with the
problem of a missing chlorine source in the stratosphere (research over the past
20 years has shown that the volcanic source is rather insignificant).

   In the fall of 1973 and early 1974 I spent some time looking for potential
anthropogenic sources of chlorine in the stratosphere. Initially my main interest was
with DDT and other pesticides. Then by the beginning of 1974 I read a paper by
James Lovelock and co-workers [53], who reported atmospheric measurements of
CFCl$_3$ (50 picomol/mol) and CCl$_4$ (71 pmol mol^{-1}) over the Atlantic. (Such
measurements had been made possible by Lovelock's invention of the electron
capture detector for gas chromatographic analysis, a major advance in the environmental sciences). Lovelock's paper gave me the first estimates of the industrial
production rates of CF$_2$Cl$_2$ and CFCl$_3$. It also stated that these compounds "are
unusually stable chemically and only slightly soluble in water and might therefore
persist and accumulate in the atmosphere… The presence of these compounds
constitutes no conceivable hazard." This statement had just aroused my curiosity
about the fate of these compounds in the atmosphere when a preprint of a paper by
M.J. Molina and F.S. Rowland with the title "Stratospheric Sink for
Chlorofluoromethanes—Chlorine Atom Catalyzed Destruction of Ozone" was sent
to me by the authors. I knew immediately that this was a very important paper and
decided to mention it briefly during a presentation on stratospheric ozone to which I
had been invited by the Royal Swedish Academy of Sciences in Stockholm. What I
did not know was that the press was likewise invited to the lecture. To my great
surprise, within a few days, an article appeared in the Swedish newspaper *Svenska
Dagbladet*, which drew attention to the topic. This article quickly attracted wide
international attention, and soon I was visited by representatives of the German

chemical company Hoechst and also by Professor Rowland, who at that time was spending a sabbatical at the Atomic Energy Agency in Vienna. This was the first time I had ever heard of Molina or Rowland, which is not surprising as they had not been active in studies on the chemistry of the atmosphere. Needless to say, I remained highly interested in the topic, and by September 1974, about 2 months after the publication of Molina and Rowland's paper [54], I presented a model analysis of the potential ozone depletion resulting from continued use of chlorofluorocarbons (CFCs) [55], which indicated the possibility of up to about 40 % ozone depletion near 40 km altitude as a result of continued use of these compounds at 1974 rates. Almost simultaneously, Cicerone et al. [56]. published a paper in which they predicted that by 1985–1990, continued use of CFCs at early 1970 levels could lead to ClOx-catalyzed ozone destruction of a similar magnitude to the natural sinks of ozone. Following Molina and Rowland's proposal, research on stratospheric chemistry further intensified, now with the emphasis on chlorine compounds.

By the summer of 1974, together with my family, I moved to Boulder, Colorado, where I assumed two halftime positions, one as a consultant at the Aeronomy Laboratory of the National Oceanic and Atmospheric Administration (NOAA) and the other at the Upper Atmosphere Project of the National Center for Atmospheric Research (NCAR). The NOAA group, which under the able direction of Dr. Eldon Ferguson had become the world leading group in the area of laboratory studies of ion/molecule reactions, had just decided to direct their considerable experimental skills to studies of stratospheric chemistry. My task was to guide them in that direction. I still feel proud to have been part of a most remarkable transformation. Together with Eldon Ferguson, scientists like Dan Albritton, Art Schmeltekopf, Fred Fehsenfeld, Paul Goldan, Carl Howard, George Reid, John Noxon, and Dieter Kley rapidly made major contributions to stratospheric research, including such activities as air sampling with balloon-borne evacuated cans, so-called "salad bowls" for later gas chromatographic analysis, optical measurements of the vertical abundances and distributions of NO_2 and NO_3 (later expanded by Susan Solomon to BrO and OClO), the design and operation of an instrument to measure extremely low water vapor mixing ratios, and laboratory simulations of important, but previously poorly known rate coefficients of important reactions. In later years the NOAA group also devoted itself to studies of tropospheric chemistry, reaching a prominent position in this research area as well. At NCAR the emphasis was more on infrared spectrographic measurements by John Gille and Bill Mankin, work that also developed into satellite-borne experiments. Another prominent activity was the analysis of the vertical distributions of less reactive gases, such as CH_4, H_2O, N_2O and the CFCs, employing the cryogenic sampling technique which had been pioneered by Ed Martell and Dieter Ehhalt.

In 1977 I took up the directorship of the Air Quality Division of NCAR, my first partially administrative position. I continued, however, my scientific work, something which many thought would be impossible. Fortunately, in Nelder Medrud I had a highly competent administrative officer. In my position as director I promoted work on both stratospheric and tropsopheric chemistry. My own research was mostly devoted to the development of photochemical models, conducted mostly with my students Jack Fishman, Susan Solomon, and Bob Chatfield. Together with Pat Zimmerman we started studies on atmosphere-biosphere interactions, especially the release of hydrocarbons from vegetation and pollutant emissions due to biomass burning in the tropics. I also tried to strengthen interactions between atmospheric chemists and meteorologists to improve the interpretation of the chemical measurements obtained during various field campaigns. To get this interdisciplinary research going was a challenge in those days.

During this period, as part of various activities in the United States and internationally, much of my research remained centered on the issue of anthropogenic, chlorine-catalyzed ozone destruction. However, because I am sure that this topic will be covered extensively by my two fellow recipients of this year's Nobel Prize, I would like to make a jump to the year 1985, when Joe Farman and his colleagues [57] of the British Antarctic Survey published their remarkable set of October total ozone column measurements from the Halley Bay station, showing a rapid depletion, on the average by more than 3 % per year, starting from the latter half of the 1970s. Although their explanation (ClO_x/NO_x interactions) was wrong, Farman et al. [57]. correctly suspected a connection with the continued increase in stratospheric chlorine (nowadays more than five times higher than natural levels). Their display of the downward trend of ozone, matching the upward trend of the chlorofluorocarbons (with the appropriate scaling) was indeed highly suggestive.

The discovery of the ozone hole came during a period in which I was heavily involved in various international studies on the potential environmental impacts of a major nuclear war between the NATO and Warsaw Pact nations, an issue to which I will briefly return to in one of the following sections. Because so many researchers became quickly involved in the "ozone hole" research, initially I stayed out of it. Then, in early 1986 I attended a scientific workshop in Boulder, Colorado, which brought me up-to-date with the various theories that had been proposed to explain the ozone hole phenomenon. Although it turned out that some of the hypotheses had elements of the truth, in particular the idea put forward by Solomon et al. [58]. of chlorine activation on the surface of stratospheric ice particles, followed by reactions (23) and (24).

$$HCl + ClONO_2 \rightarrow Cl_2 + HNO_3 \qquad (23)$$

$$Cl_2 + h\nu \rightarrow 2Cl\,(<350\,nm) \qquad (24)$$

I felt rather dissatisfied with the treatment of the chemistry in the heterogeneous phase. On my flight back to Germany to which we had moved (I hardly sleep on trans-Atlantic flights), I had good time to think it over and suddenly realized that if HNO_3 and NO_x were removed from the gas phase into the particulate phase, then an important defense against the attack of ClO_x on O_3 would be removed. The thought goes as follows: Under normal stratospheric conditions, there are strong interactions between the NO_x and ClO_x radicals, leading to the formation of $ClONO_2$ and HCl, which do not react with ozone. This leads to protection of ozone from otherwise much more severe destruction. Important examples of these are reaction (25) and the pair of reactions (26) + (27).

$$CIO + NO_2 + M \rightarrow CIONO_2 + M \qquad (25)$$

$$CIO + NO \rightarrow Cl + NO_2 \qquad (26)$$

$$Cl + CH_4 \rightarrow HCl + CH_3 \qquad (27)$$

Because of these reactions, under normal stratospheric conditions most of the inorganic chlorine is present as HCl and $CIONO_2$. Like two Mafia families, the CIO_x and NO_x thus fight each other, to the advantage of ozone. As shown in Fig. 1.9, there are plenty of complex interactions between the OX, HX, NX, and CIX families (for definitions see the legend of Fig. 1.9). Now, if the NX compounds were removed from the gas phase, reactions (25)–(27) would not occur and most chlorine may become available in the activated forms. During my return trip to Germany I started to think about this possibility. First, all NO_x compounds are converted to HNO_3 during the long polar nights by reactions (11), (28), (29), (14'), (30).

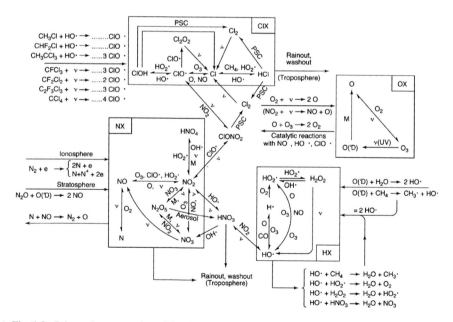

Fig. 1.9 Schematic presentation of the chemical interactions in the stratosphere. At the start of my scientific career only the OX and some of the HX reactions had been taken into account. Note that OX stands for the odd oxygen compounds, HX for H, OH, HO_2 and H_2O_2; NX for N, NO, NO_2, NO_3, N_2O_5, HNO_3 and HNO_4; and CIX for all inorganic chlorine compounds, Cl, CIO, Cl_2O_2, $CIONO_2$, HCl, OCIO and Cl_2. Not included are the bromine compounds which likewise play a significant role in stratospheric ozone depletion

$$NO + O_3 \rightarrow NO_2 + O_2 \tag{11}$$

$$NO_2 + O_3 \rightarrow NO_3 + O_2 \tag{28}$$

$$NO_3 + NO_2(+M) \rightarrow N_2O_5(+M) \tag{29}$$

$$N_2O_5 + H_2O(\text{surface}) \rightarrow 2HNO_3(\text{gas}) \tag{14'}$$

$$HNO_3(\text{gas}) \rightarrow HNO_3(\text{particles}) \tag{30}$$

As noted before, reaction (14) does not occur in the gas phase, but it readily occurs on wetted particulate surfaces. These are always present in the lower stratosphere in the form of sulfate particles, a fact which was first discovered by Christian Junge, a pioneer in atmospheric chemistry and my predecessor as director at the Max Planck Institute for Chemistry in Mainz [59]. The sulfate particles are formed by nucleation of gas phase H_2SO_4, which is formed from SO_2, following attack by OH [60, 61] [Eqs. (31)–(33)].

$$SO_2 + OH + M \rightarrow HSO_3 + M \tag{31}$$

$$HSO_3 + O_2 \rightarrow SO_3 + HO_2 \tag{32}$$

$$SO_3 + H_2O \rightarrow H_2SO_4 \tag{33}$$

The sources of stratospheric SO_2 are either direct injections by volcanic explosions [59] or oxidation of OCS, produced at the earth's surface [62], by reactions (34)–(36).

$$OCS + h\nu \rightarrow S + CO \tag{34}$$

$$S + O_2 \rightarrow SO + O \tag{35}$$

$$SO + O_2 \rightarrow SO_2 + O \tag{36}$$

The possibility of HNO_3 formation by heterogeneous reactions on sulfate particles had already been considered in a 1975 paper that I co-authored with Richard Cadle and Dieter Ehhalt [63]. Based on laboratory experiments, this reaction was for a long while thought to be unimportant, until it was discovered that the original laboratory measurements were grossly incorrect and that reaction (14) readily occurs on H2O-containing surfaces [64–66]. Earlier tropospheric measurements had, however, already indicated this [67]. The introduction of reaction (14) leads to a significant conversion of reactive NO_x to much less reactive HNO_3, thus diminishing the role of NO_x in ozone chemistry, especially in the lower stratosphere. By including reaction (14), better agreement was obtained between theory and observation [68]. The experience with reaction (14) emphasizes again the importance of high-quality measurements. It is better to have no measurements at all than bad measurements.

As soon as I returned to Mainz, I contacted Dr. Frank Arnold of the Max Planck Institute for Nuclear Physics in Heidelberg to explain my idea about NO_x removal from the gas phase to him. After about a week he had shown that under stratospheric conditions, solid nitric acid trihydrate (NAT) particles could be formed at temperatures below about 200 K, that is, at temperature about 10 K higher than that needed for water ice particle formation. The paper about our findings was published in Nature at the end of 1986 [69]. Independently, the idea had also been developed by Brian Toon, Rich Turco, and co-workers [70]. Subsequent laboratory investigations, notably by David Hanson and Konrad Mauersberger [71], then of the University of Minnesota, provided accurate information on the thermodynamic properties of NAT. Next it was also shown that the NAT particles could provide efficient surfaces to catalyze the production of ClO_x by reactions (23) and (24) [72, 73]. Finally, Molina and Molina [74] proposed a powerful catalytic reaction cycle involving ClO-dimer formation [Eqs. (21), (37), and (38)], which

$$Cl + O_3 \rightarrow ClO + O_2 (2x) \tag{21}$$

$$ClO + ClO + M \rightarrow Cl_2O_2 + M \tag{37}$$

$$Cl_2O_2 + h\nu \rightarrow Cl + ClO_2 \rightarrow 2Cl + O_2 \tag{38}$$

$$2O_3 \rightarrow 3O_2 \qquad (21) + (37) + (38)$$

could complete the chain of events causing rapid ozone depletion under cold, sunlit stratospheric conditions. Note that reaction (37) implies an ozone depletion response that is proportional to the square of the ClO concentrations. Furthermore, as chlorine activation by reaction (23) is also nonlinearly dependent on the stratospheric chlorine content, a powerful, nonlinear, positive feedback system is created, which is responsible for the accelerating loss of ozone under "ozone hole" conditions. The "ozone hole" is a drastic example of a man-made chemical instability, which developed at a location most remote from the industrial releases of the chemicals responsible for the effect.

The general validity of the chain of events leading to chlorine activation has been confirmed by both ground-based [75, 76] and airborne, in situ [77] radical observations. Especially the latter, performed by James Anderson and his colleagues of Harvard University, have been very illuminating, showing large enhancements in ClO concentrations in the cold, polar region of the lower stratosphere, coincident with a rapid decline in ozone concentrations. Together with other observations this confirms the correctness of the ozone depletion theory as outlined above. In the meanwhile the seriousness of this global problem has been recognized by all nations of the world, and international agreements have been signed to halt the production of CFCs and halons.

1.7   And Things Could Have Been Much Worse

Gradually, over a period of a century or so, stratospheric ozone should recover. However, it was a close call. Had Joe Farman and his colleagues from the British Antarctic Survey not persevered in making their measurements in the harsh Antarctic environment for all those years since the International Geophysical Year 1958/1959, the discovery of the ozone hole may have been substantially delayed, and there may have been far less urgency to reach international agreement on the phasing out of CFC production. There might thus have been a substantial risk that an ozone hole could also have developed in the higher latitudes of the northern hemisphere.

Furthermore, while the establishment of an instability in the O_x/ClO_x system requires chlorine activation by heterogeneous reactions on solid or supercooled liquid particles, this is not required for inorganic bromine, which is normally largely present in its activated forms due to gas-phase photochemical reactions. This makes bromine almost a hundred times more dangerous for ozone than chlorine on an atom to atom basis [78]. This brings up the nightmarish thought that if the chemical industry had developed organobromine compounds instead of CFCs—or alternatively, if chlorine chemistry had behaved more like that of bromine—then without any preparedness, we would have been faced with a catastrophic ozone hole everywhere and at all seasons during the 1970s, probably before the atmospheric chemists had developed the necessary knowledge to identify the problem and the appropriate techniques for the necessary critical measurements. Noting that nobody had worried about the atmospheric consequences of the release of Cl or Br before 1974. I can only conclude that we have been extremely lucky and it shows that we should always be on our guard for the potential consequences of the release of new products into the environment. Continued surveillance of the composition of the stratosphere, therefore, remains a matter of high priority for many years ahead.

In the meantime, we know that freezing of $H_2SO_4/HNO_3/H_2O$ mixtures to give NAT particle formation does not always occur and that supercooled liquid droplets can exist in the stratosphere substantially below nucleation temperatures, down to the ice freezing temperatures [79]. This can have great significance for chlorine activation [80, 81]. This issue, and its implications for heterogeneous processes, have been under intensive investigation at a number of laboratories, especially in the United States, notably by the groups headed by A.R. Ravishankara at the Aeronomy Laboratory of NOAA, Margaret Tolbert at the University of Colorado, Mario Molina at MIT, Doug Worsnop and Chuck Kolb at Aerodyne, Boston, and Dave Golden at Stanford Research Institute in Palo Alto. I am very happy that a team of young colleagues at the Max Planck Institute for Chemistry under the leadership of Dr. Thomas Peter, now at the ETH in Zurich, is likewise very successfully involved in experimental and theoretical studies of the physical and chemical properties of stratospheric particles at low temperatures. A new finding from this work was that freezing of supercooled ternary $H_2SO_4/HNO_3/H_2O$ mixtures may actually start when air parcels go through orographically induced cooling

Paul J. Crutzen with an air sampling canister in Chicago in1989. The photo was taken by Jos Lelieveld

events. Under these conditions the smaller particles originally mostly consisting of a mixture of H_2SO_4 and H_2O, will most rapidly be diluted with HNO_3 and H_2O and attain a chemical composition resembling that of a NAT aerosol, which, according to laboratory investigations, can readily freeze [82, 83].

1.8   Biomass Burning in the Tropics

By the end of the 1970s considerable attention was given to the possibility of a large net source of atmospheric CO_2 due to tropical deforestation [84]. Biomass burning is, however, not only a source of CO_2, but also of a great number of photochemically and radiatively active trace gases, such as NO_x, CO, CH_4, reactive hydrocarbons, H_2, N_2O, OCS and CH_3Cl. Furthermore, biomass burning in the tropics is not restricted to forest conversion, but is also a common activity related to agriculture, involving the burning of savanna grasses, wood, and agricultural wastes. In the summer of 1978, on our way back to Boulder from measurements of the emissions of OCS and N_2O from feedlots in Northeastern Colorado, we saw a big forest fire high up in the Rocky Mountain National Forest, which provided us with the opportunity to collect air samples from a major forest fire plume. To derive the emission ratios of the above gases relative to CO_2 could be established. Multiplying these ratios with estimates of the global extent of CO_2 production by biomass burning, estimated to be of the order of $2 \times 10^{15} - 4 \times 10^{15}$ g C per year [85], we next derived the first estimates of the global emissions of H_2, CH_4, CO, N_2O, NO_x, COS, and CH_3Cl and could show that the emissions of these gases could constitute a significant fraction of their total global emissions. These first measurements stimulated considerable international research efforts. Except for N_2O (for which our first measurements have since proved incorrect) our original findings were largely confirmed, although large uncertainties in the quantification of the various human activities contributing to biomass burning and individual trace gas releases remain [86]. Because biomass burning releases substantial quantities of reactive trace gases, such as hydrocarbons, CO, and NO_x, in photochemically very active environments, large quantities of ozone were expected to be formed in the tropics and subtropics during the dry season. Several measurement campaigns in South America and Africa, starting in 1979 and 1980 with NCAR's Quemadas expedition in Brazil, have confirmed this expectation [87–92]. The effects of biomass burning are especially noticeable in the industrially lightly polluted southern hemisphere, as is clearly shown from satellite observations of the tropospheric column amounts of CO and O_3 in Figs. 1.10 and 1.11 [93, 94].

Paul J. Crutzen with gas sampling canister in the burning rain forest in Brazil in 1978. The photo was taken by his colleague Tony Delany

Paul J. Crutzen in Brazil in 1978. Taking soil samples. The photo was taken by his colleague Patrick Zimmerman

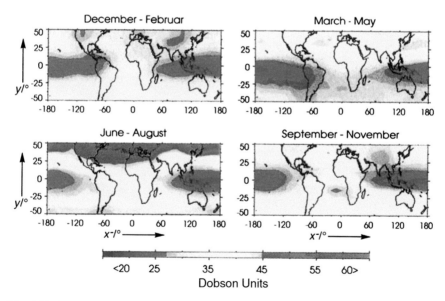

Fig. 1.10 Observed distributions of vertical column ozone in the troposphere for 4 periods from Fishman et al. [91, 93]. 1 Dobson unit represents a vertical column of $2.62 \ 10^{16}$ molecules cm^{-2}

Measurement of Air Pollution from Satellites
Carbon Monoxide Mixing Ratios in Middle Troposphere during April and October 1994

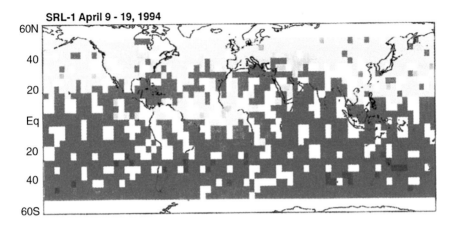

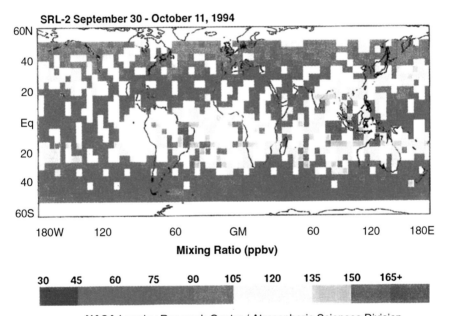

NASA Langley Research Center / Atmospheric Sciences Division

Fig. 1.11 Observed distributions of vertical column CO in the troposphere for 4 periods, measured on the space shuttle during April and October 1994. Courtesy of Drs. Vicki Connors, Hank Reichle and the MAPS team. Reference should be made to Connors et al. (94). (1 ppbv is the same as 1 nmole/mole)

Seasonal Depictions of Troposheric Ozone Distribution

Panels below depict global climatologies of tropospheric ozone (smog) developed at NASA Langley. Note high summertime values in the Northern Hemisphere and enhancements over South Atlnatic Ocean due to widespread biomass burning in Africa September–November.

1.9   "Nuclear Winter"

My research interest both into the effects of NO_x on stratospheric ozone and in biomass burning explain my involvement in the "nuclear winter" studies. When in 1981 I was asked by the editor of *Ambio* to contribute to a special issue on the environmental consequences of a major nuclear war, an issue co-edited by Dr. Joseph Rotblat, the 1995 Nobel Peace Prize awardee, the initial thought was that I would make an update on predictions of the destruction of ozone by the NO_x that would be produced and carried up by the fireballs into the stratosphere [95, 96]. Prof. John Birks of the University of Colorado, Boulder, one of the co-authors of the Johnston study on this topic [96], who spent a sabbatical year in my research division in Mainz, joined me in this study. Although the ozone depletion effects were significant, it was also clear to us that these effects could not come close to the direct impacts of the nuclear explosions. However, we then came to think about the potential climatic effects of the large amounts of sooty smoke from fires in the forests and in urban and industrial centers and oil storage facilities, which would reach the middle and higher troposphere. Our conclusion was that the absorption of sunlight by the black smoke could lead to darkness, strong cooling of the earth's continents, and a heating of the atmosphere at higher elevations, thus creating atypical meteorological and climatic conditions that would jeopardize agricultural production for a large part of the human population [97]. This idea was picked up by others, especially the so-called TTAPS (Turco, Toon, Ackerman, Pollack, Sagan) group [98], who even predicted that subfreezing temperatures could be possible over much of the earth. This was supported by detailed climate modeling [99]. A major international study of the issue, which was conducted by a group of scientists working under the auspices of SCOPE (Scientific Committee on Problems of the Environment) of the ICSU (International Council of Scientific Unions), also supported the initial hypothesis, and concluded that far more people could die because of the climatic and environmental consequences of a nuclear war than directly because of the explosions [100, 101].

Paul J. Crutzen after the announcement of the Nobel Prize award. Reception organized by employees at the Max Planck Institute for Chemistry in Mainz October 1995. Photo by Alfred Klemm (MPIC)

Although I do not count the nuclear winter idea among my greatest scientific achievements, I am convinced that, from a political point of view, it is by far the most important, because it magnifies and highlights the dangers of a nuclear war and convinces me that in the long run mankind can only escape such horrific consequences if nuclear weapons are totally abolished by international agreement. I thus wholeheartedly agree in this respect with Joseph Rotblat and the Pugwash organization, the 1995 recipient of the Nobel Prize for Peace.

1.10 Current Research Interests

Realizing the great importance of heterogeneous reactions in stratospheric chemistry, together with my Dutch students Jos Lelieveld (now my successor as director at the MPI for Chemistry in Mainz) and Frank Dentener, I have been involved in studies on the effects of reactions taking place in cloud droplets and tropospheric aerosol particles. In general, such reactions result in removal of NO_x and lower concentrations of O_3 and OH [102, 103]. Furthermore, even at high enough NO_x concentrations to allow ozone formation by reactions (16′) + (17) + (2), such reactions would be much limited within clouds, because the NO_x molecules, which are only slightly water soluble, stay in the gas phase, while the HO_2 radicals readily dissolve in the cloud droplets [Eq. (39)], where they can destroy ozone by reaction (40).

$$HO_2(gas) \rightarrow HO_2(aq) \leftrightarrow H^+ + O_2^- \qquad (39)$$

$$O_2^- + O_3 + H^+ \rightarrow OH + 2O_2 \qquad (40)$$

The role of rapid convective transport of reactive compounds from the planetary boundary layer into the upper troposphere is another topic with which I have been involved with some of my students. This may have important effects on the chemistry of the upper troposphere [104, 105].

A new project in which I have been much interested is the possibility of Cl and especially Br activation in the marine boundary layer. It is already known that Br activation can explain the near-zero O_3 concentrations, which are often found in the high-latitude, marine boundary layer during springtime [106]. In our most recent papers we discuss the possibility that Br activation may also occur in other marine regions and seasons [107, 108].

1.10.1   A Look Ahead from 1995, the Year I Was Awarded the Nobel Prize

Despite the fundamental progress that has been made over the past decades, much research will be needed to fill major gaps in our knowledge of atmospheric chemistry. I will indicate some of those research areas that I consider to be of greatest interest [109].

1.10.2   Observations of Tropospheric Ozone

Despite the great importance of tropospheric ozone in atmospheric chemistry, there are still major uncertainties concerning its budget and global concentration distribution. Everywhere, but especially in the tropics and the subtropics, there is a severe lack of data on tropospheric ozone. Considering the enormous role of tropical ozone in the oxidation efficiency of the atmosphere, the already recognized large anthropogenic impact on ozone through biomass burning, and the expected major agricultural and industrial expansion of human activities in this part of the world, this knowledge gap is very serious. At this stage it is not possible to test photochemical transport models owing to the scarcity of ozone observations, especially in the tropics and subtropics. Of critical importance in the effort to obtain data from the tropics and subtropics will be the training and long-term active participation of scientists from the developing countries. Besides ozone measurements at a number of stations and during intensive measurement campaigns, it will be important to also obtain data on reactive hydrocarbons, CH_4, CO, NO_x, and NO_y.

1.10.3 Long-Term Observations of Properties of the Atmosphere

Two major findings have demonstrated the extreme value of long-term observations of important chemical properties of the atmosphere. One example was the discovery of the rapid depletion of stratospheric ozone over Antarctica during the spring months, as discussed before. Another is the unexpected major, temporary break in the trends of CH_4 and CO. Most surprising were the changes in CO, for which Khalil and Rasmussen [110] derived a downward trend in surface concentrations by (1.4 ± 0.9) % per year in the northern hemisphere and by as much as (5.2 ± 0.7) % per year in the southern hemisphere between 1987 and 1992. Even larger downward trends, (6.1 ± 1) % per year in the northern hemisphere and (7 ± 0.6) % per year in the southern hemisphere, were reported for the period between June 1990 and June 1993 by Novelli et al. [111]. Although these trends have again reversed (P. Novelli, private communication) into the previous upward trend of +0.7 % per year for CO and almost 1 % per year for CH_4 [112, 113], the temporary break is remarkable. The reasons for this surprising behavior are not known. They may consist of a combination of: (1) variable annual emissions from biomass burning, (2) higher concentrations of OH radicals, maybe due to loss of stratospheric ozone, triggered by an increase in reactive aerosol surfaces in the stratosphere following the Pinatubo volcanic eruption in June 1991, (3) a dynamically forced global redistribution of CO, introducing a bias due to the location of the limited number of measuring sites, (4) reduced CO formation from the oxidation of natural hydrocarbons emitted by tropical forests due to globally altered precipitation and temperature patterns, or, most likely, a combination of these and other, yet unknown, factors. At this stage we can only conclude that the causes for the surprisingly rapid changes in CO trends are not known, mainly because of incomplete global coverage of the CO measurement network.

1.10.4 Intensive Measurement Campaigns

Comprehensive field programmes that have been conducted in the past with detailed observations of all factors that influence the photochemistry of the troposphere will also be much needed in the future, especially in various regions of the marine and continental tropics and subtropics, to find out whether we understand the major processes that determine the chemistry of ozone and related photo-chemically active compounds. Applications of comprehensive chemical and transport models should be an important part of these activities. Topics in which greatly improved knowledge is necessary are especially the improved quantification of the stratospheric influx of ozone; distributions, sources, and sinks of CH_4, reactive hydrocarbons, CO, NO_x and NO_y, and the quantification of natural NO emissions from lightning and soils.

1.10.5 Cloud Transport

The role of clouds as transporters of chemical constituents such as CO, NO_x, reactive hydrocarbons, and their oxidation products from the boundary layer to the middle and upper troposphere (and possibly into the lower stratosphere) should be better understood and quantified, so that they can be parameterized for inclusion in large scale photochemical models of the atmosphere. Similarly the production of NO by lightning and its vertical redistribution by convective storms should also be much better understood and quantified, both for marine and continental conditions.

1.10.6 Chemical Interactions with Hydrometeors

The interactions of chemical constituents emanating from the boundary layer with liquid and solid hydrometeors in the clouds will be of special importance. There is, for instance, the question of why strong ozone formation has not been noticed around the most convective regions in the continental tropics in which large amounts of forest-derived reactive hydrocarbon such as isoprene (C_5H_8) and their oxidation products are rapidly lifted to the middle and upper troposphere and mixed with lightning-produced NO to provide favorable conditions for photochemical ozone formation. Could significant ozone destruction take place in cloud water or on the surface of ice particles that may be partially covered by water? [114, 115]. Such questions regarding potential loss of ozone by reactions with hydrometeors may be especially relevant in connection with observations of record low O_3 volume mixing ratios often of less than 10 nmol/mol over much of the entire tropospheric column in March 1993 in an extended, heavily convective region between Fiji and Christmas Island over the Pacific Ocean [116]. Although such low ozone volume mixing ratios had been noted on several occasions in the tropical marine boundary layer and can be explained by the ozone-destroying reactions (5)–(8) in the lower troposphere, it should be ascertained whether these reactions alone suffice to explain the extremely low ozone concentrations in such a large volume of air.

1.10.7 Photolysis Rates in Cloudy Atmospheres

Regarding the photochemistry taking place in cloudy atmospheric conditions, recent observations of unexpectedly high absorption of solar radiation in cloudy atmospheres [117] point to the possibility that multiple scattering in broken cloud systems may lead to strongly enhanced photolysis rates and photochemical activity, leading, for example, to much higher O_3 destruction and OH production by reactions (5)–(8), than thought so far. The influence of clouds on the photochemically active UV radiation field is a potentially very important research topic, which should be pursued by measurements and the development of appropriate radiative transfer models.

1.10.8 Biogenic Sources of Hydrocarbons, CO, and NO

The continental biosphere is a large source of hydrocarbons. Quantification of these sources in terms of geophysical (e.g., temperature, humidity, light levels) and biogeochemical (soil physical and chemical properties, land use) parameters are urgently needed for inclusion in atmospheric models. The hydrocarbon oxidation mechanisms in the atmosphere should also be better understood, so that formation of ozone, OH, carbon monoxide, partially oxidized gaseous hydrocarbons, and organic aerosol can be better quantified. The formation of organic aerosol from hydrocarbon precursors and their capability to serve as cloud condensation nuclei are related, potentially important, subjects that have not been studied in any depth so far.

1.10.9 Potential Role of Halogens in Tropospheric Ozone Destruction

There are strong indications that tropospheric ozone can be destroyed by reactions in addition to those discussed so far. Observations of surface ozone levels during polar sunrise in the Arctic have frequently shown the occurrence of immeasurably low ozone concentrations, coinciding with high "filterable Br" [106]. Further measurements [118] identified BrO as one of the active Br compounds, which, as is well known from stratospheric measurements, may rapidly attack ozone by a series of catalytic reactions, such as:

$$2 \times (Br + O_3 \rightarrow BrO + O_2) + (BrO + BrO \rightarrow 2Br + O_2) = (2O_3 \rightarrow 3O_2)$$

or

$$(Br + O_3 \rightarrow BrO + O_2) + (BrO + HO_2 \rightarrow HOBr + O_2) + (HOBr + h\nu \rightarrow OH + Br)$$
$$+ (OH + CO + O_2 \rightarrow HO_2 + CO_2) = (CO + O_3 \rightarrow CO_2 + O_2)$$

It should be explored whether halogen activation reactions may also occur under different circumstances than indicated above [106–108].

1.10.10 Heterogeneous Reactions on Aerosol Particles

The issue of interactions between gases and atmospheric aerosol is largely unexplored and very little considered in tropospheric chemistry models. Examples are interactions of dimethysulfide-derived sulfur compounds with sea salt in the marine boundary layer and reactions of SO_2, H_2SO_4, NO_x, N_2O_5, and HNO_3 on soil dust particles, which remove these compounds from the gas phase. In the case of industrial SO_2, the neglect of such heterogeneous reactions may well have led to overestimations of the climatic cooling effects of anthropogenic aerosol, as any incorporation of sulfur in soil dust or sea salt will prevent the nucleation of new sunlight backscattering sulfate particles.

1.10.11 Ozone/Climate/Greenhouse Gas Feedbacks in the Stratosphere

Ozone is a significant greenhouse gas with an infrared absorption band in the atmospheric window region, centered at 9.6 μm. Although the amount of ozone in the troposphere is only about 10 % of that of the stratosphere, the effective long-wave optical depth of tropospheric ozone is larger. Of greatest importance would be any changes that might take place in the ozone concentrations in the tropopause regions as a result of human activities, such as those caused by H_2O, NO, SO_2, and particulate emissions from expanding fleets of civil aircraft flying in the stratosphere and upper troposphere. On one hand this may lead to increasing temperatures in the lower stratosphere. However, increased HNO_3 and H_2O concentrations in the lower stratosphere may increase the likelihood of polar stratospheric particle formation and ozone destruction. Such a course of events is also promoted by cooling of the stratosphere through increasing concentrations of CO_2. This cooling effect also increases with height in the stratosphere and mesosphere. The implications of this for the future dynamics of the stratosphere, mesosphere, and lower thermosphere is likewise a topic deserving considerable attention. Changes in chemical and radiative conditions in the lower stratosphere may, therefore, create feedbacks that we need to understand well. They include understanding their potential impact on tropopause heights and temperatures, stratospheric water vapor, lower stratospheric cloud characteristics, and the tropospheric hydrological cycle. Recent observations of increasing trends of water vapor concentrations in the lower stratosphere over Boulder emphasize this point [119].

1.10.12 Long Range Indirect Transport of NOx Embedded in PAN

Because of efficient conversion of NO_x to HNO_3, which is mostly removed by precipitation, the lifetime of NO_x is short, on the order of a day, so that long range transport of NO_x is limited. However, non-methane hydrocarbon oxidation chemistry can convert NO_x to PAN (peroxyacetyl nitrate), whose thermal stability under the cold temperature conditions in the middle and upper troposphere, especially following rapid convective transport from the boundary layer, indirectly promotes long-range transport of NO_x. Thermal decomposition of PAN after it is mixed down into the warmer planetary boundary layer releases NO_x from the PAN reservoir again. Aircraft-borne measurements led by H.B. Singh and co-workers confirmed the important role of PAN [120, 121].

1.10.13 Field Campaign in Suriname

In contrast to the studies which took place in polluted atmospheric environments, a group of scientists from the MPI and the universities of Utrecht and Innsbruck, a field program was conducted in February–March 1998 in the largely still pristine tropical rain forests of Surinam, which concentrated on measurements of organic compounds emitted by the forests, such as isoprene, and their oxidation products. During this field program for the first time the proton transfer mass spectrometer, developed by the late Werner Lindinger and his colleagues at the University of Innsbruck, was successfully employed on board of a Cessna Citation aircraft of the Technical University of Delft, the Netherlands [122–124].

1.10.14 The Indian Ocean Expedition (INDOEX) and the Asian Brown Clouds ABC Project

The postulated effects of air pollution arising from biomass and fossil fuel burning on regional and global atmospheric chemistry and climate in the developing world led to the Indian Ocean Expedition (INDOEX) of 1999, organized principally by V. Ramanathan of Scripps Institution of Oceanography of the University of California, San Diego, with me serving as co-chief scientist. One main finding of INDOEX was the high concentration of smoke particles originating from the burning of fossil and biomass fuels, containing substantial amounts (10–15 %) of black carbon, causing low visibility, heating of the atmosphere and cooling of the earth's surface, thus reducing evaporation over the Indian Ocean and increasing the thermal stability of the earth's boundary layer over large tracks of tropical and subtropical Asia [125–128]. Similar effects occur during the dry season over Africa and South America.

The important findings of INDOEX has further led to a new major international research program initiative, ABC (Atmospheric Brown Clouds), with participation of scientists from India, China, Japan, Korea, the Maldives, Thailand, USA and Europe. Like INDOEX, the ABC program is headed by V. Ramanathan with me serving as co-chairman until 2005.

Major disturbances of the chemistry and radiation balance of the troposphere in the Mediterranean region were also revealed during the MINOS campaign led by Jos Lelieveld in the summer of 2001 [129].

1.10.15 Halogen Chemistry in the Troposphere

After the discovery of ozone-poor air masses near the surface and in the lower troposphere at polar sunrise, in a paper co-authored with Len Barrie, Jan Bottenheim, Russ Schnell and Rei Rasmussen, I showed that the ozone loss was due to chemical reactions, involving BrOx radicals as catalysts.

Similar reactions, involving bromine chemistry [106], derived photochemically from sea salt, also take place in the "background" marine troposphere. As calculated by Roland von Glasow and me, catalytic BrO_x chemistry may lead to global O_3 and DMS (dimethysulfide) loss by up to 20 and 50 % respectively [106]. The DMS loss reduces CCN (cloud condensation nuclei) production. Over the past 15 years, this research was extended to other regions of the troposphere, such as salt lakes, and other halogen species (I, Cl) by Crutzen and collaborators, in particular Rolf Sander, Roland von Glasow and Rainer Vogt [130–136]. In the meanwhile much research has been devoted to the complicated interaction between NO_y and halogen chemistry, as summarized for instance by Sander et al. [137]. and in a 64-page review by von Glasow and Crutzen [138]. Of particular interest have been the observations of halogen radicals by the DOAS technique by Platt and colleagues of the University of Heidelberg [139]. In a recent publication [140] many of the theoretical predictions, derived from models, were confirmed by field campaigns conducted at a site on Cape Verde.

1.10.16 Geo-engineering

In 2006 I published a paper which opened an intense debate on geoengineering, proposing that it should be studied whether the injection of sulfur gases (H_2S, SO_2) in the stratosphere, leading to the formation of a sunlight-reflecting sulfuric acid particle layer can cool the planet, thereby counteracting the global warming caused by the increase in the atmospheric concentrations of carbon dioxide and other greenhouse gases. In a modeling effort with Phil Rasch and D. Coleman we showed that such an operation indeed produced the anticipated lower atmosphere cooling. However, it must be mentioned that there may be negative side effects, such as depletion of stratospheric ozone.

1.10.17 Biofuels—Climate Effects of N_2O Emissions

Together with Arvin Mosier, Keith Smith, and Wilfried Winiwarter I showed in 2008 that currently used methods of biofuel production (bio-ethanol and bio-diesel) could cause the atmospheric release of N_2O in an amount that can wipe out any advantages provided by savings in the emissions of fossil fuel derived CO_2, This is possible due to the large global warming potential of N_2O, a product of fertilizer application. Earlier studies severely underestimated the importance of the N_2O production.

References

1. 1972: "Sweden's Case Study for the United Nations Conference on the Human Environment 1972", Air Pollution Across National Boundaries. The Impact on the Environment of Sulfur in Air and Precipitation, Stockholm.
2. Chapman, S., 1930: "A Theory of Upper Atmospheric Ozone", in: *Memoirs of the Royal Meteorological Society,* 3: 103–125.
3. Benson, S.W.; Axworthy, A.E., 1965: "Reconsiderations of the Rate Constants from the Thermal Decomposition of Ozone", in: *The Journal of Chemical Physics*, 42: 2614.
4. Bates, D.R.; Nicolet, M., 1950: "The Photochemistry of Atmospheric Water Vapour", in: *Journal of Geophysical Research,* 55: 301.
5. McGrath, W.D.; Norrish, R.G.W., 1960: "Studies of the Reaction of Excited Oxygen Atoms and Molecules Produced in the Flash Photolysis of Ozone", in: *Proceedings of the Royal Society of London A*, 254: 317.
6. Norrish, R.G.W.; Wayne, R.P., 1965: "The Photolysis of Ozone by Ultraviolet Radiation. The Photolysis of Ozone Mixed with Certain Hydrogen-Containing Substances", in: *Proceedings of the Royal Society of London A*, 288: 361.
7. Hampson, J., 1965: "Chemiluminescent Emission Observed in the Stratosphere and Mesosphere", in: *Les problèmes météorologiques de la stratosphère et de la mesosphère* (Presses universitaires de France, Paris): 393.
8. Hunt, B.G., 1966: "Photochemistry of Ozone in a Moist Atmosphere", in: *Journal of Geophysical Research,* 71: 1385.

9. Crutzen, P.J., 1969: "Determination of Parameters Appearing in the "Dry" and "Wet" Photochemical Theories for Ozone in the Stratosphere", in: *Tellus,* 21: 368–388.

10. Murcray, D.G.; Kyle, T.G.; Murcray, F.H.; Williams, W.J., 1968: "Nitric Acid and Nitric Oxide in the Lower Stratosphere", in: *Nature,* 218: 78.

11. Rhine, P.E.; Tubbs, L.D.; Williams, D., 1969: "Nitric Acid Vapor Above 19 km in the Earth's Atmosphere", in: *Applied Optics,* 8: 1501.

12. Crutzen, P.J., 1970: "The Influence of Nitrogen Oxides on the Atmospheric Ozone Content", in: *Quarterly Journal of the Royal Meteorological Society,* 96: 320–325.

13. Bates, D.R.; Hays, P.B., 1967: "Atmospheric Nitrous Oxide", in: *Planetary and Space Science,* 15: 189.

14. Greenberg, R.I.; Heicklen, J., 1970: "Reaction of $O(^1D)$ with N_2O", in: *International Journal of Chemical Kinetics,* 2: 185.

15. Crutzen, P.J., 1971: "Ozone Production Rates in an Oxygen-Hydrogen-Nitrogen Oxide Atmosphere", in: *Journal of Geophysical Research,* 76: 7311.

16. McElroy, M.B.; McConnell, J.C., 1971: "Nitrous Oxide. A Natural Source of Stratospheric NO", in: *Journal of the Atmospheric Sciences,* 28: 1085.

17. Davis, D.D.; et al., 1973: "Recent Kinetic Measurements on the Reactions of $O(^3P)$, Hand HO_2", in: DOT-TSC-OST-73-4, p 126.

18. Yoshino, K.; et al., 1988: "Improved Absorption Cross-Sections of Oxygen in the Wavelength Region 205–240 nm of the Herzberg Continuum", in: *Planetary and Space Science,* 36: 1469.

19. Frederick, J.E.; Mentall, J.E., 1982: "Solar Irradiance in the Stratosphere: Implications for the Herzberg Continuum Absorption of 0_2", in: *Geophysical Research Letters,* 9: 461.

20. Nicolet, M., 1981: "The Solar Spectral Irridiance and its Action in the Atmospheric Photodissociation Processes", in: *Planetary and Space Science,* 29: 951.

21. Weiss, R.F., 1981: "The Temporal and Spatial Distribution of Tropospheric Nitrous Oxide", in: *Journal of Geophysical Research,* 86: 7185.

22. SCEP (Study on Critical Environmental Problems), 1970: *Man's Impact on the Global Environment. Assessment and Recommendations for Action* (Cambridge: The MIT Press).

23. Nicolet, M., 1965: "Nitrogen Oxides in the Chemosphere", in: *Journal of Geophysical Research,* 70: 679.

24. Harrison, H.S., 1970: "Stratospheric Ozone with Added Water Vapour: Influence of High Altitude Aircraft", in: *Science,* 170: 734.

25. Johnston, H.S., Crosby, H.J., 1954: "Kinetics of the Fast Gas Phase Reaction Between Ozone and Nitric Oxide", in: *The Journal of Chemical Physics,* 22: 689.

26. Johnston, H.S., Garvin, D., 1972: "Working Papers for a Survey of Rate Data for Chemical Reactions in the Stratosphere", in: *NBS Report,* 10931.

27. Johnston, H.S., 1971: "Reduction of Stratospheric Ozone by Nitrogen Oxide Catalysts from Supersonic Transport Exhaust", in: *Science,* 173: 517.

28. Johnston, H.S., 1992: "Atmospheric Ozone", in: *The Annual Review of Physical Chemistry,* 43: 1.

29. Johnston, H.S.; Graham, R.A., 1974: "Photochemistry of NO_x and HNO_x compounds", in: *Canadian Journal of Chemistry,* 52: 1415.

30. CIAP (Climate Impact Assessment Program), 1974: *Report of Findings: The Effects of Stratospheric Pollution by Aircraft* (US Department of Transportation Washington, DC, DOT-TSC-75-50).

31. COMESA (The Report of the Committee on Meteorological Effects of Stratospheric Aircraft), 1975 (Bracknell, England: UK Meteorological Office).

32. COVOS (Comité d'Etudes sur les Conséquences des Vols Stratosphériques), 1976 (Boulogne, France: Société Météorologique de France).

33. National Academy of Sciences (Ed.), 1975: *Environmental Impact of Stratospheric Flight* (Washington, DC).

34. Crutzen P.J.; Isaksen, I.S.A.; Reid, G.C., 1975: "Solar Proton Events: Stratospheric sources of nitric oxide", in: *Science,* 189: 457.

35. Heath, D.F.; Krueger, A.J.; Crutzen, P.J., 1977: "Solar PROTON EVENT: INFLUENCE on Stratospheric Ozone", in: *Science,* 197: 886.
36. Solomon, S.; Crutzen, P.J., 1981: "Analysis of the August 1972 Solar Proton Event, Including Chlorine Chemistry", in: *Journal of Geophysical Research*, 86: 1140.
37. Levy, H., 1971: "Normal Atmosphere: Large Radical and Formaldehyde Concentrations Predicted", in: *Science*, 173: 141.
38. McConnell, J.C.; McElroy, M.B.; Wofsy, S.C., 1971: "Natural Sources of Atmospheric CO", in: *Nature,* 233: 187.
39. Prinn, R.G.; Weiss, R.F.; Miller B.R.; Huang, J.; Alyea, F.N.; Cunnold, D.H.; Fraser, P.J.; Hartley, D.E.; Simmonds, P.G., 1995: "Atmospheric Trends and Lifetime of Trichloroethane and Global Average Hydroxyl Radical Concentrations Based on 1978–1994 ALE/GAGE measurements", in: *Science*, 269: 187.
40. Levy III, H., 1972: "Photochemistry of the Lower Troposphere", in: *Planetary and Space Science*, 20: 919.
41. Crutzen, P.J., 1973: "A Discussion of the Chemistry of Some Minor Constituents in the Stratosphere and Troposphere", in: *Pure and Applied Geophysics*, 106–108: 1385.
42. Crutzen, P.J., 1974: "Photochemical Reactions Initiated by an Influencing Ozone in Unpolluted Tropospheric Air", in: *Tellus*, 26: 47.
43. Chameides, W.L.; Walker, J.C.G., 1973: "A Photochemical Theory of Tropospheric Ozone", in: *Journal of Geophysical Research*, 78: 8751.
44. Fishman, J.; Crutzen, P.J., 1978: "The Origin of Ozone in the Troposphere", in: *Nature*, 274: 855.
45. Fishman, J.; Solomon, S.; Crutzen, P.J., 1979: "Observational and Theoretical Evidence in Support of a Significant In-Situ Photochemical Source of Tropospheric Ozone", in: *Tellus*, 31: 432.
46. Howard, C.J.; Evenson, K.M., 1977: "Kinetics of the Reaction of HO_2 Radicals with NO", in: *Geophysical* Research *Letters*, 4: 437.
47. Crutzen, P.J.; Howard, C.J., 1978: "The Effect of the HO_2 + NO Reaction Rate Constant on One-Dimensional Model Calculations of Stratospheric Ozone Depletions", in: *Pure and Applied Geophysics,* 116: 497.
48. Crutzen, P.J.; Zimmermann, P.H., 1991: "The Changing Photochemistry of the Troposphere:, in: *Tellus*, 43 A/B:136.
49. Houghton J.T.; et al., 1990: "Intergovernmental Panel on Climate Change", in: *Climate Change: The IPCC Scientific Assessment* (Cambridge: Cambridge University Press), p 365.
50. Stolarski, R.S.; Cicerone, R.J., 1974: "Stratospheric Chlorine: A possible Sink for Ozone", in: *Canadian Journal of Chemistry*, 52: 1610.
51. Wofsy, S.C.; McElroy, M.B., 1974: "HO_x NO_x and ClO_x: Their role in atmospheric photochemistry", in: *Canadian Journal of Chemistry*, 52: 1582.
52. Crutzen, P.J., 1974: "A Review of Upper Atmospheric Photochemistry", in: *Canadian Journal of Chemistry*, 52: 1569.
53. Lovelock, J.E.; Maggs, R.J.; Wade, R.J., 1973: "Halogenated Hydrocarbons in and over the Atlantic", in: *Nature*, 241: 194.
54. Molina, M.J.; Rowland, F.S., 1974: "Stratospheric Sink of Chlorofluoromethanes: Chlorine Atom-Catalyzed Destruction of Ozone", in: *Nature*, 249: 810.
55. Crutzen, P.J., 1974: "Estimates of Possible Future Ozone Reductions from Continued Use of Fluorochloromethanes (CF_2Cl_2 $CFCl_3$)", in: *Geophysical Research Letters*, 1: 205.
56. Cicerone, R.J.; Stolarski, R.S.; Walters, S., 1974: "Stratospheric Ozone Destruction by Man-Made Chlorofluoromethanes", in: *Science*, 185: 1165.
57. Farman, J.C.; Gardiner, B.G.; Shanklin, J.D., 1985: "Large Losses of Total Ozone in Antarctica Reveal Seasonal ClO_x/NO_x Interaction", in: *Nature*, 315: 201.
58. Solomon, S.; Garcia, R.R.; Rowland, F.S.; Wuebbles, D.J., 1986: "On the Depletion of Antarctic Ozone", in: *Nature,* 321: 755.
59. Junge, C.E.; Chagnon, C.W.; Manson, J.E., 1961: "Stratospheric Aerosols", in: *Journal of Meteorology*, 18: 81.

60. Davis, D.D.; Ravishankara, A.R.; Fischer, S., 1979: "SO_2 Oxidation Via the Hydroxyl Radical: Atmospheric Fate of the HSO_x radicals", in: *Geophysical Research Letters*, 6: 113.
61. Stockwell, W.R.; Calvert, J.G., 1983: "The Mechanism of the $HO-SO_2$ Reaction", in: *Atmospheric Environment*, 17: 2231.
62. Crutzen, P.J., 1976: "The Possible Importance of CSO for the Sulfate Layer of the Stratosphere", in: *Geophysical Research Letters*, 3: 73.
63. Cadle, R.D.; Crutzen, P.J.; Ehhalt, D.H., 1975: "Heterogeneous Chemical Reactions in the Stratosphere", in: *Journal of Geophysical Research*, 80: 3381.
64. Mozurkewich, M.; Calvert, J.G., 1988: "Reaction Probabilities of N_2O_5 on Aqueous Aerosols", in: *Journal of Geophysical Research*, 93: 15889.
65. Hanson, D.R.; Ravishankara, A.R., 1991: "The Reaction Probabilities of $ClONO_2$ and N_2O_5 on 40 to 75 Percent Sulfuric Acid Solutions", in: *Journal of Geophysical Research*, 96: 17307.
66. Van Doren J.M.; Watson, L.R.; Davidovits, P.; Worsnop, D.R.; Zahniser, M.S.; Kolb, C.E., 1990: "Temperature Dependence of the Uptake Coefficients of HNO_3 HCl, and N_2O_5 by Water Droplets", in: *The Journal of Chemical Physics*, 94: 3265.
67. Platt, U.; Perner, D.; Winer, A.M.; Harris, G.W.; Pitts, J.N., 1980: "Detection of NO_3 in the Polluted Troposphere by Differential Optical Absorption", in: *Geophysical Research Letters*, 7: 89.
68. Pommereau, J.F.; Goutail, F., 1988: "Stratospheric O_3 and NO_2 Observations at the Southern Polar Circle in Summer and Fall 1988", in: *Geophysical Research Letters*, 15: 895.
69. Crutzen, P.J.; Arnold, F., 1986: "Nitric Acid Cloud Formation in the Cold Antarctic Stratosphere: A Major Cause for the Springtime 'Ozone Hole'", in: *Nature*, 324: 651.
70. Toon, O.B.; Hamill, P.; Turco R.P.; Pinto J., 1986: "Condensation of HNO_3 and HCl in the Winter Polar Stratosphere", in: *Geophysical Research Letters*, 13: 1284.
71. Hanson, D.R.; Mauersberger, K., 1988: "Vapor Pressures of HNO_3/H_2O Solutions At Low Temperatures", in: *The Journal of Physics and Chemistry*, 92: 6167.
72. Molina, M.J.; Tso, T.L.; Molina, L.T.; Wang, F.C.Y., 1987: "Antarctic Stratospheric Chemistry of Chlorine Nitrate, Hydrogen Chloride and Ice", in: *Science*, 238: 1253.
73. Tolbert, M.A.; Rossi, M.J.; Malhotra, R.; Golden, D.M., 1987: "Reaction of Chlorine Nitrate with Hydrogen Chloride and Water at Antarctic Stratospheric Temperatures", in: *Science*, 238: 1258.
74. Molina, L.T.; Molina, M.J., 1987: "Production of Cl_2O_2 from the Self-reaction of the ClO Radical", in: *The Journal of Physics and Chemistry*, 91: 433.
75. de Zafra, R.L.; Jaramillo, M.; Parrish, A.; Solomon, P.M.; Connor, P.M.; Barrett, J., 1987: "High Concentration of Chlorine Monoxide at Low Altitudes in the Antarctic Spring Stratosphere, I. Diurnal variation", in: *Nature*, 328: 408.
76. Solomon, S.; Mount, G.H.; Sanders, R.W.; Schmeltekopf, A.L., 1987: "Visible Spectroscopy at McMurdo Station, Antarctica: Observations of OCIO", in: *Journal of Geophysical Research*, 92: 8329.
77. Anderson, J.G.; Brune, W.H.; Proffitt, M.H., 1989: "Ozone Destruction by Chlorine Radicals Within the Antarctic vortex: The Spatial and Temporal Evolution of $ClO-O_3$ Anticorrelation Based on In Situ ER-2 Data", in: *Journal of Geophysical Research*, 94: 11465.
78. Wofsy, S.C.; McElroy, M.B.; Yung, Y.L., 1975: "The Chemistry of Atmospheric Bromine", in: *Geophysical Research Letters*, 2: 215.
79. Dye, J.E.; Baumgardner, D.; Gandrud, B.W.; Kawa, S.R.; Kelly, K.K.; Loewenstein, M.; Ferry, G.V.; Chan, K.R.; Gary, B.L., 1992: "Particle Size Distribution in Arctic Polar Stratospheric Clouds, Growth and Freezing of Sulfuric Acid Droplets, and Implications for Cloud Formation", in: *Journal of Geophysical Research*, 97: 8015.
80. Cox, R.A.; MacKenzie, A.R.; Müller, R.; Peter, T.; Crutzen, P.J., 1994: "Activation of Stratospheric Chlorine by Reactions in Liquid Sulphuric Acid", in: *Geophysical Research Letters*, 21: 1439.
81. Hanson, D.R.; Ravishankara, A.R., Solomon, S., 1994: "Heterogeneous Reactions in Sulfuric Acid Aerosol: A Framework for Model Calculations", in: *Journal of Geophysical Research*, 99: 3615.

82. Meilinger, S.K.; Koop, T.; Luo, B.P.; Huthwelker T; Carslaw, K.S.; Crutzen, P.J.; Peter, T., 1995: "Size-dependent Stratospheric Droplet Composition in lee Wave Temperature Fluctuations and Their Potential Role in PSC Freezing", in: *Geophysical Research Letters*, 22: 3031.

83. Koop, T.; Luo, B.P.; Biermann, U.M.; Crutzen, P.J.; Peter, T.: "Freezing of $HNO_3/H_2SO_4/H_2O$ Solutions at Stratospheric Temperatures: Nucleation Statistics and Experiments", in: noch unveröffentlicht.

84. Woodwell, G.M.; Whittaker, R.H.; Reiners, W.A.; Likens, G.E.; Delwiche, C.C.; Botkin, D. B., 1980: "The Biota and the World Carbon Budget", in: *Science*, 199: 141.

85. Seiler, W.; Crutzen, P.J., 1980: "Estimates of Gross and Net Fluxes of Carbon Between the Biosphere and the Atmosphere from Biomass Burning", in: *Climatic Change*, 2: 207.

86. Crutzen, P.J.; Andreae, M.O., 1990: "Biomass Burning in the Tropics: Impact on Atmospheric Chemistry and Biogeochemical Cycles", in: *Science*, 250: 1669.

87. Crutzen, P.J.; Delany, A.C.; Greenberg, J.; Haagenson, P.; Heidt, L.; Lueb, R.; Pollock, W.; Seiler, W.; Wartburg, A.; Zimmerman, P., 1985: "Tropospheric Chemical Composition Measurements in Brazil During the Dry Season", in: *The Journal of Atmospheric Chemistry*, 2: 233.

88. Amazon Boundary Layer Experiment (ABLE 2A): Dry season 1985, Collection of 24 papers, 1988: *Journal of Geophysical Research*, 93 (D2), 1349–1624).

89. Andreae, M.O.; et al., 1988: "Biomass Burning Emissions and Associated Haze Layers over Amazonia", in: *Journal of Geophysical Research*, 93: 1509.

90. Andreae, M.O.; et al., 1992: "Ozone and Aitken Nuclei over Equatorial Africa: Airborne Observations During DECAFE 88", in: *Journal of Geophysical Research*, 97: 6137.

91. Fishman, J.; Fakhruzzaman, K.; Cros, B.; Nyanga, D., 1991: "Identification of Widespread Pollution in the Southern Hemisphere Deduced from Satellite Analyses", in: *Science*, 252: 1693.

92. FOS/DECAFE 91 Experiment, Collection of 13 papers 1995: *The Journal of Atmospheric Chemistry*, 22, 1–239).

93. Fishman, J., 1991: "Probing Planetary Pollution from Space", in: *Environmental Science and Technology*, 25: 612.

94. Connors, V.; Flood, M.T.; Jones, T.; Gormsen, B.; Nolt, S.; Reichle, H., 1996: "Global Distribution of Biomass Burning and Carbon Monoxide in the Middle Troposphere During Early April and October 1994", in: *Biomass Burning and Global Change* (Ed.: 1. Levine, MIT Press, in press).

95. Foley, H.M.; Ruderman, M.A., 1973: "Stratospheric NO Production from Past Nuclear Explosions", in: *Journal of Geophysical Research*, 78: 4441.

96. Johnston, H.S.; Whitten, G.; Birks, J.W., 1973: "Effects of Nuclear Explosions on Stratospheric Nitric Oxide and Ozone", in: *Journal of Geophysical Research*, 78: 6107.

97. Crutzen, P.J.; Birks, J., 1982: "The Atmosphere After a Nuclear War: Twilight at noon", in: *Ambio*, 12: 114.

98. Turco, R.P.; Toon, O.B.; Ackerman, R.P.; Pollack, H.B.; Sagan, C., 1983: "Nuclear Winter: Global Consequences of Multiple Nuclear Explosion", in: *Science*, 222: 1283.

99. Thompson, S.L.; Alexandrov, V.; Stenchikov, G.L.; Schneider, S.H.; Covey, C.; Chervin, R. M., 1984: "Global Climatic Consequences of Nuclear War: Simulations with Three-Dimensional Models", in: *Ambio*, 13: 236.

100. Pittock, A.B.; Ackerman, T.P.; Crutzen, P.J.; MacCracken, M.C.; Shapiro, C.S.; Turco, R.P., 1986: *Environmental Consequences of Nuclear War, Volume 1: Physical and Atmospheric Effects, SCOPE 28* (New York: Wiley).

101. Harwell, M.A.; Hutchinson, T.C., 1985: *Environmental Consequences of Nuclear War, Volume 11: Ecological and Agricultural Effects, SCOPE 28* (New York: Wiley).

102. Lelieveld, J.; Crutzen, P.J., 1990: "Influences of Cloud Photochemical Processes on Tropospheric Ozone", in: *Nature*, 343: 227.

103. Dentener, F.J.; Crutzen, P.J., 1993: "Reaction of N_2O_5 on Tropospheric Aerosols: Impact on the Global Distributions of NO_x, O_3, and OH", in: *Journal of Geophysical Research*, 98 (DU): 7149.
104. Chatfield, R.; Crutzen, P.J., 1984: "Sulfur dioxide in Remote Oceanic Air: Cloud Transport of Reactive Precursors", in: *Journal of Geophysical Research*, 89 (D5): 711.
105. Lelieveld, J.; Crutzen, P.J., 1994: "Role of Deep Convection in the Ozone Budget of the Troposphere": *Science*, 264: 1759.
106. Barrie, L.A.; Bottenheim, J.W.; Schnell, R.C.; Crutzen, P.J.; Rasmussen, R.A., 1988: "Ozone Destruction and Photochemical Reactions at Polar Sunrise in the Lower Arctic Atmosphere", in: *Nature*, 334: 138.
107. Sander, R.; Crutzen, P.J., 1996: "Model study Indicating Halogen Activation and Ozone Destruction in Polluted Air Masses Transported to the Sea", in: *Journal of Geophysical Research*, 101: 9121.
108. Vogt, R.; Crutzen, P.J.; Sander, R., in press: "A New Mechanism for Bromine and Chlorine Release from Sea Salt Aerosol in the Unpolluted Marine Boundary Layer", in: *Nature*.
109. Crutzen, P.J., 1995: "Overview of Tropospheric Chemistry: Developments During the Past Quarter Century and a Look Ahead", in: *Faraday Discuss*, 100: 1.
110. Khalil, M.A.K.; Rasmussen, R.A., 1993: "Global Decrease of Atmospheric Carbon Monoxide", in: *Nature*, 370: 639.
111. Novelli, P.C.; Masario, K.A; Tans, P.P.; Lang, P.M.: "Recent Changes in Atmospheric Carbon Monoxide", in: *Science*, 263: 1587.
112. Zander, R.; Demoulin, P.; Ehhalt, D.H.; Schmidt, U.; Rinsland, C.P., 1989: "Secular Increases in the Total Vertical Abundances of Carbon Monoxide Above Central Europe since 1950", in: *Journal of Geophysical Research*, 94: 11021.
113. Zander, R.; Demoulin, P.; Ehhalt, D.H.; Schmidt, U., 1989: "Secular Increases of the Vertical Abundance of Methane Derived from IR Solar Spectra Recorded at the Jungfraujoch Station", in: *Journal of Geophysical Research*, 94: 11029.
114. Crutzen, P.J., 1994: "Global Tropospheric Chemistry", in: Moortgat, G.K.; et al. (Eds.): *Low-Temperature Chemistry of the Atmosphere* (Springer, Berlin): 467–498.
115. Crutzen, P.J., 1995: "Ozone in the Troposphere": in: Singh, H.B. (Ed.): *Composition. Chemistry and Climate of the Atmosphere* (Von Nostrand Reinhold, New York): 349–393.
116. Kley, D.; Crutzen, P.J.; Smit, H.G.J.; Vömel, H.; Oltmans, S.J.; Grassl, H.; Ramanathan, V., 1996: "Observations of Near-Zero Ozone Concentrations over the Convective Pacific: Effects on Air Chemistry", in: *Science*, 274: 230–233.
117. Ramanathan, V.; Subasilar, B.; Zhang, G.J.; Conant, W.; Cess, R.D.; Kiehl, J.T.; Grassl, H.; Shi, L., 1995: "Warm Pool Heat Budget and Shortwave Cloud Forcing: A Missing Physics?", in: *Science*, 267: 499.
118. Hausmann, M.; Platt, U., 1992: "Spectroscopic Measurement of Bromine Oxide and Ozone in the High Arctic during Polar Sunrise Experiment 1992", in: *Journal of Geophysical Research*, 99: 25399.
119. Oltmans, S.J.; Hofmann, D.J., 1995: "Increase in Lower-Stratospheric Water Vapour At A Mid-Latitude Northern Hemisphere Site from 1981 to 1994", in: *Nature*, 374: 146.
120. Singh, H.B.; O'Hara, D.; Herlth, D.; Bradshaw, J.D.; Sandholm, S.T.; Gregory, G.L.; Sachse, G.W.; Blake, D.R.; Crutzen, P.J.; Kanakidou, M., 1992: "Atmospheric Measurements of PAN and Other Organic Nitrates At High Latitudes: Possible Sources and Sinks", in: *Journal of Geophysical Research*, 97: 16511–16522.
121. Singh, H.B.; Herlth, D.; Zahnle, K.; O'Hara, D.; Bradshaw, J.; Sandholm, S.T.; Talbot, R.; Crutzen, P.J.; Kanakidou, M., 1992: "Relationship of PAN to Active and Total Odd Nitrogen At Northern High Latitudes: Possible Influence of Reservoir Species on NO_x and O_3", in: *Journal of Geophysical Research*, 97: 16523–16530.
122. Warneke, C.; Holzinger, R.; Hansel, A.; Jordan, A.; Lindinger, W.; Pöschl, U.; Williams, J.; Hoor, P.; Fischer, H.; Crutzen, P.J.; Scheeren, H.A.; Lelieveld, J., 2001: "Isoprene and its Oxidation Products Methyl Vinyl Ketone, Methacrolein, and Isoprene Related Peroxides

Measured Online over the Tropical Rain Forest of Surinam in March 1998", in: *The Journal of Atmospheric Chemistry*, 38: 167–185.
123. Williams, J.; Fischer, H.; Hoor, P.; Pöschl, U.; Crutzen, P.J.; Andreae, M.O.; Lelieveld, J., 2001: "The Influence of the Tropical Rainforest on Atmospheric CO and CO_2 As Measured by Aircraft over Surinam, South America", in: *Chemosphere*, 3: 157–170.
124. Williams, J.; Pöschl, U.; Crutzen, P.J.; Hansel, A.; Holzinger, R.; Warneke, C.; Lindinger, W.; Lelieveld, J., 2001: "An Atmospheric Chemistry Interpretation of Mass Scans Obtained from a Proton Transfer Mass Spectrometer Flown over the Tropical Rainforest of Surinam", in: *The Journal of Atmospheric Chemistry*, 38: 133–166.
125. Crutzen, P.J.; Ramanathan, V., 2001: "Foreword, INDOEX special issue", in: *Journal of Geophysical Research*, 106 (D22): 28369–28370.
126. Lelieveld, J.; Crutzen, P.J.; Ramanathan, V.; Andreae, M.O.; Brenninkmeijer, C.A.M.; Campos, T.; Cass, G.R.; Dickerson, R.R.; Fischer, H.; de Gouw, J.A.; Hansel A.; Jefferson, A.; Kley, D.; de Laat, A.T.J.; Lal, S.; Lawrence, M.G.; Lobert, J.M.; Mayol-Bracero, O.L.; Mitra, A.P.; Novakov, T.; Oltmans, S.J.; Prather, K.A.; Reiner, T.; Rodhe, H.; Scheeren, H. A.; Sikka, D.; Williams, J., 2001: "The Indian Ocean Experiment: Widespread Air Pollution from South to Southeast Asia", in: *Science*, 291: 1031–1036.
127. Ramanathan, V.; Crutzen, P.J.; Mitra, A.P.; Sikka, D., 2002: "The Indian Ocean Experiment and the Asian Brown Cloud", in: *Current Science,* 83: 947–955.
128. Ramanathan, V.; Crutzen, P.J., 2003: "New Directions: Atmospheric Brown "Clouds"", in: *Atmospheric Environment*, 37: 4033–4035.
129. Lelieveld, J.; Berresheim, H.; Borrmann, S.; Crutzen, P.J.; Dentener, F.J.; Fischer, H.; Feichter, J.; Flatau, P.F.; Heland, J.; Holzinger, R.; Korrmann, R.; Lawrence, M.G.; Levin, Z.; Markowicz, K.M.; Mihalopoulos, N.; Minikin, A.; Ramanathan, V.; de Reus, M.; Roelofs, G.J.; Scheeren, H.A.; Sciare, J.; Schlager, H.; Schultz, M.; Siegmund, P.; Steil, B.; Stephanou, E.; Stier, P.; Schlager, H.; Schultz, M.; Siegmund, P.; Steil, B.; Stephanou, E.G.; Stier, P.; Traub, M.; Warneke, C.; Williams, J.; Ziereis, H., 2002: " Global Air Pollution Crossroads over the Mediterranean", in: *Science*, 298: 794–799.
130. Vogt, R.; Crutzen, P.J.; Sander, R., 1996: "A Mechanism for Halogen Release Form Sea-Salt Aerosol in the Remote Marine Boundary Layer", in: *Nature*, 383: 327–330.
131. Sander, R.; Vogt, R.; Harris, G.W.; Crutzen, P.J., 1997: "Modeling the Chemistry of Ozone, Halogen Compounds, and Hydrocarbons in the Arctic Troposphere During Spring", in: *Tellus*, 49B: 522–532.
132. Dickerson, R.R.; Rhoads, K.P.; Carsey, T.P.; Oltmans, S.J.; Burrows, J.P.; Crutzen, P.J., 1999: "Ozone in the Remote Marine Boundary Layer: A Possible Role for Halogens", in: *Journal of Geophysical Research*, 104(D17),21: 385–21, 395.
133. Vogt, R; Sander, R.; von Glasow, R.; Crutzen, P.J., 1999: "Iodine Chemisty and its Role in Halogen Activation and Ozone Loss in the Marine Boundary Layer: A Model Study", in: *The Journal of Atmospheric Chemistry*, 32: 375–395.
134. von Glasow, R.; Sander, R.; Bott, A.; Crutzen, P.J., 2002: "Modeling Halogen Chemistry in the Marine Boundary Layer. 1. Cloud-free MBL", in: *Journal of Geophysical Research*, 107D: 4341, doi:10.1029/2001JD000942.
135. von Glasow, R.; Sander, R.; Bott, A.; Crutzen, P.J., 2002: "Modeling Halogen Chemistry in the Marine Boundary Layer. 2. Interactions with Sulfur and the Cloud-Covered MBL", in: *Journal of Geophysical Research*, 107D: 4323, doi:10.1029/2001JD000943.
136. von Glasow, R.; Crutzen, P.J., 2007: "Tropospheric Halogen Chemistry", in: Holland, H.D.; Turekian, K.K. (Eds.): *Treatise on Geochemistry Update,* 4,02: 1–67.
137. Sander, R.; Keene, W.C.; Pszenny, A.A.P.; Arimoto, R.; Ayers, G.P.; Baboukas, E.; Cainey, J.M.; Crutzen, P.J.; Duce, R.A.; Hönninger, G.; Huebert, B.J.; Maenhaut, W.; Mihalopoulos, N.; Turekian, V.C.; Van Dingenen, R., 2003: "Inorganic Bromine in the Marine Boundary Layer: A Critical Review", in: *Atmospheric Chemistry and Physics*, 3: 1301–1336.

138. von Glasow, R.; Crutzen, P.J., 2003: "Tropospheric Halogen Chemistry", in: Holland, HD; Turekian, KK; Keeling, RF (Eds.): *Treatise on Geochemistry (Elsevier Pergamon)*: 21–64.
139. Platt, U., 2000: "Reactive Halogen Species in the mid-Latitude Troposphere - Recent discoveries", in: *Water, Air, Soil & Pollution*, 123: 229–224.
140. Read, K.A.; Mahajan, A.S.; Carpenter, L.J.; Evans, M.J.; Faria, B.V.E.; Heard, D.E.; Hopkins, J.R.; Lee, J.D.; Moller, S.J.; Lewis, A.C.; Mendes, L.; McQuaid, J.B.; Oetjen, H.; Saiz-Lopez, A.; Pilling, M.J.; Plane, J.M.C., 2008: "Extensive Halogen-Mediated Ozone Destruction over the Tropical Atlantic Ocean", in: *Nature*, 453: 1232–1235.

Chapter 2
Complete Bibliography of the Writings of Paul J. Crutzen (1965–2015)

Paul J. Crutzen

2.1 Books

1. Crutzen, P.J.; Hahn, J., 1985: *Schwarzer Himmel* (S. Fischer Verlag): 240 pp.
2. Pittock, A.B.; Ackerman, T.P.; Crutzen, P.J.; MacCracken, M.C.; Shapiro, C.S.; Turco, R.P., 1986: *Environmental Consequences of Nuclear War, SCOPE 28, Volume I: Physical and Atmospheric Effects* (Chichester: Wiley): 359 pp; 2nd edition 1989.
3. Crutzen, P.J.; Müller, M., 1989: *Das Ende des blauen Planeten?* (C.H. Beck Verlag): 271 pp.
4. Graedel, T.E.; Crutzen, P.J., 1993: *Atmospheric Change: An Earth System Perspective* (New York: W.H. Freeman): 446 pp.
5. Graedel, T.E.; Crutzen, P.J., 1995: *Atmosphere, Climate, and Change* (New York: W.H. Freeman): 208 pp.
6. Graedel, T.E.; Crutzen, P.J., 1996: *Atmosphäre im Wandel. Die empfindliche Lufthülle unseres Planeten* (Heidelberg: Spektrum Akademischer Verlag): 221 pp.
7. Enquete Commission "Preventive Measures to Protect the Earth's Atmosphere", 1989: *Interim Report: Protecting the Earth's Atmosphere: An International Challenge* (Bonn: Deutscher Bundestag, Referat Öffentlichkeitsarbeit): 592 pp.

© The Author(s) 2016
P.J. Crutzen and H.G. Brauch (eds.), *Paul J. Crutzen: A Pioneer on Atmospheric Chemistry and Climate Change in the Anthropocene*, Nobel Laureates 50, DOI 10.1007/978-3-319-27460-7_2

8. Enquete Commission "Preventive Measures to Protect the Earth's Atmosphere", 1990: *Protecting the Tropical Forests: A High-Priority International Task* (Bonn: Deutscher Bundestag, Referat Öffentlichkeitsarbeit): 968 pp.

9. Enquete Commission "Preventive Measures to Protect the Earth's Atmosphere", 1991: *Protecting the Earth: A Status Report with Recommendations for a New Energy Policy* (Bonn: Deutscher Bundestag, Referat Öffentlichkeitsarbeit): 2 Volumes.

10. Crutzen, P.J.; Gerard, J.-C.; Zander, R. (Eds.), 1989: "Our Changing Atmosphere", Proceedings of the 28th Liège International Astrophysical Colloquium June 26–30, 1989 (Belgium: Université de Liège, Cointe-Ougree): 534 pp.

11. Crutzen, P.J.; Goldammer, J.G., 1993: *Fire in the Environment: The Ecological, Atmospheric, and Climatic Importance of Vegetation Fires.* Dahlem Konferenz (15–20 March 1992, Berlin), ES13 (Chichester: Wiley): 400 pp.

12. Graedel, T.E.; Crutzen, P.J., 1994: *Chemie der Atmosphäre. Bedeutung für Klima und Umwelt* (Heidelberg: Spektrum Akademischer Verlag): 511 pp.

13. Crutzen, Paul J.; Ramanathan, Veerabhadran (Eds.), 1996: *Clouds, Chemistry and Climate, Nato ASI Subseries I*

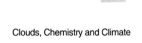

Clouds, Chemistry and Climate

Edited by
Paul J. Crutzen and Veerabhadran Ramanathan

NATO ASI Series

Series I: Global Environmental Change, Vol. 35

14. Crutzen, P.J.; Komen, G.; Verbeek, K.; van Dorland, R., 2004: *Veranderingen in het klimaat* (De Bilt, The Netherlands: Koninklijk Nederlands Meteorologisch Instituut): 16 pp (in Dutch); at: http://www.dbnl.org/basisbibliotheek/.

15. Schellnhuber, H.J.; Crutzen, P.J.; Clark, W.C.; Claussen, M.; Held, H. (Eds.), 2004: *Earth System Analysis for Sustainability.* Dahlem Workshop Reports (MIT Press).

2.2   Special Publications

1. Crutzen, P.J., 1996: "Mein Leben mit O_3, NO_x und anderen YZO_x-Verbindungen (Nobel-Vortrag)", in: *Angewandte Chemie*, 108: 1878–1898.

2. Crutzen, P.J., 1996: "My Life with O_3, NO_x, and Other YZO_x Compounds (Nobel Lecture)", in: *Angewandte Chemie International Edition in English*, 35: 1758–1777.

3. Crutzen, P.J., 1997: "Die Beobachtung atmosphärisch-chemischer Veränderungen: Ursachen und Folgen für Umwelt und Klima. (Festvortrag

anläßlich der Hauptversammlung der Max-Planck-Gesellschaft in Bremen am 6. Juni 1997)", in: *Max-Planck-Gesellschft. Jahrbuch 1997. Ed. Generalverwaltung der Max-Planck-Gesellschaft München* (Göttingen: Van den Hoek & Ruprecht): 51–71.
4. Crutzen, P.J., 1998: "The BULLETIN Interviews. Professor Paul Josef Crutzen", in: *WMO Bulletin*, 47,2: 3–15.
5. Crutzen, P.J., 1999: "The Nuclear Winter", in: Kornacki, J.; Budzynski, R.; Kotomycki, J. (Eds.): *Proceedings of the international conference "The discovery of Polonium and Radium", Warsaw, Poland, 17–20 September 1998, Warsawska Drukornia Nankowa PAN, 1998, 85–104*.
6. Crutzen, P.J., 2004: *How I Became a Scientist?* (Trieste, Italy: The Abdus Salam Centre for Theoretical Physics).
7. Crutzen, P.J., 2004: "A Late Change to the Programme—How an Engineer Became Hooked on Atmospheric Chemistry", in: *Nature*, 429: 349.
8. Crutzen, P.J., 2005: *Benvenuti nell'Antropocene!* (Milano, Italy: Arnoldo Modadori Editore S.p.A.) (in Italian).

2.3   Journal Articles (Refereed)

1. Blankenship, J.R.; Crutzen, P.J., 1965: "A Photochemical Model for the Space-Time Variations of the Oxygen Allotropes in the 20 to 100 km Layer", in: *Tellus*, 18: 160–175.
2. Crutzen, P.J., 1969: "Determination of Parameters Appearing in the "Dry" and the "Wet" Photochemical Theories for Ozone in the Stratos", in: *Tellus*, 21: 368–388.
3. Crutzen, P.J., 1969: "Determination of Parameters Appearing in the Oxygen-Hydrogen Atmosphere", in: *Annals of Géophysics*, 25: 275–279.
4. Crutzen, P.J., 1970: "The Influence of Nitrogen Oxides on the Atmospheric Ozone Content", in: *Quarterly Journal of the Royal Meteorological Society*, 96: 320–325.
5. Crutzen, P.J., 1970: "Comments on Absorption and Emission by Carbon Dioxide in the Mesosphere", in: *Quarterly Journal of the Royal Meteorological Society*, 96: 767–769.
6. Crutzen, P.J., 1971: "Energy Conversions and Mean Vertical Motions in the High Latitude Summer Mesosphere and Lower Thermosphere", in: Fiocco, G. (Ed.): *Mesospheric Models and Related Experiments* (Dordrecht, Holland: D. Reidel Publ. Co.): 78–88.
7. Crutzen, P.J.; Jones, I.T.N.; Wayne, R.P., 1971: "Calculation of O_2 ($1\Delta g$) in the Atmosphere Using New Laboratory Data", in: *Journal of Geophysical Research*, 76: 1490–1497.

8. Crutzen, P.J., 1971: "Ozone Production Rates in an Oxygen-Hydrogen-Nitrogen Oxide Atmosphere", in: *Journal of Geophysical Research*, 76: 7311–7327.

9. Crutzen, P.J., 1972: "SST's—A Threat to the Earth's Ozone Shield", in: *Ambio*, 1: 41–51.

10. Crutzen, P.J., 1973: "A Discussion of the Chemistry of Some Minor Constituents in the Stratosphere and Troposphere", in: *Pure and Applied Geophysics*, 106–108: 1385–1399.

11. Crutzen, P.J., 1973: "Gas-Phase Nitrogen and Methane Chemistry in the Atmosphere", in: McCormac, B.M. (Ed.): *Physics and Chemistry of Upper Atmospheres* (Dordrecht, Holland: Reidel): 110–124.

12. Crutzen, P.J., 1974: "A Review of Upper Atmospheric Photochemistry", in: *Canadian Journal of Chemistry*, 52: 1569–1581.

13. Crutzen, P.J., 1974: "Estimates of Possible Future Ozone Reductions from Continued Use of Fluorochloromethanes (CF_2Cl_2, $CFCl_3$)", in: *Geophysical Research Letters*, 1: 205–208.

14. Crutzen, P.J., 1974: "Estimates of Possible Variations in Total Ozone due to Natural Causes and Human Activities", in: *Ambio*, 3: 201–210.

15. Crutzen, P.J., 1974: "Photochemical Reactions Initiated by and Influencing Ozone in Unpolluted Tropospheric Air:, in: *Tellus*, 26: 48–57.

16. Cadle, R.D.; Crutzen, P.J.; Ehhalt, D.H., 1975: "Heterogeneous Chemical Reactions in the Stratosphere", in: *Journal of Geophysical Research*, 80: 3381–3385.

17. Crutzen, P.J.; Isaksen, I.S.A.; Reid, G.C., 1975: "Solar Proton Events: Stratospheric Sources of Nitric Oxide", in: *Science*, 189: 457–459.

18. Johnston, H.S.; Garvin, D.; Corrin, M.L.; Crutzen, P.J.; Cvetanovic, R.J.; Davis, D.D.; Domalski, E.S.; Ferguson, E.E.; Hampson, R.F.; Hudson, R.D.; Kieffer, L.J.; Schiff, H.I.; Taylor, R.L.; Wagman, D.D.; Watson, R.T., 1975: *Chemistry in the Stratosphere, Chapter 5, CIAP Monograph 1. The Natural Stratosphere of 1974*, DOT-TST-75-51, U.S. Department of Transportation, Climate Impact Assessment Program.

19. Schmeltekopf, A.L.; Goldan, P.D.; Henderson, W.R.; Harrop, W.J.; Thompson, T.L.; Fehsenfeld, F.C.; Schiff, H.I.; Crutzen, P.J.; Isaksen, I.S.A.; Ferguson, E.E., 1975: "Measurements of Stratospheric $CFCl_3$, CF_2Cl_2, and N_2O", in: *Geophysical Research Letters*, 2: 393–396.

20. Zerefos, C.S.; Crutzen, P.J., 1975: "Stratospheric Thickness Variations over the Northern Hemisphere and Their Possible Relation to Solar Activity", in: *Journal of Geophysical Research*, 80: 5041–5043.

21. Crutzen, P.J., 1976: "Upper Limits on Atmospheric Ozone Reductions Following Increased Application of Fixed Nitrogen to the Soil", in: *Geophysical Research Letters*, 3: 169–172.

22. Crutzen, P.J., 1976: "The Possible Importance of CSO for the Sulfate Layer of the Stratosphere", in: *Geophysical Research Letters*, 3: 73–76.

23. Crutzen, P.J.; Reid, G.C., 1976: "Comments on Biotic Extinctions by Solar Flares", in: *Nature*, 263: 259.

24. Fehsenfeld, F.C.; Crutzen, P.J.; Schmeltekopf, A.L.; Howard, C.J.; Albritton, D.L.; Ferguson, E.E.; Davidson, J.A.; Schiff, H.I., 1976: "Ion Chemistry of Chlorine Compounds in the Troposphere and Stratosphere", in: *Journal of Geophysical Research*, 81: 4454–4460.

25. Reid, G.C.; Isaksen, I.S.A.; Holzer, T.E.; Crutzen, P.J., 1976: "Influence of Ancient Solar Proton Events on the Evolution of Life", in: *Nature*, 259: 177–179.

26. Crutzen, P.J.; Ehhalt, D.H., 1977: "Effects of Nitrogen Fertilizers and Combustion on the Stratospheric Ozone Layer", in: *Ambio*, 6: 1–3, 112–117.

27. Crutzen, P.J.; Fishman, J., 1977: "Average Concentrations of OH in the Troposphere, and the Budgets of CH_4, CO, H_2 and CH_3CCl_3", in: *Geophysical Research Letters*, 4: 321–324.

28. Fishman, J.; Crutzen, P.J., 1977: "A Numerical Study of Tropospheric Photochemistry Using a One-Dimensional Model", in: *Journal of Geophysical Research*, 82: 5897–5906.

29. Heath, D.F.; Krueger, A.J.; Crutzen, P.J., 1977: "Solar Proton Event: Influence on Stratospheric Ozone", in: *Science*, 197: 886–889.

30. Hidalgo, H.; Crutzen, P.J., 1977: "The Tropospheric and Stratospheric Composition Perturbed by NO_x Emissions of High Altitude Aircraft", in: *Journal of Geophysical Research*, 82: 5833–5866.

31. Isaksen, I.S.A.; Crutzen, P.J., 1977: "Uncertainties in Calculated Hydroxyl Radical Densities in the Troposphere and Stratosphere", in: *Geophysica Norvegica*, 31,4: 1–10.

32. Isaksen, I.S.A.; Midtbøe, K.H.; Sunde, J.; Crutzen, P.J., 1977: "A Simplified Method to Include Molecular Scattering and Reflection in Calculations of Photon Fluxes and Photo-dissociation Rates", in: *Geophysica Norvegica*, 31: 11–26.

33. Schmeltekopf, A.L.; Albritton, D.L.; Crutzen, P.J.; Goldan, D.; Harrop, W.J.; Henderson, W.R.; McAfee, J.R.; McFarland, M.; Schiff, H.I.; Thompson, T.L.; Hofmann, D.J.; Kjome, N.T., 1977: "Stratospheric Nitrous Oxide Altitude Profiles at Various Latitudes", in: *Journal of the Atmospheric Sciences*, 34: 729–736.

34. Crutzen, P.J.; Howard, C.J., 1978: "The Effect of the HO_2 + NO Reaction Rate Constant on One-Dimensional Model Calculations of Stratospheric Ozone Perturbations", in: *Applied Geophysics*, 116: 487–510.

35. Crutzen, P.J.; Isaksen, I.S.A.; McAfee, J.R., 1978: "The Impact of the Chlorocarbon Industry on the Ozone Layer", in: *Journal of Geophysical Research*, 83: 345–363.

36. Fishman, J.; Crutzen, P.J., 1978: "The Origin of Ozone in the Troposphere", in: *Nature*, 274: 855–858.

37. Reid, G.C.; McAfee, J.R.; Crutzen, P.J., 1978: "Effects of Intense Stratospheric Ionization Events", in: *Nature*, 257: 489–492.

38. Zimmerman, P.R.; Chatfield, R.B.; Fishman, J.; Crutzen, P.J.; Hanst, P.L., 1978: "Estimates on the Production of CO and H_2 from the Oxidation of

Hydrocarbon Emissions from Vegetation", in: *Geophysical Research Letters*, 5: 679–682.

39. Crutzen, P.J., 1979: "The Role of NO and NO_2 in the Chemistry of the Troposphere and Stratosphere", in: *Annual Review of Earth and Planetary Sciences*, 7: 443–472.

40. Crutzen, P.J., 1979: "Chlorofluoromethanes: Threats to the Ozone Layer", in: *Reviews of Geophysics and Space Physics*, 17: 1824–1832.

41. Crutzen, P.J.; Heidt, L.E.; Krasnec, J.P.; Pollock, W.H.; Seiler, W., 1979: "Biomass Burning as a Source of Atmospheric Gases CO, H_2, N_2O, NO, CH_3Cl and COS", in: *Nature*, 282: 253–256.

42. Dickerson, R.R.; Stedman, D.H.; Chameides, W.L.; Crutzen, P.J.; Fishman, J., 1979: "Actinometric Measurements and Theoretical Calculations of $J(O_3)$, the Rate of Photolysis of Ozone to $O(^1D)$", in: *Geophysical Research Letters*, 6: 833–836.

43. Fishman, J.; Ramanathan, V.; Crutzen, P.J.; Liu, S.C., 1979: "Tropospheric Ozone and Climate", in: *Nature*, 282: 818–820.

44. Fishman, J.; Solomon, S.; Crutzen, P.J., 1979: "Observational and Theoretical Evidence in Support of a Significant In Situ Photochemical Source of Tropospheric Ozone", in: *Tellus*, 31: 432–446.

45. Berg, W.W.; Crutzen, P.J.; Grahek, F.E.; Gitlin, S.N.; Sedlacek, W.A., 1980: "First Measurements of Total Chlorine and Bromine in the Lower Stratosphere", in: *Geophysical Research Letters*, 7: 937–940

46. Crutzen, P.J.; Solomon, S., 1980: "Response of Mesospheric Ozone to Particle Precipitation", in: *Planetary and Space Science*, 28: 1147–1153.

47. Heidt, L.E.; Krasnec, J.P.; Lueb, R.A.; Pollock, W.H.; Henry, B.E.; Crutzen, P.J., 1980: "Latitudinal Distributions of CO and CH_4 over the Pacific", in: *Journal of Geophysical Research*, 85: 7329–7336.

48. Seiler, W.; Crutzen, P.J., 1980: "Estimates of Gross and Net Fluxes of Carbon Between the Biosphere and the Atmosphere from Biomass Burning", in: *Climatic Change*, 2: 207–247.

49. Thomas, G.E.; Barth, C.A.; Hansen, E.R.; Hord, C.W.; Lawrence, G.M.; Mount, G.H.; Rottman, G.J.; Rusch, D.W.; Stewart, A.I.; Thomas, R.J.; London, J.; Bailey, P.L.; Crutzen, P.J.; Dickinson, R.E.; Gille, J.C.; Liu, S.C.; Noxon, J.F.; Farmer, C.B., 1980: "Scientific Objectives of the Solar Mesosphere Explorer Mission", in: *Pure and Applied Geophysics*, 118, 591–615.

50. Rodhe, H.; Crutzen, P.J.; Vanderpol, A., 1981: "Formation of Sulfuric Acid in the Atmosphere During Long Range Transport", in: *Tellus*, 33: 132–141.

51. Rusch, D.W.; Gérard, J.C.; Solomon, S.; Crutzen, P.J.; Reid, G.C., 1981: "The Effects of Particle Precipitation Events on the Neutral and Ion Chemistry of the Middle Atmosphere—I. Odd Nitrogen", in: *Planetary and Space Science*, 29: 767–774.

52. Solomon, S.; Crutzen, P.J., 1981: "Analysis of the August 1972 Solar Proton Event Including Chlorine Chemistry", in: *Journal of Geophysical Research*, 86: 1140–1146.

53. Solomon, S.; Rusch, D.W.; Gérard, J.C.; Reid, G.C.; Crutzen, P.J., 1981: "The Effect of Particle Precipitation Events on the Neutral and Ion Chemistry of the Middle Atmosphere—II. Odd Hydrogen", in: *Planetary and Space Science*, 29: 885–892.

54. Baulch, D.L.; Cox, R.A.; Crutzen, P.J.; Hampson Jr., R.F.; Kerr, J.A.; Troe, J., 1982: "Evaluated Kinetic and Photochemical Data for Atmospheric Chemistry: Supplement I", in: *Journal of Physics and Chemistry Reference Data*, 11: 327–496.

55. Crutzen, P.J., 1982: "The Global Distribution of Hydroxyl", in: Goldberg, E.D. (Ed.): *Atmospheric Chemistry*. Dahlem Konferenzen 1982 (Berlin, Heidelberg, New York: Springer): 313–328.

56. Crutzen, P.J.; Birks, J.W., 1982: "The Atmosphere After a Nuclear War: Twilight at Noon", in: *Ambio*, 2&3: 114–125.

57. Hahn, J.; Crutzen, P.J., 1982: "The Role of Fixed Nitrogen in Atmospheric Photochemistry", in: *Philosophical Transactions of the Royal Society of London*, B 296: 521–541.

58. Solomon, S.; Crutzen, P.J.; Roble, R.G., 1982: "Photochemical Coupling Between the Thermosphere and the Lower Atmosphere. 1. Odd Nitrogen from 50–120 km", in: *Journal of Geophysical Research*, 87: 7206–7220.

59. Solomon, S.; Ferguson, E.E.; Fahey, D.W.; Crutzen, P.J., 1982: "On the Chemistry of H_2O, H_2 and Meteoritic Ions in the Mesosphere and Lower Thermosphere", in: *Planetary and Space Science*, 30: 1117–1126.

60. Solomon, S.; Reid, G.C.; Roble, R.G.; Crutzen, P.J., 1982: "Photochemical Coupling Between the Thermosphere and the Lower Atmosphere. 2. D-Region Ion Chemistry and the Winter Anomaly", in: *Journal of Geophysical Research*, 87: 7221–7227.

61. Zimmerman, P.R.; Greenberg, J.P.; Wandiga, S.O.; Crutzen, P.J., 1982: "Termites: A Potentially Large Source of Atmospheric Methane, Carbon Dioxide, and Molecular Hydrogen", in: *Science*, 218: 563–565.

62. Bolin, B.; Crutzen, P.J.; Vitousek, P.M.; Woodmansee, R.G.; Goldberg, E.D.; Cook, R.B., 1983: "Interactions of Biochemical Cycles", in: Bolin, B.; Cook, R.B. (Eds.): *The Major Biochemical Cycles and Their Interactions*, SCOPE 21 (Chichester: Wiley): 1–40.

63. Crutzen, P.J., 1983: "Atmospheric Interactions—Homogeneous Gas Reactions of C, N, and S Containing Compounds", in: Bolin, B.; Cook, R.B. (Eds.): *The Major Biochemical Cycles and Their Interactions*, SCOPE 21 (Chichester: Wiley): 67–114.

64. Crutzen, P.J.; Gidel, L.T., 1983: "A Two-Dimensional Photochemical Model of the Atmosphere. 2. The Tropospheric Budgets of the Anthropogenic Chlorocarbons, CO, CH_4, CH_3Cl and the Effect of Various NO_x Sources on Tropospheric Ozone", in: *Journal of Geophysical Research*, 88: 6641–6661.

65. Crutzen, P.J.; Schmailzl, U., 1983: "Chemical Budgets of the Stratosphere", in: *Planetary and Space Science*, 31: 1009–1032.

66. Frederick, J.E.; Abrams, R.B.; Crutzen, P.J., 1983: "The Delta Band Dissociation of Nitric Oxide: A Potential Mechanism for Coupling Thermospheric Variations to the Mesosphere and Stratosphere", in: *Journal of Geophysical Research*, 88: 3829–3835.
67. Gidel, L.T.; Crutzen, P.J.; Fishman, J., 1983: "A Two-Dimensional Photochemical Model of the Atmosphere. 1: Chlorocarbon Emissions and Their Effect on Stratospheric Ozone", in: *Journal of Geophysical Research*, 88: 6622–6640.
68. Chatfield, R.B.; Crutzen, P.J., 1984: "Sulfur Dioxide in Remote Oceanic Air: Cloud Transport of Reactive Precursors", in: *Journal of Geophysical Research*, 89,D5: 7111–7132.
69. Crutzen, P.J.; Galbally, I.E.; Brühl, C., 1984: "Atmospheric Effects from Postnuclear Fires", in: *Climatic Change*, 6: 323–364.
70. Crutzen, P.J.; Coffey, M.T.; Delany, A.C.; Greenberg, J.; Haagenson, P.; Heidt, L.; Heidt, R.; Lueb, L.; Mankin, W.G.; Pollock, W.; Seiler, W.; Wartburg, A.; Zimmerman, P., 1985: "Observations of Air Composition in Brazil Between the Equator and 20°S During the Dry Season", in: *Acta Amazonica, Manaus*, 15,1–2: 77–119.
71. Crutzen, P.J.; Delany, A.C.; Greenberg, J.; Haagenson, P.; Heidt, L.; Lueb, R.; Pollock, W.; Seiler, W.; Wartburg, A.; Zimmerman, P., 1985: "Tropospheric Chemical Composition Measurements in Brazil During the Dry Season", in *Journal of Atmospheric Chemistry*, 2: 233–256.
72. Crutzen, P.J.; Whelpdale, D.M.; Kley, D.; Barrie, L.A., 1985: "The Cycling of Sulfur and Nitrogen in the Remote Atmosphere", in: Galloway, J.N.; Charlson, R.J.; Andreae, M.O.; Rodhe, H. (Eds.): *The Biogeochemical Cycling of Sulfur and Nitrogen in the Remote Atmosphere*, NATO ASI Series C 158 (Dordrecht, Holland: Reidel): 203–212.
73. Delany, A.C.; Haagenson, P.; Walters, S.; Wartburg, A.F.; Crutzen, P.J., 1985: "Photochemically Produced Ozone in the Emission of Large Scale Tropical Vegetation Fires", in: *Journal of Geophysical Research*, 90,D1: 2425–2429.
74. Crutzen, P.J.; Arnold, F., 1986: "Nitric Acid Cloud Formation in the Cold Antarctic Stratosphere: A Major Cause for the Springtime "ozone hole"", in: *Nature*, 324: 651–655.
75. Crutzen, P.J.; Aselmann, I.; Seiler, W., 1986: "Methane Production by Domestic Animals, Wild Ruminants, Other Herbivorous Fauna, and Humans", in: *Tellus*, 38B: 271–284.
76. Crutzen, P.J.; Graedel, T.E., 1986: "The Role of Atmospheric Chemistry in Environment-Development Interactions", in: Clark, W.C.; Munn, R.E. (Eds.): *Sustainable Development of the Environment* (Cambridge: Cambridge University Press): 213–251.
77. Pittock, A.B.; Ackerman, T.P.; Crutzen, P.J.; MacCracken, J.C.; Shapiro, C.S.; Turco, R.P., 1986: "Scenarios for a Nuclear Exchange", in: *Environmental Consequences of Nuclear War Volume 1: Physical and Atmospheric Effects*, SCOPE (New York: Wiley): 25–37.

78. Pittock, A.B.; Ackerman, T.P.; Crutzen, P.J.; MacCracken, J.C.; Shapiro, C.S.; Turco, R.P., 1986: "Atmospheric Processes", in: *Environmental Consequences of Nuclear War Volume 1: Physical and Atmospheric Effects*, SCOPE (New York: Wiley): 105–147.

79. Bingemer, H.G.; Crutzen, P.J., 1987: "The Production of Methane from Solid Wastes", in: *Journal of Geophysical Research*, 92,D2: 2181–2187.

80. Crutzen, P.J., 1987: "Role of the Tropics in the Atmospheric Chemistry", in: Dickinson, R. (Ed.): *Geophysiology of the Amazon* (Chichester–New York: Wiley): 107–131.

81. Crutzen, P.J., 1987: "Acid Rain at the K/T Boundary", in: *Nature*, 330: 108–109.

82. Barrie, L.A.; Bottenheim, J.W.; Schnell, R.C.; Crutzen, P.J.; Rasmussen, R.A., 1988: "Ozone Destruction and Photochemical Reactions at Polar Sunrise in the Lower Arctic Atmosphere", in: *Nature*, 334: 138–141.

83. Brühl, C.; Crutzen, P.J., 1988: "Scenarios of Possible Changes in Atmospheric Temperatures and Ozone Concentrations due to Man's Activities as Estimated with a One-Dimensional Coupled Photochemical Climate Model", in: *Climate Dynamics*, 2: 173–203.

84. Crutzen, P.J., 1988: "Tropospheric Ozone: An Overview", in: Isaksen, I.S.A. (Ed.): *Tropospheric Ozone* (Dordrecht: Reidel): 3–32.

85. Crutzen, P.J., 1988: "Variability in Atmospheric-Chemical Systems", in: Rosswall, T.; Woodmansee, R.G.; Risser, P.G. (Eds.): *Scales and Global Change*, SCOPE 35 (Chichester: Wiley): 81–108.

86. Crutzen, P.J.; Brühl, C.; Schmailzl, U.; Arnold, F., 1988: "Nitric Acid Haze Formation in the Lower Stratosphere: A Major Contribution Factor to the Development of the Antarctic "Ozone Hole"", in: McCormick, M.P.; Hobbs, P.V. (Eds.): *Aerosols and Climate* (Hampton, Virginia, USA: A. Deepak Publ.): 287–304.

87. Hao, W.M.; Scharffe, D.; Sanhueza, E.; Crutzen, P.J., 1988: "Production of N_2O, CH_4, and CO_2 from Soils in the Tropical Savanna During the Dry Season", in: *Journal of Atmospheric Chemistry*, 7: 93–105.

88. Horowitz, A.; von Helden, G.; Schneider, W.; Crutzen, P.J.; Moortgat, G.K., 1988: "Ozone Generation in the 214 nm Photolysis of Oxygen at 25 °C", in: *Journal of Physics and Chemistry*, 92, 4956-4960.

89. Liu, S.C.; Cox, R.A.; Crutzen, P.J.; Ehhalt, D.H.; Guicherit, R.; Hofzumahaus, A.; Kley, D.; Penkett, S.A.; Phillips, L.F.; Poppe, D.; Rowland, F.S., 1988: "Group Report: Oxidizing Capacity of the Atmosphere", in: Rowland, F.S.; Isaksen, I.S.A. (Eds.): *The Changing Atmosphere* (Chichester: Wiley): 219–232.

90. Wilson, S.R.; Crutzen, P.J.; Schuster, G.; Griffith, D.W.T.; Helas, G., 1988: "Phosgene Measurements in the Upper Troposphere and Lower Stratosphere", in: *Nature*, 334: 689–691.

91. Aselmann, I.; Crutzen, P.J., 1989: "Global Distribution of Natural Freshwater Wetlands and Rice Paddies, Their Net Primary Productivity, Seasonality and

Possible Methane Emissions", in: *Journal of Atmospheric Chemistry*, 8: 307–358.

92. Brühl, C.; Crutzen, P.J., 1989: "On the Disproportionate Role of Tropospheric Ozone as a Filter Against Solar UV-B Radiation", in: *Geophysical Research Letters*, 16: 703–706.

93. Crutzen, P.J.; Brühl, C., 1989: "The Impact of Observed Changes in Atmospheric Composition on Global Atmospheric Chemistry and Climate", in: Oeschger, H.; Langway, C.C. (Eds.): *The Environmental Record in Glaciers and Ice Sheets*, Dahlem Konferenzen 1988 (Chichester: Wiley): 249–266.

94. Pearman, G.I.; Charlson, R.J.; Class, T.; Clausen, H.B.; Crutzen, P.J.; Hughes, T.; Peel, D.A.; Rahn, K.A.; Rudolph, J.; Siegenthaler, U.; Zardini, D. S., 1989: "Group Report: What Anthropogenic Impacts are Recorded in Glaciers?", in: Oeschger, H.; Langway, C.C. (Eds.): *Dahlem Workshop Reports: The Environmental Record in Glaciers and Ice Sheets*, Dahlem Konferenzen (Chichester: Wiley): 269–286.

95. Robertson, R.P.; Andreae, M.O.; Bingemer, H.G.; Crutzen, P.J.; Delmas, R.A.; Duizer, J.H.; Fung, I.; Harriss, R.C.; Kanakidou, M.; Keller, M.; Melillo, J.M.; Zavarzin, G.A., 1989: "Group Report: Trace Gas Exchange and the Chemical and Physical Climate: Critical Interactions", in: Andreae, M.O.; Schimel, D.S. (Eds.): *Dahlem Workshop Reports: Exchange of Trace Gases between Terrestrial Ecosystems and the Atmosphere*, Life Sciences Research Report 47 (Chichester: Wiley): 303–320.

96. Simon, F.G.; Burrows, J.P.; Schneider, W.; Moortgat, G.K.; Crutzen, P.J., 1989: "Study of the Reaction $ClO + CH_3O_2 \rightarrow$ Products at 300 K", in: *Journal of Physics and Chemistry*, 93: 7807–7813.

97. Zimmermann, P.H.; Feichter, H.; Rath, H.K.; Crutzen, P.J.; Weiss, W., 1989: "A Global Three-Dimensional Source-Receptor Model Investigating Kr^{85}", in: *Atmospheric Environment*, 23: 25–35.

98. Brühl, C.; Crutzen, P.J., 1990: "Ozone and Climate Changes in the Light of the Montreal Protocol, a Model Study", in: *Ambio*, 19: 293–301.

99. Chatfield, R.B.; Crutzen, P.J., 1990: "Are There Interactions of Iodine and Sulfur Species in Marine Air Photochemistry?", in: *Journal of Geophysical Research*, 95: 22319–22341.

100. Crutzen, P.J.; Andreae, M.O., 1990: "Biomass Burning in the Tropics: Impact on Atmospheric Chemistry and Biogeochemical Cycles", in: *Science*, 250: 1669–1678.

101. Feichter, J.; Crutzen, P.J., 1990: "Parameterization of Vertical Tracer Transport due to Deep Cumulus Convection in a Global Transport Model and Its Evaluation with 222 Radon Measurements", in: *Tellus*, 42B: 100–117.

102. Graedel, T.E.; Crutzen, P.J., 1990: "Atmospheric Trace Constituents", in: Turner II, B.L.; et al. (Eds.): *The Earth as Transformed by Human Action* (Cambridge: Cambridge University Press): 295–311.

103. Hao, W.M.; Liu, M.H.; Crutzen, P.J., 1990: "Estimates of Annual and Regional Releases of CO_2 and Other Trace Gases to the Atmosphere from

Fires in the Tropics, Based on the FAO Statistics for the Period 1975–1980", in: Goldammer, J.G. (Ed.): *Fire in the Tropical Biota*, Ecological Studies, 84 (Berlin: Springer): 440–462.

104. Lelieveld, J.; Crutzen, P.J., 1990: "Influences of Cloud and Photochemical Processes on Tropospheric Ozone", in: *Nature*, 343: 227–233.

105. Lobert, J.M.; Scharffe, D.H.; Hao, W.M.; Crutzen, P.J., 1990: "Importance of Biomass Burning in the Atmospheric Budgets of Nitrogen-Containing Gases", in: *Nature*, 346: 552–554.

106. Sanhueza, E.; Hao, W.M.; Scharffe, D.; Donoso, L.; Crutzen, P.J., 1990: "N_2O and NO Emissions from Soils of the Northern Part of the Guayana Shield, Venezuela", in: *Journal of Geophysical Research*, 95,D13: 22481–22488.

107. Scharffe, D.; Hao, W.M.; Donoso, L.; Crutzen, P.J.; Sanhueza, E., 1990: "Soil Fluxes and Atmospheric Concentrations of CO and CH_4 in the Northern Part of the Guayana Shield, Venezuela", in: *Journal of Geophysical Research*, 95, D13: 22475–22480.

108. Crutzen, P.J.; Zimmermann, P.H., 1991: "The Changing Photochemistry of the Troposphere", in: *Tellus*, 43 A/B: 136–151.

109. Hao, W.M.; Scharffe, D.; Lobert, J.M.; Crutzen, P.J., 1991: "Emissions of N_2O from the Burning of Biomass in an Experimental System", in: *Geophysical Research Letters*, 18: 999–1002.

110. Kanakidou, M.; Singh, H.B.; Valentin, K.M.; Crutzen, P.J., 1991: "A 2-D Study of Ethane and Propane Oxidation in the Troposphere", in: *Journal of Geophysical Research*, 96: 15395–15413.

111. Kuhlbusch, A.T.; Lobert, J.M.; Crutzen, P.J.; Warneck, P., 1991: "Molecular Nitrogen Emissions from Denitrification During Biomass Burning", in: *Nature*, 351: 135–137.

112. Lelieveld, J.; Crutzen, P.J., 1991: "The Role of Clouds in Tropospheric Photochemistry", in: *Journal of Atmospheric Chemistry*, 12: 229–267.

113. Lobert, J.M.; Scharffe, D.H.; Hao, W.M.; Kuhlbusch, T.A.; Seuwen, R.; Warneck, P.; Crutzen, P.J., 1991: "Experimental Evaluation of Biomass Burning Emissions: Nitrogen and Carbon Containing Compounds", in: Levine, J.S. (Ed.): *Global Biomass Burning: Atmospheric, Climatic and Biosphere Implications* (Cambridge, MA: MIT Press): 122–125.

114. Peter, T.; Brühl, C.; Crutzen, P.J., 1991: "Increase in the PSC-Formation Probability Caused by High-Flying Aircraft", in: *Geophysical Research Letters*, 18: 1465–1468.

115. Crutzen, P.J.; Golitsyn, G.S., 1992: "Linkages Between Global Warming, Ozone Depletion and Other Aspects of Global Environmental Change", in: Mintzer, I.M. (Ed.): *Confronting Climate Change* (Cambridge: Cambridge University Press): 15–32.

116. Crutzen, P.J.; Müller, R.; Brühl, C.; Peter, T., 1992: "On the Potential Importance of the Gas Phase Reaction $CH_3O_2 + ClO \rightarrow ClOO + CH_3O$ and the Hererogeneous Reaction $HOCl + HCl \rightarrow H_2O + Cl_2$ in "Ozone Hole" Chemistry", in: *Geophysical Research Letters*, 19: 1113–1116.

117. Kanakidou, M.; Crutzen, P.J.; Zimmermann, P.H.; Bonsang, B., 1992: "A 3-Dimensional Global Study of the Photochemistry of Ethane and Propane in the Troposphere: Production and Transport of Organic Nitrogen Compounds, in: van Dop, H.; Kallos, G. (Eds.): *Air Pollution Modeling and its Application IX* (New York: Plenum Press): 415–426.

118. Langner, J.; Rodhe, H.; Crutzen, P.J.; Zimmermann, P., 1992: "Anthropogenic Influence on the Distribution of Tropospheric Sulphate Aerosol", in: *Nature*, 359: 712–715.

119. Lelieveld, J.; Crutzen, P.J., 1992: "Indirect Chemical Effects of Methane on Climate Warming", in: *Nature*, 355: 339–342.

120. Luo, B.P.; Peter, T.; Crutzen, P.J., 1992: "Maximum Supercooling of H_2SO_4 Acid Aerosol Droplets", in: *Berichte der Bunsengesellschaft für physikalische Chemie*, 96: 334–338.

121. Singh, H.B.; O'Hara, D.; Herlth, D.; Bradshaw, J.D.; Sandholm, S.T.; Gregory, G.L.; Sachse, G.W.; Blake, D.R.; Crutzen, P.J.; Kanakidou, M., 1992: "Atmospheric Measurements of PAN and Other Organic Nitrates at High Latitudes: Possible Sources and Sinks", in: *Journal of Geophysical Research*, 97: 16511–16522.

122. Singh, H.B.; Herlth, D.; Zahnle, K.; O'Hara, D.; Bradshaw, J.; Sandholm, S.T.; Talbot, R.; Crutzen, P.J.; Kanakidou, M., 1992: "Relationship of PAN to Active and Total Odd Nitrogen at Northern High Latitudes: Possible Influence of Reservoir Species on NO_x and O_3", in: *Journal of Geophysical Research*, 97: 16523–16530.

123. Berges, M.G.M.; Hofmann, R.M.; Scharffe, D.; Crutzen, P.J., 1993: "Nitrous Oxide Emissions from Motor Vehicles in Tunnels", in: *Journal of Geophysical Research*, 98: 18527–18531.

124. Crutzen, P.J.; Brühl, C., 1993: "A Model Study of Atmospheric Temperatures and the Concentrations of Ozone, Hydroxyl, and Some Other Photochemically Active Gases During the Glacial, the Preindustrial Holocene and the Present", in: *Geophysical Research Letters*, 20: 1047–1050.

125. Crutzen, P.J.; Carmichael, G.R., 1993: "Modeling the Influence of Fires on Atmospheric Chemistry", in: Crutzen, P.J.; Goldammer, J.G. (Eds.): *Fire in the Environment: The Ecological, Atmospheric, and Climatic Importance of Vegetation Fires*, op. cit., 90–105.

126. Dentener, F.; Crutzen, P.J., 1993: "Reaction of N_2O_5 on Tropospheric Aerosols: Impact on the Global Distributions of NO_x, O_3 and OH", in: *Journal of Geophysical Research*, 98: 7149–7163.

127. Goldammer, J.G.; Crutzen, P.J., 1993: "Fire in the Environment: Scientific Rationale and Summary of Results of the Dahlem Workshop", in: Crutzen, P.J.; Goldammer, J.G. (Eds.): *Fire in the Environment: The Ecological, Atmospheric and Climatic Importance*, op. cit., 1–14.

128. Kanakidou, M.; Crutzen, P.J., 1993: "Scale Problems in Global Tropospheric Chemistry Modeling: Comparison of Results Obtained with a Three-Dimensional Model, Adopting Longitudinally Uniform and Varying Emissions of NO_x and NMHC", in: *Chemosphere*, 26: 787–801.

129. Lelieveld, J.; Crutzen, P.J.; Brühl, C., 1993: "Climate Effects of Atmospheric Methane", in: *Chemosphere*, 26: 739–768.

130. Müller, R.; Crutzen, P.J., 1993: "A Possible Role of Galactic Cosmic Rays in Chlorine Activation During Polar Night", in: *Journal of Geophysical Research*, 98: 20483–20490.

131. Peter, T.; Crutzen, P.J., 1993: "The Role of Stratospheric Cloud Particles in Polar Ozone Depletion. An Overview", in: *Journal of Aerosol Science*, 24, Suppl. 1: 119–120.

132. Russell III, J.M.; Tuck, A.F.; Gordley, L.L.; Park, J.H.; Drayson, S.R.; Harries, J.E.; Cicerone, R.J.; Crutzen, P.J., 1993: "HALOE Antarctic Observations in the Spring of 1991", in: *Geophysical Research Letters*, 20: 719–722.

133. Schupp, M.; Bergamaschi, P.; Harris, G.W.; Crutzen, P.J., 1993: "Development of a Tunable Diode Laser Absorption Spectrometer for Measurements of the $^{13}C/^{12}C$ Ratio in Methane", in: *Chemosphere*, 26: 13–22.

134. Carslaw, K.S.; Luo, B.P.; Clegg, S.L.; Peter, T.; Brimblecombe, P.; Crutzen, P.J., 1994: "Stratospheric Aerosol Growth and HNO_3 Gas Phase Depletion from Coupled HNO_3 and Water Uptake by Liquid Particles", in: *Geophysical Research Letters*, 21: 2479–2482.

135. Chen, J.-P.; Crutzen, P.J., 1994: "Solute Effects on the Evaporation of Ice Particles", in: *Journal of Geophysical Research*, 99: 18847–18859.

136. Cox, R.A.; MacKenzie, A.R.; Müller, R.; Peter, T.; Crutzen, P.J., 1994: "Activation of Stratospheric Chlorine by Reactions in Liquid Sulphuric Acid", in: *Geophysical Research Letters*, 21: 1439–1442.

137. Crowley, J.N.; Helleis, F.; Müller, R.; Moortgat, G.K.; Crutzen, P.J., 1994: "CH_3OCl: UV/Visible Absorption Cross Sections, J Values and Atmospheric Significance", in: *Journal of Geophysical Research*, 99: 20683–20688.

138. Crutzen, P.J., 1994: "Global Tropospheric Chemistry", in: Moortgat, G.K.; et al. (Eds.): *Proceedings of the NATO Advanced Study Institute on Low Temperature Chemistry of the Atmosphere, Maratea, Italy, August 29–September 11, 1993*, NATO ASI Series I, 21 (Heidelberg: Springer): 465–498.

139. Crutzen, P.J., 1994: "Global Budgets for Non-CO_2 Greenhouse Gases", in: *Environmental Monitoring and Assessement*, 31: 1–15.

140. Crutzen, P.J.; Lelieveld, J.; Brühl, C., 1994: "Oxidation Processes in the Atmosphere and the Role of Human Activities: Observations and Model Results", in: Nriagu, J.O.; Simmons, M.S. (Eds.): *Environmental Oxidants*, Vol. 28 in: *Advances in Environmental Science and Technology* (Chichester, Wiley): 63–93.

141. Dentener, F.J.; Crutzen, P.J., 1994: "A Three Dimensional Model of the Global Ammonia Cycle", in: *Journal of Atmospheric Chemistry*, 19: 331–369.

142. Deshler, T.; Peter, T.; Müller, R.; Crutzen, P.J., 1994: "The Lifetime of Leewave-Induced Ice Particles in the Arctic Stratosphere: I. Balloonborne Observations", in: *Geophysical Research Letters*, 21: 1327–1330.

143. Lelieveld, J.; Crutzen, P.J., 1994: "Role of Deep Cloud Convection in the Ozone Budget of the Troposphere", in: *Science*, 264: 1759–1761.

144. Luo, B.P.; Peter, T.; Crutzen, P.J., 1994: "Freezing of Stratospheric Aerosol Droplets", in: *Geophysical Research Letters*, 21: 1447–1450.

145. Luo, B.P.; Clegg, S.L.; Peter, T.; Müller, R.; Crutzen, P.J., 1994: "HCl Solubility and Liquid Diffusion in Aqueous Sulfuric Acid Under Stratospheric Conditions", in: *Geophysical Research Letters*, 21: 49–52.

146. Müller, R.; Peter, T.; Crutzen, P.J.; Oelhaf, H.; Adrian, G.; Clarman, T.V.; Wegner, A.; Schmidt, U.; Lary, D., 1994: "Chlorine Chemistry and the Potential for Ozone Depletion in the Arctic Stratosphere in the Winter of 1991/92", in: *Geophysical Research Letters*, 21: 1427–1430.

147. Peter, T.; Crutzen, P.J., 1994: "Modelling the Chemistry and Micro-physics of the Could Stratosphere", in: Moortgat, G.K.; et al. (Eds.): *Proceedings of the NATO Advanced Study Institute on Low Temperature Chemistry of the Atmosphere, Maratea, Italy, August 29–September 11, 1993, NATO ASI Series I, 21* (Heidelberg: Springer): 499–530.

148. Peter, T.; Crutzen, P.J.; Müller, R.; Deshler, T., 1994: "The Lifetime of Leewave-Induced Particles in the Artic Stratosphere: II. Stabilization due to NAT-Coating", in: *Geophysical Research Letters*, 21: 1331–1334.

149. Sanhueza, E.; Donoso, L.; Scharffe, D.; Crutzen, P.J., 1994: "Carbon Monoxide Fluxes from Natural, Managed, or Cultivated Savannah Grasslands", in: *Journal of Geophysical Research*, 99: 16421–16425.

150. Sassen, K.; Peter, T.; Luo, B.P.; Crutzen, P.J., 1994: "Volcanic Bishop's Ring: Evidence for a Sulfuric Acid Tetrahydrate Particle Aerosol", in: *Applied Optics*, 33: 4602–4606.

151. Singh, H.B.; O'Hara, D.; Herlth, D.; Sachse, W.; Blake, D.R.; Bradshaw, J.D.; Kanakidou, M.; Crutzen, P.J., 1994: "Acetone in the Atmosphere: Distribution, Sources and Sinks", in: *Journal of Geophysical Research*, 99: 1805–1819.

152. Crutzen, P.J., 1995: "On the Role of CH_4 in Atmospheric Chemistry: Sources, Sinks and Possible Reductions in Anthropogenic Sources", in: *Ambio*, 24: 52–55.

153. Crutzen, P.J., 1995: "Introductory Lecture: Overview of Tropospheric Chemistry: Developments During the Past Quarter Century and a Look Ahead", in: *Faraday Discussion*, 100: 1–21.

154. Crutzen, P.J., 1995: "Ozone in the Troposphere", in: Singh, H.B. (Ed.): *Composition, Chemistry, and Climate of the Atmosphere* (New York: Van Nostrand Reinhold Publ.): 349–393.

155. Crutzen, P.J., 1995: "The Role of Methane in Atmospheric Chemistry and Climate", in: Engelhardt, W.V.; Leonhard-Marek, S.; Breves, G.; Giesecke, D. (Eds.): *Ruminant Physiology: Digestion, Metabolism, Growth and Reproduction: Proceedings of the Eighth International Symposium on Ruminant Physiology* (Stuttgart: Ferdinand Enke Verlag): 291–315.

156. Crutzen, P.J.; Grooß, J.-U.; Brühl, C.; Müller, R.; Russell III, J.M., 1995: "A Reevaluation of the Ozone Budget with HALOE UARS Data: No Evidence for the Ozone Deficit", in: *Science*, 268: 705–708.

157. Finkbeiner, M.; Crowley, J.N.; Horie, O.; Müller, R.; Moortgat, G.K.; Crutzen, P.J., 1995: "Reaction Between HO_2 and ClO: Product Formation Between 210 and 300 K", in: *Journal of Physics and Chemistry*, 99: 16264–16275.

158. Kanakidou, M.; Dentener, F.J.; Crutzen, P.J., 1995: "A Global Three-Dimensional Study of the Fate of HCFCs and HFC-134a in the Troposphere", in: *Journal of Geophysical Research*, 100: 18781–18801.

159. Koop, T.; Biermann, U.M.; Raber, W.; Luo, B.P.; Crutzen, P.J.; Peter, T., 1995: "Do Stratospheric Aerosol Droplets Freeze Above the Ice Frost Point?", in: *Journal of Geophysical Research*, 22: 917–920.

160. Kuhlbusch, T.A.J.; Crutzen, P.J., 1995: "Toward a Global Estimate of Black Carbon in Residues of Vegetation Fires Representing a Sink of Atmospheric CO_2 and a Source of O_2", in: *Global Biogeochemical Cycles*, 4: 491–501.

161. Meilinger, S.K.; Koop, T.; Luo, B.P.; Huthwelker, T.; Carslaw, K.S.; Crutzen, P.J.; Peter, T., 1995: "Size-Dependent Stratospheric Droplet Composition in Lee Wave Temperature Fluctuations and Their Potential Role in PSC Freezing", in: *Geophysical Research Letters*, 22: 3031–3034.

162. Rodhe, H.; Crutzen, P.J., 1995: "Climate and CCN", in: *Nature*, 375: 111.

163. Sander, R.; Lelieveld, J.; Crutzen, P.J., 1995: "Modelling of the Nighttime Nitrogen and Sulfur Chemistry in Size Resolved Droplets of an Orographic Cloud", in: *Journal of Atmospheric Chemistry*, 20: 89–116.

164. Schade, G.W.; Crutzen, P.J., 1995: "Emission of Aliphatic Amines from Animal Husbandry and Their Reactions: Potential Source of N_2O and HCN", in: *Journal of Atmospheric Chemistry*, 22: 319–346.

165. Singh, H.B.; Kanakidou, M.; Crutzen, P.J.; Jacob, D.J., 1995: "High Concentrations and Photochemical Fate of Oxygenated Hydrocarbons in the Global Troposphere", in: *Nature*, 378: 50–54.

166. Vömel, H.; Oltmans, S.J.; Kley, D.; Crutzen, P.J., 1995: "New Evidence for the Stratospheric Dehydration Mechanism in the Aquatorial Pacific", in: *Geophysical Research Letters*, 22: 3235–3238.

167. Wang, C.; Crutzen, P.J., 1995: "Impact of a Simulated Severe Local Storm on the Redistribution of Sulfur Dioxide", in: *Journal of Geophysical Research*, 100: 11357–11367.

168. Wang, C.; Crutzen, P.J.; Ramanathan, V.; Williams, S.F., 1995: "The Role of a Deep Convective Storm over the Tropical Pacific Ocean in the Redistribution of Atmospheric Chemical Species", in: *Journal of Geophysical Research*, 100: 11509–11516.

169. Wayne, R.P.; Poulet, G.; Biggs, P.; Burrows, J.P.; Cox, R.A.; Crutzen, P.J.; Hayman, G.D.; Jenkin, M.E.; Le Bras, G.; Moortgat, G.K.; Platt, U.; Schindler, R.N., 1995: "Halogen Oxides: Radicals, Sources and Reservoirs in the Laboratory and in the Atmosphere", in: *Atmospheric Environment*, 29: 2677–2881 (special issue).

170. Bergamaschi, P.; Brühl, C.; Brenninkmeijer, C.A.M.; Saueressig, G.; Crowley, J.N.; Grooß, J.U.; Fischer, H.; Crutzen, P.J., 1996: "Implications of the Large Carbon Kinetic Isotope Effect in the Reaction CH_4 + Cl for the $^{13}C/^{12}C$ Ratio of Stratospheric CH_4", in: *Geophysical Research Letters*, 23,17: 2227–2230.

171. Berges, M.G.M.; Crutzen, J., 1996: "Estimates of Global N_2O Emissions from Cattle, Pig and Chicken Manure, Including a Discussion of CH_4 Emissions", in: *Journal of Atmospheric Chemistry*, 24: 241–269.

172. Biermann, U.M.; Presper, T.; Mößinger, J.; Crutzen, P.J.; Peter, T., 1996: "The Unsuitability of Meteoritic and Other Nuclei for Polar Stratospheric Cloud Freezing", in: *Geophysical Research Letters*, 23: 1693–1696.

173. Brenninkmeijer, C.A.M.; Müller, R.; Crutzen, P.J.; Lowe, D.C.; Manning, M.R.; Sparks, R.J.; van Velthoven, P.F.J., 1996: "A Large ^{13}CO Deficit in the Lower Antarctic Stratosphere due to "Ozone Hole" Chemistry: Part I, Observations", in: *Geophysical Research Letters*, 23,16: 2125–2128.

174. Brühl, C.; Drayson, S.R.; Russell III, J.M.; Crutzen, P.J.; McInerney, J.M.; Purcell, P.N.; Claude, H.; Gernandt, H.; McGee, T.J.; McDermid, I.S.; Gunson, M.R., 1996: "Halogen Occultation Experiment Ozone Channel Validation", in: *Journal of Geophysical Research*, 101: 10217–10240.

175. Chen, J.-P.; Crutzen, P.J., 1996: "Reply", in: *Journal of Geophysical Research*, 101,D17: 23037–23038.

176. Crutzen, P.J.; Brühl, C., 1996: "Mass Extinctions and Supernova Explosions", in: *Proceedings of the National Academy of Sciences of the United States of America*, 93: 1582–1584.

177. Crutzen, P.J.; Golitsyn, G.S.; Elanskii, N.F.; Brenninkmeijer, C.A.M.; Scharffe, D.; Belikov, I.B.; Elokhov, A.S., 1996: "Observations of Minor Impurities in the Atmosphere over the Russian Territory with the Application of a Railroad Laboratory Car", in: *Transaction of the Russian Academy of Sciences/Earth Science Sections*, 351: 1289–1293 (Translated from *Doklady Akademii Nauk*, 350: 819–823, 1996).

178. Dentener, F.J.; Carmichael, G.R.; Zhang, Y.; Lelieveld, J.; Crutzen, P.J., 1996: "Role of Mineral Aerosol as a Reactive Surface in the Globel Troposphere", in: *Journal of Geophysical Research*, 101,D17: 22869–22889.

179. Kley, D.; Crutzen, P.J.; Smit, H.G.J.; Vömel, H.; Oltmans, S.; Grassl, H.; Ramanathan, V., 1996: "Observations of Near-Zero Ozone Concentrations over the Convective Pacific: Effects on Air Chemistry", in: *Science*, 274: 230–233.

180. Kuhlbusch, T.A.J.; Andreae, M.O.; Cachier, H.; Goldammer, J.G.; Lacaux, J.-P.; Shea, R.; Crutzen, P.J., 1996: "Black Carbon Formation by Savannah Fires: Measurements and Implications for the Global Carbeon Cycle", in: *Journal of Geophysical Research*, 101: 23651–23665.

181. Müller, R.; Brenninkmeijer, C.A.M.; Crutzen, P.J., 1996: "A Large ^{13}CO Deficit in the Lower Antarctic Stratosphere due to "Ozone Hole" Chemistry: Part II, Modeling", in: *Geophysical Research Letters*, 23,16: 2129–2132.

182. Müller, R.; Crutzen, P.J.; Grooß, J.-U.; Brühl, C.; Russell III, J.M.; Tuck, A.F., 1996: "Chlorine Activation and Ozone Depletion in the Arctic Vortex: Observations by the Halogen Occultation Experiment on the Upper Atmosphere Research Satellite", in: *Journal of Geophysical Research*, 101: 12531–12554.

183. Sander, R.; Crutzen, P.J., 1996: "Model Study Indicating Halogen Activation and Ozone Destruction in Polluted Air Masses Transported to the Sea", in: *Journal of Geophysical Research*, 101D: 9121–9138.

184. Shorter, J.H.; McManus, J.B.; Kolb, C.E.; Allewine, E.J.; Lamb, B.K.; Mosher, B.W.; Harriss, R.C.; Parchatka, U.; Fischer, H.; Harris, G.W.; Crutzen, P.J.; Karbach, H.-J., 1996: "Methane Emissions in Urban Aereas in Eastern Germany", in: *Journal of Atmospheric Chemistry*, 24: 121–140.

185. Singh, H.B.; Herlth, D.; Kolyer, R.; Salas, L.; Bradshaw, J.D.; Sandholm, S.T.; Davis, D.D.; Crawford, J.; Kondo, Y.; Koike, M.; Talbot, R.; Gregory, G.L.; Sachse, G.W.; Browell, E.; Blake, D.R.; Rowland, F.S.; Newell, R.; Merrill, J.; Heikes, B.; Liu, S.C.; Crutzen, P.J.; Kanakidou, M., 1996: "Reactive Nitrogen and Ozone over the Western Pacific: Distribution, Partitioning, and Sources", in: *Journal of Geophysical Research*, 101: 1793–1808.

186. Vogt, R.; Crutzen, P.J.; Sander, R., 1996: "A Mechanism for Halogen Release form Sea-Salt Aerosol in the Remote Marine Boundary Layer", in: *Nature*, 383: 327–330.

187. Andreae, M.O.; Crutzen, P.J., 1997: "Atmospheric Aerosols: Biogeochemical Sources and Role in Atmospheric Chemistry", in: *Science*, 276: 1052–1058.

188. Fischer, H.; Waibel, A.E.; Welling, M.; Wienhold, F.G.; Zenker, T.; Crutzen, P.J.; Arnold, F.; Bürger, V.; Schneider, J.; Bregman, A.; Lelieveld, J.; Siegmund, P.C., 1997: "Observations of High Concentration of Total Reactive Nitrogen (NO_y) and Nitric Acid (HNO_3) in the Lower Arctic Stratosphere During the Stratosphere-Troposphere Experiment by Aircraft Measurements (STREAM) II Campaign in February 1995", in: *Journal of Geophysical Research*, 102,D19: 23559–23571.

189. Grooss, J.-U.; Pierce, R.B.; Crutzen, P.J.; Grose, W.L.; Russell III, J.M., 1997: "Re-formation of Chlorine Reservoirs in Southern Hemisphere Polar Spring", in: *Journal of Geophysical Research*, 102: 13141–13152.

190. Hein, R.; Crutzen, P.J., 1997: "An Inverse Modeling Approach to Investigate the Global Atmospheric Methane Cycle", in: *Global Biogeochemical Cycles*, 11: 43–76.

191. Kley, D.; Smit, H.G.J.; Vömel, H.; Grassl, H.; Ramanathan, V.; Crutzen, P.J.; Williams, S.; Meywerk, J.; Oltmans, S.J., 1997: "Tropospheric Water-Vapour and Ozone Cross-Sections in a Zonal Plane over the Central Equatorial Pacific Ocean", in: *Quarterly Journal of the Royal Meteorological Society*, 123: 2009–2040.

192. Koop, T.; Luo, B.P.; Biermann, U.-M.; Crutzen, P.J.; Peter, T., 1997: "Freezing of $HNO_3/H_2SO_4/H_2O$ Solutions at Stratospheric Temperatures: Nucleation Statistics and Experiments", in: *Journal of Physics and Chemistry*, 101: 1117–1133.

193. Lelieveld, J.; Bregman, A.; Arnold, F.; Bürger, V.; Crutzen, P.J.; Fischer, H.; Waibel, A.; Siegmund, P.; van Velthoven, P.F.J., 1997: "Chemical Perturbation of the Lowermost Stratosphere Through Exchange with the Troposphere", in: *Geophysical Research Letters*, 24: 603–606.

194. Müller, R.; Crutzen, P.J.; Grooß, J.-U.; Brühl, C.; Russell III, J.M.; Gernandt, H.; McKenna, D.S.; Tuck, A.F., 1997: "Severe Chemical Ozone Loss in the Arctic During the Winter of 1995–96", in: *Nature*, 389: 709–712.

195. Müller, R.; Grooß, J.-U.; McKenna, D.S.; Crutzen, P.J.; Brühl, C.; Russell III, J.M.; Tuck, A.F., 1997: "HALOE Observations of the Vertical Structure of Chemical Ozone Depletion in the Arctic Vortex During Winter and Early Spring 1996–1997", in: *Geophysical Research Letters*, 24: 2717–2720.

196. Pierce, R.B.; Grooss, J.-U.; Grose, W.L.; Russell III, J.M.; Crutzen, P.J.; Fairlie, T.D.; Lingenfelser, G., 1997: "Photochemical Calculations Along Air Mass Trajectories During ASHOE/MAESA", in: *Journal of Geophysical Research*, 102: 13153–13167.

197. Roehl, C.M.; Burkholder, J.B.; Moortgat, G.K.; Ravishankara, A.R.; Crutzen, P.J., 1997: "Temperature Dependence of UV Absorption Cross Sections and Atmospheric Implications of Several Alkyl Iodides", in: *Journal of Geophysical Research*, 102,D11: 12819–12829.

198. Sander, R.; Vogt, R.; Harris, G.W.; Crutzen, P.J., 1997: "Modeling the Chemistry of Ozone, Halogen Compounds, and Hydrocarbons in the Arctic Troposphere During Spring", in: *Tellus*, 49B: 522–532.

199. Bergamaschi, P.; Brenninkmeijer, C.A.M.; Hahn, M.; Röckmann, T.; Scharffe, D.H.; Crutzen, P.J.; Elansky, N.F.; Belikov, I.B.; Trivett, N.B.A.; Worthy, D.E.J., 1998: "Isotope Analysis Based Source Identification for Atmospheric CH_4 and CO Sampled Across Russia Using the Trans-Siberian Railroad", in: *Journal of Geophysical Research*, 103,D7: 8227–8235.

200. Biermann, U.M.; Crowley, J.N.; Huthwelker, T.; Moortgat, G.K.; Crutzen, P.J.; Peter, T., 1998: "FTIR Studies on Lifetime Prolongation of Stratospheric Ice Particles due to NAT Coating", in: *Geophysical Research Letters*, 25: 3939–3942.

201. Brühl, C., Crutzen, P.J.; Grooß, J.-U., 1998: "High-Latitude, Summertime NO_x Activation and Seasonal Ozone Decline in the Lower Stratosphere: Model Calculations Based on Observations by HALOE on UARS", in: *Journal of Geophysical Research*, 103,D3: 3587–3597.

202. Crutzen, P.J.; Elansky, N.F.; Hahn, M.; Golitsyn, G.S.; Benninkmeijer, C.A.M.; Scharffe, D.H.; Belikov, I.B.; Maiss, M.; Bergamaschi, P.; Röckmann, T.; Grisenko, A.M.; Sevostyanov, V.M., 1998: "Trace Gas Measurements Between Moscow and Vladivistok Using the Trans-Siberian Railroad", in: *Journal of Atmospheric Chemistry*, 29: 179–194.

203. Dong, Y.; Scharffe, D.; Lobert, J.M.; Crutzen, P.J.; Sanhueza, E., 1998: "Fluxes of CO_2, CH_4 and N_2O from a Temperate Forest Soil: The Effects of Leaves and Humus Layers", in: *Tellus*, 50B: 243–252.

204. Hegels, E.; Crutzen, P.J.; Klüpfel, T.; Perner, D.; Burrows, J.P., 1998: "Global Distribution of Atmospheric Bromine-Monoxide from GOME on

Earth Observing Satellite ERS-2", in: *Geophysical Research Letters*, 25: 3127–3130.

205. Keene, W.C.; Sander, R.; Pszenny, A.A.P.; Vogt, R.; Crutzen, P.J.; Galloway, J.N., 1998: "Aerosol pH in the Marine Boundary Layer: A Review and Model Evaluation", in: *Journal of Aerosol Science*, 29: 339–356.

206. Landgraf, J.; Crutzen, P.J., 1998: "An Efficient Method for Online Calculations of Photolysis and Heating Rates", in: *Journal of the Atmospheric Sciences*, 55: 863–878.

207. Lawrence, M.G.; Crutzen, P.J., 1998: "The Impact of Cloud Particle Gravitational Settling on Soluble Trace Gas Distributions", in: *Tellus*, 50B: 263–289.

208. Lelieveld, J.; Crutzen, P.J.; Dentener, F.J., 1998: "Changing Concentration, Lifetime and Climate Forcing of Atmospheric Methane", in: *Tellus*, 50B: 128–150.

209. Röckmann, T.; Brenninkmeijer, C.A.M.; Neeb, P.; Crutzen, P.J., 1998: "Ozonolysis of Nonmethane Hydrocarbons as a Source of the Observed Mass Independent Oxygen Isotope Enrichment in Tropospheric CO", in: *Journal of Geophysical Research*, 103,D1: 1463–1470.

210. Röckmann, T.; Brenninkmeijer, C.A.M.; Saueressig, G.; Bergamaschi, P.; Crowley, J.N.; Fischer, H.; Crutzen, P.J., 1998: "Mass-Independent Oxygen Isotope Fractionation in Atmospheric CO as a Result of the Reaction CO + OH", in: *Science*, 281: 544–546.

211. Sanhueza, E.; Crutzen, P.J., 1998: "Budgets of Fixed Nitrogen in the Orinoco Savannah Region: Role of Pyrodenitrification", in: *Global Biogeochemical Cycles*, 12: 653–666.

212. Sanhueza, E.; Dong, Y.; Scharffe, D.; Lobert, J.M.; Crutzen, P.J., 1998: "Carbon Monoxide Uptake by Temperature Forest Soils: The Effects of Leaves and Humus Layers", in: *Tellus*, 50B: 51–58.

213. Steil, B.; Dameris, M.; Brühl, C.; Crutzen, P.J.; Grewe, V.; Ponater, M.; Sausen, R., 1998: "Development of a Chemistry Module for GCMs: First Results of a Multiannual Integration", in: *Annals of Geophysics*, 16: 205–228.

214. Brenninkmeijer, C.A.M.; Crutzen, P.J.; Fischer, H.; Güsten, H.; Hans, W.; Heinrich, G.; Heintzenberg, J.; Hermann, M.; Immelmann, T.; Kersting, D.; Maiss, M.; Nolle, M.; Pitscheider, A.; Pohlkamp, H.; Scharffe, D.; Specht, K.; Wiedensohler, A., 1999: "Caribic—Civil Aircraft for Global Measurement of Trace Gases and Aerosols in the Tropopause Region", in: *Journal of Atmospheric and Oceanic Technology*, 16; 1373–1383.

215. Brühl, C.; Crutzen, P.J., 1999: "Reductions in the Anthropogenic Emissions of CO and Their Effect on CH₄", in: *Chemosphere*, 1: 249–254.

216. Crutzen, P.J.; Lawrence, M.; Pöschl, U., 1999: "On the Background Photochemistry of Tropospheric Ozone", in: *Tellus*, 51 A-B: 123–146.

217. Crutzen, P.J.; Fall, R.; Galbally, I.; Lindinger, W., 1999: "Parameters for Global Ecosystem Models (Comment on "Effect of Inter-Annual Climate Variability on Carbon Storage in Amazonian ecosystems" by Tian et al.)", in: *Nature*, 399: 535.

218. Dickerson, R.R.; Rhoads, K.P.; Carsey, T.P.; Oltmans, S.J.; Burrows, J.P.; Crutzen, P.J., 1999: "Ozone in the Remote Marine Boundary Layer: A Possible Role for Halogens", in: *Journal of Geophysical Research*, 104,D17: 21, 385–395.
219. Grooß, J.U.; Müller, R.; Becker, G.; McKenna, D.S.; Crutzen, P.J., 1999: "The Upper Stratospheric Ozone Budget: An Update of Calculations Based on HALOE Data", in: *Journal of Atmospheric Chemistry*, 34: 171–183.
220. Holzinger, R.; Warneke, C.; Hansel, A.; Jordan, A.; Lindinger, W.; Scharffe, D.; Schade, G.; Crutzen, P.J., 1999: "Biomass Burning as a Source of Formaldehyde, Acetaldehyde, Methanol, Acetone, Acetonitrile and Hydrogen Cyanide", in: *Geophysical Research Letters*, 26: 1161–1164.
221. Ingham, T.; Bauer, D.; Sander, R.; Crutzen, P.J.; Crowley, J.N., 1999: "Kinetics and Products of the Reactions BrO + DMS and Br + DMS at 298 K", in: *Journal of Physical Chemistry*, 103: 7199–7209.
222. Kanakidou, M.; Crutzen, P.J., 1999: "The Photochemical Source of Carbon Monoxide: Importance, Uncertainties and Feedbacks", in: *Chemosphere*, 1: 91–109.
223. Lawrence, M.G.; Crutzen, P.J.; Rasch, P.J., 1999: "Analysis of the CEPEX Ozone Data Using a 3D Chemistry-Meteorology Model", in: *Quarterly Journal of the Royal Meteorological Society*, 125: 2987–3009.
224. Lawrence, M.G.; Crutzen, P.J., 1999: "Influence of NO_x Emissions from Ships on Tropospheric Photochemistry and Climate", in: *Nature*, 402: 167–170.
225. Lawrence, M.G.; Crutzen, P.J.; Rasch, P.J.; Eaton, B.E.; Mahowald, N.M., 1999: "A Model for Studies of Tropospheric Photochemistry: Description, Global Distributions and Evaluation", in: *Journal of Geophysical Research*, 104,D21: 26245–26277.
226. Müller, R.; Grooß, J.-U.; McKenna, D.S.; Crutzen, P.J.; Brühl, C.; Russell III, J.M.; Gordley, L.L.; Burrows, J.P.; Tuck, A.F., 1999: "Chemical Ozone Loss in the Arctic Vortex in the Winter 1995–96: HALOE Measurements in Conjunction with Other Observations", in: *Annals of Geophysicae*, 17: 101–114.
227. Röckmann, T.; Brenninkmeijer, C.A.M.; Crutzen, P.J., 1999: "Short-Term Variations in the $^{13}C/^{12}C$ Ratio of CO as a Measure of Cl Activation During Tropospheric Ozone Depletion Events in the Arctic", in: *Journal of Geophysical Research*, 104,D21: 1691–1697.
228. Sander, R.; Rudich, Y.; von Glasow, R.; Crutzen, P.J., 1999: "The Role of $BrNO_3$ in Marine Tropospheric Chemistry: A Model Study", in: *Geophysical Research Letters*, 26,18: 2857–2860.
229. Sanhueza, E.; Crutzen, P.J.; Fernández, E., 1999: "Production of Boundary Layer Ozone from Tropical American Savannah Biomass Burning Emissions", in: *Atmospheric Environment*, 33: 4969–4975.
230. Schade, G.; Hofmann, R.-M.; Crutzen, P.J., 1999: "CO Emissions from Degrading Plant Matter. Part I: Measurements", in: *Tellus*, 51B: 889–908.

231. Schade, G.; Crutzen, P.J., 1999: "CO Emissions from Degrading Plant Matter. Part II: Estimate of a Global Source Strength", in: *Tellus*, 51B: 909–918.

232. Trautmann, T.; Podgorny, I.; Landgraf, J.; Crutzen, P.J., 1999: "Actinic Fluxes and Photodissociation Coefficients in Cloud Fields Embedded in Realistic Atmospheres", in: *Journal of Geophysical Research*, 104,D23: 30173–30192.

233. Vogt, R.; Sander, R.; von Glasow, R.; Crutzen, P.J., 1999: "Iodine Chemisty and Its Role in Halogen Activation and Ozone Loss in the Marine Boundary Layer: A Model Study", in: *Journal of Atmospheric Chemistry*, 32: 375–395.

234. Waibel, A.E.; Fischer, H.; Wienhold, F.G.; Siegmund, P.C.; Lee, B.; Ström, J.; Lelieveld, J.; Crutzen, P.J., 1999: "Highly Elevated Carbon Monoxide Concentrations in the Upper Troposphere and Lowermost Stratosphere at Northern Midlatitudes During the STREAM II Summer Campaign in 1994", in: *Chemosphere*, 1: 233–248.

235. Waibel, A.E.; Peter, T.; Carslaw, K.S.; Oelhaf, H.; Wetzel, G.; Crutzen, P.J.; Pöschl, U.; Tsias, A.; Reimer, E.; Fischer, H., 1999: "Arctic Ozone Loss due to Denitrification", in: *Science*, 283: 2064–2069.

236. Warneke, C.; Karl, T.; Judmaier, H.; Hansel, A.; Jordan, A.; Lindinger, W.; Crutzen, P.J., 1999: "Acetone, Methanol, and Other Partially Oxidized Volatile Organic Emissions from Dead Plant Matter by Abiological Processes: Significance for Atmospheric HO_x Chemistry", in: *Global Biogeochemial Cycles*, 13: 9–17.

237. Ariya, P.A.; Sander, R.; Crutzen, P.J., 2000: "Significance of HO_x and Peroxides Production due to Alkene Ozonolysis During Fall and Winter: A Modeling Study", in: *Journal of Geophysical Research*, 105,D14: 17721–17738.

238. Bergamaschi, P.; Hein, R.; Heimann, M.; Crutzen, P.J., 2000: "Inverse Modeling of the Global CO Cycle: 1. Inversion of CO Mixing Ratios", in: *Journal of Geophysical Research*, 105,D2: 1909–1927.

239. Bergamaschi, P.; Hein, R.; Brenninkmeijer, C.A.M.; Crutzen, P.J., 2000: "Inverse Modeling of the Global CO Cycle: 2. Inversion of $^{13}C/^{12}C$ and $^{18}O/^{16}O$ Isotope Ratios", in: *Journal of Geophysical Research*, 105,D2: 1929–1945.

240. Brühl, C.; Crutzen, P.J., 2000: "NO_x-Catalyzed Ozone Destruction and NO_x Activation at Mid to High Latitudes as the Main Cause of the Spring to Fall Ozone Decline in the Northern Hemisphere", in: *Journal of Geophysical Research*, 105,D10: 12163–12168.

241. Brühl, C.; Pöschl, U.; Crutzen, P.J.; Steil, B., 2000: "Acetone and PAN in the Upper Troposphere: Impact on Ozone Production from Aircraft Emissions", in: *Atmospheric Environment*, 34: 3931–3938.

242. Crutzen, P.J.; Lawrence, M.G., 2000: "The Impact of Precipitation Scavenging on the Transport of Trace Gases: A 3-Dimensional Model Sensitivity Study", in: *Journal of Atmospheric Chemistry*, 37: 81–112.

243. Crutzen, P.J.; Ramanathan, V., 2000: "The Ascent of Atmospheric Sciences", in: *Science*, 290: 299–304.

244. Crutzen, P.J.; Williams, J.; Pöschl, U.; Hoor, P.; Fischer, H.; Warneke, C.; Holzinger, R.; Hansel, A.; Lindinger, W.; Scheeren, B.; Lelieveld, J., 2000: "High Spatial and Temporal Resolution Measurements of Primary Organics and Their Oxidation Products over the Tropical Forests of Surinam", in: *Atmospheric Environment*, 34,8: 1161–1165.

245. Holzinger, R.; Sandoval-Soto, L.; Rottenberger, S.; Crutzen, P.J.; Kesselmeier, J., 2000: "Emissions of Volatile Organic Compounds from *Quercus ilex* L. Measured by Proton Transfer Reaction Mass Spectrometry Under Different Environmental Conditions", in: *Journal of Geophysical Research*, 105,D16: 20573–20579.

246. Kanakidou, M.; Tsigaridis, K.; Dentener, F.J.; Crutzen, P.J., 2000: "Human-Activity-Enhanced Formation of Organic Aerosols by Biogenic Hydrocarbon Oxidation", in: *Journal of Geophysical Research*, 105,D7: 9243–9254.

247. Law, K.S.; Plantevin, P.-H.; Thouret, V.; Marenco, A.; Asman, W.A.H.; Lawrence, M.; Crutzen, P.J.; Muller, J.-F.; Hauglustaine, D.A.; Kanakidou, M., 2000: "Comparison Between Global Chemistry Transport Model Results and Measurement of Ozone and Water Vapor by Airbus In-Service Aircraft (MOZAIC) Data", in: *Journal of Geophysical Research*, 105,D1: 1503–1525.

248. Pöschl, U.; von Kuhlmann, R.; Poisson, N.; Crutzen, P.J., 2000: "Development and Intercomparison of Condensed Isoprene Oxidation Mechanisms for Global Atmospheric Modeling", in: *Journal of Atmospheric Chemistry*, 37: 29–52.

249. Pöschl, U.; Lawrence, M.G.; von Kuhlmann, R.; Crutzen, P.J., 2000: "Comment on Methane Photooxidation in the Atmosphere: Contrast Between Two Methods of Analysis by Harold Johnston and Douglas Kinnison", in: *Journal of Geophysical Research*, 105,D1: 1431–1433.

250. Poisson, N.; Kanakidou, M.; Crutzen, P.J., 2000: "Impact of Non-methane Hydrocarbons on Tropospheric Chemistry and the Oxidizing Power of the Global Troposphere: 3-Dimensional Modelling Results", in: *Journal of Atmospheric Chemistry*, 36: 157–230.

251. Röckmann, T.; Brenninkmeijer, C.A.M.; Wollenhaupt, M.; Crowley, J.N.; Crutzen, P.J., 2000: "Measurement of the Isotopic Fractionation of $^{15}N^{14}N^{16}O$, $^{14}N^{15}N^{16}O$ and $^{14}N^{14}N^{18}O$ in the UV Photolysis of Nitrous Oxide", in: *Geophysical Research Letters*, 27,9: 1399–1402.

252. Sander, R.; Crutzen, P.J., 2000: "Comment on A Chemical Aqueous Phase Radical Mechanism for Tropospheric Chemistry by Herrmann et al.", in: *Chemosphere*, 41: 631–632.

253. Williams, J.; Fischer, H.; Harris, G.W.; Crutzen, P.J.; Hoor, P.; Hansel, A.; Holzinger, R.; Warnecke, C.; Lindinger, W.; Scheeren, B.; Lelieveld, J., 2000: "Variability-Lifetime Relationship for Organic Trace Gases: A Novel Aid to Compound Identification and Estimation of HO Concentrations", in: *Journal of Geophysical Research*, 105,D16: 20473–20486.

254. Heintzenberg; Wiedensohler, A.; Güsten, H.; G. Heinrich, Fischer, H.; Cuijpers, J.W.M.; van Velthoven, P.F.J., 2000: "Identification of Extratropical Two-Way Troposphere-Stratosphere Mixing Based on CARIBIC Measurements of O_3, CO, and Ultrafine Particles", in: *Journal of Geophysical Research*, 105,D1: 1527–1535.

255. Crutzen, P.J.; Brühl, C., 2001: "Catalysis by NO_x as the Main Cause of the Spring to Fall Stratospheric Ozone Decline in the Northern Hemisphere", in: *Journal of Physics and Chemistry*, 105: 1579–1582.

256. Crutzen, P.J.; Lelieveld, J., 2001: "Human Impacts on Atmospheric Chemistry", in: *Annual Review of Earth and Planetary Sciences*, 29: 17–45.

257. Crutzen, P.J.; Ramanathan, V., 2001: Foreword, INDOEX special issue. *Journal of Geophysical Research*, 106,D22: 28369–28370.

258. Jöckel, P.; Kuhlmann, R.V.; Lawrence, M.G.; Steil, B.; Brenninkmeijer, C.A.M.; Crutzen, P.J.; Rasch, P.J.; Eaton, B., 2001: "On a Fundamental Problem in Implementing Flux-Form Advection Schemes for Tracer Transport in 3-Dimensional General Circulation and Chemistry Transport Models", in: *Quarterly Journal of the Royal Meteorological Society*, 127: 1035–1052.

259. Karl, T.; Fall, R.; Crutzen, P.J.; Jordan, A.; Lindinger, W., 2001: "High Concentrations of Reactive Biogenic VOCs at a High Altitude Site in Late Autumn", in: *Geophysical Research Letters*, 28: 507–510.

260. Karl, T.; Crutzen, P.J.; Mandl, M.; Staudinger, M.; Guenther, A.; Jordan, A.; Fall, R.; Lindinger, W., 2001: "Variability-Lifetime Relationship of VOCs Observed at the Sonnblick Observatory 1999-Estimation of HO-Densities", in: *Atmospheric Environment*, 35: 5287–5300.

261. Lelieveld, J.; Crutzen, P.J.; Ramanathan, V.; Andreae, M.O.; Brenninkmeijer, C.A.M.; Campos, T.; Cass, G.R.; Dickerson, R.R.; Fischer, H.; de Gouw, J.A.; Hansel, A.; Jefferson, A.; Kley, D.; de Laat, A.T.J.; Lal, S.; Lawrence, M.G.; Lobert, J.M.; Mayol-Bracero, O.L.; Mitra, A.P.; Novakov, T.; Oltmans, S.J.; Prather, K.A.; Reiner, T.; Rodhe, H.; Scheeren, H.A.; Sikka, D.; Williams, J., 2001: "The Indian Ocean Experiment: Widespread Air Pollution from South to Southeast Asia", in: *Science*, 291: 1031–1036.

262. Pöschl, U.; Williams, J.; Hoor, P.; Fischer, H.; Crutzen, P.J.; Warneke, C.; Holzinger, R.; Hansel, A.; Jordan, A.; Lindinger, W.; Scheeren, H.A.; Peters, W.; Lelieveld, J., 2001: "High Acetone Concentrations Throughout the 0–12 km Altitude Range over the Tropical Rainforest in Surinam", in: *Journal of Atmospheric Chemistry*, 38: 115–132.

263. Quesada, J.; Grossmann, D.; Fernández, E.; Romero, J.; Sanhueza, E.; Moortgat, G.; Crutzen, P.J., 2001: "Ground Based Gas Phase Measurements in Surinam during the LBA-Claire 98 Experiment", in: *Journal of Atmospheric Chemistry*, 39: 15–36.

264. Ramanathan, V.; Crutzen, P.J.; Kiehl, J.T.; Rosenfeld, D., 2001: "Aerosols, Climate, and the Hydrological Cycle", in: *Science*, 294: 2119–2124.

265. Ramanathan, V.; Crutzen, P.J.; Lelieveld, J.; Mitra, A.P.; Althausen, D.; Anderson, J.; Andreae, M.O.; Cantrell, W.; Cass, G.R.; Chung, C.E.; Clarke, A.D.; Coakley, J.A.; Collins, W.D.; Conant, W.C.; Dulac, F.; Heintzenberg, J.;

Heymsfield, A.J.; Holben, B.; Howell, S.; Hudson, J.; Jayaraman, A.; Kiehl, J. T.; Krishnamurti, T.N.; Lubin, D.; McFarquhar, G.; Novakov, T.; Ogren, J.A.; Podgorny, I.A.; Prather, K.; Priestley, K.; Prospero, J.M.; Quinn, P.K.; Rajeev, K.; Rasch, P.; Rupert, S.; Sadourny, R.; Satheesh, S.K.; Shaw, G.E.; Sheridan, P.; Valero, F.P.J., 2001: "Indian Ocean Experiment: An Integrated Analysis of the Climate Forcing and Effects of the Great Indo-Asian Haze", in: *Journal of Geophysical Research*, 106,D22: 28371–28398.

266. Röckmann, T.; Kaiser, J.; Crowley, J.N.; Brenninkmeijer, C.A.M.; Crutzen, P.J., 2001: "The Origin of the Anomalous or "Mass-Independent" Oxygen Isotope Fractionation in Tropospheric N_2O", in: *Geophysical Research Letters*, 28,3: 503–506.

267. Röckmann, T.; Kaiser, J.; Brenninkmeijer, C.A.M.; Crowley, J.N.; Borchers, R.; Brand, W.A.; Crutzen, P.J., 2001: "Isotopic Enrichment of Nitrous Oxide ($^{15}N^{14}NO$, $^{14}N^{15}NO$, $^{14}N^{14}N^{18}O$) in the Stratosphere and in the Laboratory", in: *Journal of Geophysical Research*, 106,D10: 10403–10410.

268. Röckmann, T.; Kaiser, J.; Brenninkmeijer, C.A.M.; Brand, W.A.; Borchers, R.; Crowley, J.N.; Wollenhaupt, M.; Crutzen, P.J., 2001: "The Position Dependent ^{15}N Enrichment of Nitrous Oxide in the Stratosphere", in: *Isotopes in Environmental and Health Studies*, 37: 91–95.

269. Warneke, C.; Holzinger, R.; Hansel, A.; Jordan, A.; Lindinger, W.; Pöschl, U.; Williams, J.; Hoor, P.; Fischer, H.; Crutzen, P.J.; Scheeren, H.A.; Lelieveld, J., 2001: "Isoprene and Its Oxidation Products Methyl Vinyl Ketone, Methacrolein, and Isoprene Related Peroxides Measured Online over the Tropical Rain Forest of Surinam in March 1998", in: *Journal of Atmospheric Chemistry*, 38: 167–185.

270. Williams, J.; Fischer, H.; Hoor, P.; Pöschl, U.; Crutzen, P.J.; Andreae, M.O.; Lelieveld, J., 2001: "The Influence of the Tropical Rainforest on Atmospheric CO and CO_2 as Measured by Aircraft over Surinam, South America", in: *Chemosphere*, 3: 157–170.

271. Williams, J.; Pöschl, U.; Crutzen, P.J.; Hansel, A.; Holzinger, R.; Warneke, C.; Lindinger, W.; Lelieveld, J., 2001: "An Atmospheric Chemistry Interpretation of Mass Scans Obtained from a Proton Transfer Mass Spectrometer Flown over the Tropical Rainforest of Surinam", in: *Journal of Atmospheric Chemistry*, 38: 133–166.

272. Crutzen, P.J.; Ramanathan, V., 2002: "The Ascent of Atmospheric Sciences", in: Amato, I. (Ed.): *Science—Pathways of Discovery* (New York: Wiley): 175–188.

273. Gabriel, R.; von Glasow, R.; Sander, R.; Andreae, M.O.; Crutzen, P.J., 2002: "Bromide Content of Sea-Salt Aerosol Particles Collected over the Indian Ocean During INDOEX 1999", in: *Journal of Geophysical Research*, 107D: 8032, doi:10.1029/2001JD001133.

274. von Glasow, R.; Sander, R.; Bott, A.; Crutzen, P.J., 2002: "Modeling Halogen Chemistry in the Marine Boundary Layer. 1. Cloud-Free MBL", in: *Journal of Geophysical Research*, 107D: 4341, doi:10.1029/2001JD000942.

275. von Glasow, R.; Sander, R.; Bott, A.; Crutzen, P.J., 2002: "Modeling Halogen Chemistry in the Marine Boundary Layer. 2. Interactions with Sulfur and the Cloud-Covered MBL", in: *Journal of Geophysical Research*, 107D: 4323, doi:10.1029/2001JD000943.

276. Holzinger, R.; Sanhueza, E.; von Kuhlmann, R.; Kleiss, B.; Donoso, L.; Crutzen, P.J., 2002: "Diurnal Cycles and Seasonal Variation of Isoprene and Its Oxidation Products in the Tropical Savanna Atmosphere", in: *Global Biogeochemical Cycles*, 16,4: 1074, doi:10.1029/2001GB001421.

277. Lelieveld, J.; Berresheim, H.; Borrmann, S.; Crutzen, P.J.; Dentener, F.J.; Fischer, H.; Feichter, J.; Flatau, P.F.; Heland, J.; Holzinger, R.; Korrmann, R.; Lawrence, M.G.; Levin, Z.; Markowicz, K.M.; Mihalopoulos, N.; Minikin, A.; Ramanathan, V.; de Reus, M.; Roelofs, G.J.; Scheeren, H.A.; Sciare, J.; Schlager, H.; Schultz, M.; Siegmund, P.; Steil, B.; Stephanou, E.G.; Stier, P.; Schlager, H.; Schultz, M.; Siegmund, P.; Steil, B.; Stephanou, E.G.; Stier, P.; Traub, M.; Warneke, C.; Williams, J.; Ziereis, H., 2002: "Global Air Pollution Crossroads over the Mediterranean", in: *Science*, 298: 794–799.

278. Markowicz, K.M.; Flatau, P.J.; Ramana, M.V.; Crutzen, P.J., 2002: "Absorbing Mediterranean Aerosols Lead to a Large Reduction in the Solar Radiation at the Surface", in: *Geophysical Research Letters*, 29, doi:10.1029/2002GL015767.

279. Mühle, J.; Zahn, A.; Brenninkmeijer, C.A.M.; Gros, V.; Crutzen, P.J., 2002: "Air Mass Classifictation During the INDOEX R/V Ronald Brown Cruise Using Measurements of Nonmethane Hydrocarbons, CH_4, CO_2, CO, ^{14}CO, and $\delta^{18}O(CO)$", in: *Journal of Geophysical Research*, 107, doi:10.1029/2001JD000730.

280. Oberlander, E.A.; Brenninkmeijer, C.A.M.; Crutzen, P.J.; Elansky, N.F.; Golitsyn, G.S.; Granberg, I.G.; Scharffe, D.H.; Hofmann, R.; Belikov, I.B.; Paretzke, H.G.; van Velthoven, P.F.J., 2002: "Trace Gas Measurements Along the Trans-Siberian Railroad: The TROICA 5 Expedition", in: *Journal of Geophysical Research*, 107.

281. Ramanathan, V.; Crutzen, P.J.; Mitra, A.P.; Sikka, D., 2002: "The Indian Ocean Experiment and the Asian Brown Cloud", in: *Current Science*, 83: 947–955.

282. Wagner, V.; von Glasow, R.; Fischer, H.; Crutzen, P.J., 2002: "Are CH_2O Measurements in the Marine Boundary Layer Suitable for Testing the Current Understanding of CH_4 Photooxidation?: A Model Study", in: *Journal of Geophysical Research*, 107,D3, doi:10.1029/2001JD000722.

283. Williams, J.; Fischer, H.; Wong, S.; Crutzen, P.J.; Scheele, M.P.; Lelieveld, J., 2002: "Near Equatorial CO and O_3 Profiles over the Indian Ocean During the Winter Monsoon: High O_3 Levels in the Middle Troposphere and Interhemispheric Exchange", in: *Journal of Geophysical Research*, 107, doi:10.1029/2001JD001126.

284. Wisthaler, A.; Hansel, A.; Dickerson, R.R.; Crutzen, P.J., 2002: "Organic Trace Gas Measurements by PTR-MS During INDOEX 1999", in: *Journal of Geophysical Research*, 107,D19, doi:10.1029/2001JD000576.

285. Wuebbles, D.J.; Brasseur, G.P.; Rodhe, H.; Barrie, L.A.; Crutzen, P.J.; Delmas, R.J.; Jacob, D.J.; Kolb, C.; Pszenny, A.; Steffen, W.; Weiss, R.F., 2002, "Changes in the Chemical Composition of the Atmosphere and Potential Impacts", in: Brasseur, G.P.; Prinn, R.G.; Pszenny, A.A.P. (Eds.): *Atmospheric Chemistry in a Changing World* (New York: Springer): 1–17.

286. Zahn, A.; Brenninkmeijer, C.A.M.; Asman, W.A.H.; Crutzen, P.J.; Heinrich, G.; Fischer, H.; Cuijpers, J.W.M.; van Velthoven, P.F.J., 2002: "Budgets of O_3 and CO in the Upper Troposphere: CARIBIC Passenger Aircraft Results 1997–2001", in: *Journal of Geophysical Research*, 107.

287. Zahn, A.; Brenninkmeijer, C.A.M.; Crutzen, P.J.; Parrish, D.D.; Sueper, D.; Heinrich, G.; Güsten, H.; Fischer, H.; Hermann, M.; Heintzenberg, J., 2002: "Electrical Discharge Source for Tropospheric Ozone-Rich Transients", in: *Journal of Geophysical Research*, 107.

288. Asman, W.A.H.; Lawrence, M.G.; Brenninkmeijer, C.A.M.; Crutzen, P.J.; Cuijpers, J.W.M.; Nédélec, P., 2003: "Rarity of Upper-Tropospheric Low O_3 Concentration Events During MOZAIC Flights", in: *Atmospheric Chemistry and Physics*, 3: 1541–1549.

289. Christian, T.J.; Kleiss, B.; Yokelson, R.J.; Holzinger, R.; Crutzen, P.J.; Hao, W.M.; Saharjo, B.H.; Ward, D.E., 2003: "Comprehensive Laboratory Measurements of Biomass-Burning Emissions: 1. Emissions from Indonesian, African, and Other Fuels", in: *Journal of Geophysical Research*, 108.

290. Crutzen, P.J.; Lelieveld, J., 2003: Comment on the paper by C. G. Roberts et al. "Cloud Condensation Nuclei in the Amazon Basin: "Marine" Conditions over a Continent?", in: *Geophysical Research Letters*, 30, doi:10.1029/2002GL015206.

291. Crutzen, P.J.; Ramanathan, V., 2003: "The Parasol Effect on Climate", in: *Science*, 302: 1679–1680.

292. Crutzen, P.J.; Steffen, W., 2003: "How Long Have You Been in the Anthropocene Era? An Editorial Comment", in: *Climatic Change*, 61: 251–257.

293. von Glasow, R.; Crutzen, P.J.; 2003: "Tropospheric Halogen Chemistry', in: Holland, H.D.; Turekian, K.K.; Keeling, R.F. (Eds.): *Treatise on Geochemistry* (Elsevier Pergamon): 21–64.

294. von Glasow, R.; Lawrence, M.G.; Sander, R.; Crutzen, P.J., 2003: "Modeling the Chemical Effects of Ship Exhaust in the Cloud-Free Marine Boundary Layer", in: *Atmospheric Chemistry and Physics*, 3: 233–250.

295. Guazzotti, S.A.; Suess, D.T.; Coffee, K.R.; Quinn, P.K.; Bates, T.S.; Wisthaler, A.; Hansel, A.; Ball, W.P.; Dickerson, R.R.; Neusüß, C.; Crutzen, P.J.; Prather, K.A., 2003: "Characterization of Carbonaceous Aerosols Outflow from India and Arabia: Biomass/Biofuel Burning and Fossil Fuel Combustion", in: *Journal of Geophysical Research*, 108, doi:10.1029/2002JD003277.

296. Jöckel, P.; Brenninkmeijer, C.A.M.; Crutzen, P.J., 2003: "A Discussion on the Determination of Atmospheric OH and Its Trends", in: *Atmospheric Chemistry and Physics*, 3: 107–118.

297. Kaiser, J.; Brenninkmeijer, C.A.M.; Röckmann, T.; Crutzen, P.J., 2003: "Wavelength Dependence of Isotope Fractionation in N_2O Photolysis", in: *Atmospheric Chemistry and Physics*, 3: 303–313.

298. von Kuhlmann, R.; Lawrence, M.G.; Crutzen, P.J.; Rasch, P.J., 2003: "A Model for Studies of Tropospheric Ozone and Non-methane Hydrocarbons: Model Description and Ozone Results", in: *Journal of Geophysical Research*, 108, doi:10.1029/2002JD002893.

299. von Kuhlmann, R.; Lawrence, M.G.; Crutzen, P.J.; Rasch, P.J., 2003: "A Model for Studies of Tropospheric Ozone and Nonmethane Hydrocarbons: Model Evaluation of Ozone-Related Species", in: *Journal of Geophysical Research*, 108, doi:10.1029/2002JD003348.

300. Lawrence, M.G.; Rasch, P.J.; von Kuhlmann, R.; Williams, J.; Fischer, H.; de Reus, M.; Lelieveld, J.; Crutzen, P.J.; Schultz, M.; Stier, P.; Huntrieser, H.; Helans, J.; Stohl, A.; Forster, C.; Elbern, H.; Jakobs, H.; Dickerson, R.R., 2003: "Global Chemical Weather Forecasts for Field Campaign Planning: Predictions and Observations of Large-Scale Features During MINOS, CONTRACE, and INDOEX", in: *Atmospheric Chemistry and Physics*, 3: 267–289.

301. Ramanathan, V.; Crutzen, P.J., 2003: "New Directions: Atmospheric Brown "Clouds"", in: *Atmospheric Environment*, 37: 4033–4035.

302. Sander, R.; Keene, W.C.; Pszenny, A.A.P.; Arimoto, R.; Ayers, G.P.; Baboukas, E.; Cainey, J.M.; Crutzen, P.J.; Duce, R.A.; Hönninger, G.; Huebert, B.J.; Maenhaut, W.; Mihalopoulos, N.; Turekian, V.C.; Van Dingenen, R., 2003: "Inorganic Bromine in the Marine Boundary Layer: A Critical Review", in: *Atmospheric Chemistry and Physics*, 3: 1301–1336.

303. Steil, B.; Brühl, C.; Manzini, E.; Crutzen, P.J.; Lelieveld, J.; Rasch, P.J.; Roeckner, E.; Krüger, K., 2003: "A New Interactive Chemistry-Climate Model: 1. Present-Day Climatology and Interannual Variability of the Middle Atmosphere Using the Model and 9 years of HALOE/UARS Data", in: *Journal of Geophysical Research*, 108, doi:10.1029/2002JD002971.

304. van Aalst, M.K.; van den Broek, M.M.P.; Bregman, A.; Brühl, C.; Steil, B.; Toon, G.C.; Garcelon, S.; Hansford, G.M.; Jones, R.L.; Gardiner, T.D.; Roelofs, G.J.; Lelieveld, J.; Crutzen, P.J., 2004: "Trace Gas Transport in the 1999/2000 Arctic Winter: Comparison of Nudged GCM Runs with Observations", in: *Atmospheric Chemistry and Physics*, 4: 81–93.

305. Christian, T.J.; Kleiss, B.; Yokelson, R.J.; Holzinger, R.; Crutzen, P.J.; Hao, W.M.; Shirai, T.; Blake, D.R., 2004: "Comprehensive Laboratory Measurements of Biomass-Burning Emissions: 2. First Intercomparison of Open-Path FTIR, PTR-MS, GC-MS/FID/ECD", in: *Journal of Geophysical Research*, 109, doi:10.1029/2003JD003874.

306. Clark W.C.; Crutzen, P.J.; Schellnhuber, H.J., 2004: "Science for Global Sustainability", in: Schellnhuber, H.J.; Crutzen, P.J.; Clark, W.C.;

Claussen, M.; Held, H. (Eds.): *Earth System Analysis for Sustainablility*. Dahlem Workshop Report (Cambridge, USA: MIT Press): 1–28

307. Crutzen, P.J., 2004: "New Directions: The Growing Urban Heat and Pollution "Island" Effect—Impact on Chemistry and Climate", in: *Atmospheric Environment*, 38: 3539–3540.

308. Crutzen, P.J.; Ramanathan, V., 2004: "Atmospheric Chemistry and Climate in the Anthropocene. Where are we Heading?", in: Schellnhuber, H.J.; Crutzen, P.J.; Clark, W.C.; Claussen, M.; Held, H. (Eds.): *Earth System Analysis for Sustainablility*. Dahlem Workshop Report (Cambridge, USA: MIT Press): 265–292.

309. Hurst, D.F.; Romashkin, P.A.; Elkins, J.W.; Oberlander, E.A.; Elansky, N.F.; Belikov, I.B.; Granberg, I.G.; Golitsyn, G.S.; Grisenko, A.M.; Brenninkmeijer, C.A.M.; Crutzen, P.J., 2004: "Emissions of Ozone-Depleting Substances in Russia During 2001", in: *Journal of Geophysical Research*, 109, doi:10.1029/2004JD004633.

310. Gabrielli, P.; Barbante, C.; Plane, J.M.C.; Varga, A.; Hong, S.; Cozzi, G.; Gaspari, V.; Planchon, F.A.M.; Cairns, W.; Ferrari, C.; Crutzen, P.; Cescon, P.; Boutron, C.F., 2004: "Meteoric Smoke Fallout over the Holocene Epoch Revealed by Iridium and Platiunum in Greenland Ice", in: *Nature*, 432: 1011–1014.

311. von Glasow, R.; Crutzen, P.J., 2004: "Model Study of Multiphase DMS Oxidation with a Focus on Halogens", in: *Atmospheric Chemistry and Physics*, 4: 589–608.

312. von Glasow, R.; von Kuhlmann, R.; Lawrence, M.G.; Platt, U.; Crutzen, P.J., 2004: "Impact of Reactive Bromine Chemistry in the Troposphere", in: *Atmospheric Chemistry and Physics*, 4: 2481–2497.

313. von Kuhlmann, R.; Lawrence, M.G.; Pöschl, U.; Crutzen, P.J., 2004: "Sensitivities in Global Scale Modeling of Isoprene", in: *Atmospheric Chemistry and Physics*, 4: 1–17.

314. Peters, W.; Krol, M.C.; Fortuin, J.P.F.; Kelder, H.M.; Thompson, A.M.; Becker, C.R.; Lelieveld, J.; Crutzen, P.J., 2004: "Tropospheric Ozone over a Tropical Atlantic Station in the Northern Hemisphere: Paramaribo, Surinam (6 N, 55 W)", in: *Tellus B*, 56,1: 21–34.

315. Pszenny, A.P.; Moldanová, J.; Keene, W.C.; Sander, R.; Maben, J.R.; Martinez, M.; Crutzen, P.J.; Perner, D.; Prinn, R.G., 2004: "Halogen Cycling and Aerosol pH in the Hawaiian Marine Boundary Layer", in: *Atmospheric Chemistry and Physics*, 4: 147–168.

316. Richter, A.; Eyring, V.; Burrows, J.P.; Bovensmann, H.; Lauer, A.; Sierk, B.; Crutzen, P.J., 2004: "Satellite Measurements of NO_2 from International Shipping Emissions", in: *Geophysical Research Letters*, 31, doi:10.1029/2004GL020822.

317. Sander, R.; Crutzen, P.J.; von Glasow, R., 2004: "Comment on Reactions at Interfaces as a Source of Sulfate Formation in Sea-Salt Particles (II)", in: *Science*, 303: 628.

318. Sanhueza, E.; Holzinger, R.; Kleiss, B.; Donoso, L.; Crutzen, P.J., 2004: "New Insights in the Global Cycle of Acetonitrile: Release from the Ocean and Dry Deposition in the Tropical Savanna of Venezuela", in: *Atmospheric Chemistry and Physics*, 4: 275–280.

319. Steffen, W.; Andreae, M.O.; Bolin, B.; Cox, P.M.; Crutzen, P.J.; Cubasch, U.; Held, H.; Nakićenović, N.; Scholes, R.J.; Talaue-McManus, L.; Turner II, B.L., 2004: "Abrupt Changes: The Achilles' Heels of the Earth System", in: *Environment*, 46: 8–20 (also published as IIASA rapport RR-04-006, June 2004).

320. Steffen, W.; Andreae, M.O.; Cox, P.M.; Crutzen, P.J.; Cubasch, U.; Held, H.; Nakicenovic, N.; Talaue-McManus, L.; Turner II, B.L., 2004: "Group Report: Earth System Dynamics in the Anthropocene", in: Schellnhuber, H.J.; Crutzen, P.J.; Clark, W.C.; Claussen, M.; Held, H. (Eds.): *Earth System Analysis for Sustainablility*. Dahlem Workshop Report (Cambridge, USA: MIT Press): 313–340.

321. Vrekoussis, M.; Kanakidou, M.; Mihalopoulos, N.; Crutzen, P.J.; Lelieveld, J.; Perner, D.; Berresheim, H.; Baboukas, E., 2004: "Role of NO_3 Radicals in Oxidation Processes in the Eastern Mediterranean Troposphere During the MINOS Campaign", in: *Atmospheric Chemistry and Physics*, 4: 169–182.

322. Wallström, M.; Bolin, B.; Crutzen, P.J.; Steffen, W., 2004: "The Earth's Life-Support System is in Peril", in: *Global Change NewsLetter*, 57: 22–23.

323. Brenninkmeijer, C.A.M.; Slemr, F.; Koeppel, C.; Scharffe, D.S.; Pupek, M.; Lelieveld, J.; Crutzen, P.; Zahn, A.; Sprung, D.; Fischer, H.; Hermann, M.; Reichelt, M.; Heintzenberg, J.; Schlager, H.; Ziereis, H.; Schumann, U.; Dix, B.; Platt, U.; Ebinghaus, R.; Martinsson, B.; Ciais, P.; Flippi, D.; Leuenberger, M.; Oram, D.; Penkett, S.; van Velthoven, P.; Waibel, A., 2005: "Analyzing Atmospheric Trace Gases and Aerosols Using Passenger Aircraft", in: *EOS*, 86,8: 77, 82, 83.

324. Fishman, J.; Creilson, J.K.; Wozniak, A.E.; Crutzen, P.J., 2005: "Interannual Variability of Stratospheric and Tropospheric Ozone Determined from Satellite Measurements", in: *Journal of Geophysical Research*, 110, doi:10.1029/2005JD005868.

325. Gabrielli, P.; Plane, J.M.C.; Boutron, C.F.; Hong, S.; Cozzi, G.; Cescon, P.; Ferrari, C.; Crutzen, P.J.; Petit, J.R.; Lipenkov, V.Y.; Barbante, C., 2006: "A Climatic Control on the Accretion of Meteoric and Super-Chondritic Iridium-Platinum to the Antarctic Ice Cap", in: *Earth and Planetary Science Letters*, 250: 459–469.

326. Holzinger, R.; Williams, J.; Salisbury, G.; Klüpfel, T.; de Reus, M.; Traub, M.; Crutzen, P.J.; Lelieveld, J., 2005: "Oxygenated Compounds in Aged Biomass Burning Plumes over the Eastern Mediterranean: Evidence for Strong Secondary Production of Methanol and Acetone", in: *Atmospheric Chemistry and Physics*, 5: 39–46.

327. Schellnhuber, H.J.; Crutzen, P.J.; Clark, W.C.; Hunt, J., 2005: "Earth System Analysis for Sustainability", in: *Environment*, 47: 11–25.

328. Crutzen, P.J., 2006: "Albedo Enhancement by Stratospheric Sulfur Injections: A Contribution to Resolve a Policy Dilemma?", in: *Climatic Change*, doi:10. 1007/s10584-006-9101-y (online published).

329. Keene, W.C.; Lobert, J.M.; Crutzen, P.J.; Maben, J.R.; Scharffe, D.H.; Landmann, T.; Hély, C.; Brain, C., 2006: "Emissions of Major Gaseous and Particulate Species During Experimental Burns of Southern African Biomass", in: *Journal of Geophysical Research*, 111, doi:10.1029/ 2005JD006319.

330. Vrekoussis, M.; Liakakou, E.; Mihalopoulos, N.; Kanakidou, M.; Crutzen, P. J.; Lelieveld, J., 2006: "Formation of HNO_3 and NO_3 in the Anthropogenically-Influenced Eastern Mediterranean Marine Boundary Layer", in: *Geophysical Research Letters*, 33, doi:10.1029/2005GL025069.

331. Birks, J.W.: Crutzen, P.J.; Roble, R.G., 2007: "Frequent Ozone Depletion Resulting from Impacts of Asteroids and Comets", in: Bobrowsky, P.; Rickman, H. (Eds.): *Comet/Asteroid Impacts and Human Society* (Berlin: Springer): 225–245, doi:10.1007/978-3-540-32711-0.

332. Brenninkmeijer, C.A.M.; Crutzen, P.; Boumard, F.; Dauer, T.; Dix, B.; Ebinghaus, R.; Filippi, D.; Fischer, H.; Franke, H.; Frieß, U.; Heintzenberg, J.; Helleis, F.; Hermann, M.; Kock, H.H.; Koeppel, C.; Lelieveld, J.; Leuenberger, M.; Martinsson, B.G.; Miemczyk, S.; Moret, H.P.; Nguyen, H. N.; Nyfeler, P.; Oram, D.; O'Sullivan, D.; Penkett, S.; Platt, U.; Pupek, M.; Ramonet, M.; Randa, B.; Reichelt, M.; Rhee, T.S.; Rohwer, J.; Rosenfeld, K.; Scharffe, D.; Schlager, H.; Schumann, U.; Slemr, F.; Sprung, D.; Stock, P.; Thaler, R.; Valentino, F.; van Velthoven, P.; Waibel, A.; Wandel, A.; Waschitschek, K.; Wiedensohler, A.; Xueref-Remy, I.; Zahn, A.; Zech, U.; Ziereis, H., 2007: "Civil Aircraft for the Regular Investigation of the Atmosphere Based on an Instrumented Container: The New CARIBIC System", in: *Atmospheric Chemistry and Physics*, 7: 4953–4976.

333. von Glasow, R.; Crutzen, P.J., 2007: "Tropospheric Halogen Chemistry", in: Holland, H.D.; Turekian K.K. (Eds.): *Treatise on Geochemistry Update1*, Vol. 4.02, 1–67.

334. Lelieveld, J.; Brühl, C.; Jöckel, P.; Steil, B.; Crutzen, P.J.; Fischer, H.; Giorgetta, M.A.; Hoor, P.; Lawrence, M.G.; Sausen, R.; Tost, H., 2007: "Stratospheric Dryness: Model Simulations and Satellite Observations", in: *Atmospheric Chemistry and Physics*, 7: 1313–1332.

335. Steffen, W.; Crutzen, P.J.; McNeill, J.R., 2007: "The Anthropocene: Are Humans Now Overwhelming the Great Forces of Nature?", in: *Ambio*, 36: 614–621.

336. Vreskoussis, M.; Mihalopoulos, N.; Gerasopoulos, E.; Kanakidou, M.; Crutzen, P.J.; Lelieveld, J., 2007: "Two-Years of NO_3 Radical Observations in the Boundary Layer over the Eastern Mediterranean", in: *Atmospheric Chemistry and Physics*, 7: 315–327.

337. Crutzen, P.J.; Oppenheimer, M., 2008: "Learning About Ozone Depletion", in: *Climatic Change*, 89: 143–154, doi:10.1007/s10584-008-9400-6.

338. Crutzen, P.J.; Mosier, A.R.; Smith, K.A.; Winiwarter, W., 2008: "N_2O Release from Agro-biofuel Production Negates Global Warming Reduction by Replacing Fossil Fuels", in: *Atmospheric Chemistry* and *Physics*, 8: 389–395.

339. O'Neill, B.C.; Crutzen, P.; Grübler, A.; Duong, M.H.; Keller, K.; Kolstad, C.; Koomey, J.; Lange, A.; Obersteiner, M.; Oppenheimer, M.; Pepper, W.; Sanderson, W.; Schlesinger, M.; Treich, N.; Ulph, A.; Webster, M.; Wilson, C., 2008: "Learning and Climate Change", in: *Climate Policy*, 6: 585–589.

340. Rasch, P.J.; Crutzen, P.J.; Coleman, D.B., 2008: "Exploring the Geoengineering of Climate Using Stratospheric Sulfate Aerosols: The Role of Particle Size", in: *Geophysical Research Letters*, 35, doi:10.1029/2007GL032179.

341. Mosier, A.R.; Crutzen, P.J.; Smith, K.A.; Winiwarter, W., 2009: "Nitrous Oxide's Impact on Net Greenhouse Gas Savings from Biofuels: Life-Cycle Analysis Comparison", in: *International Journal of Biotechnology*, 11: 60–74.

342. Rockström, J.; Steffen, W.; Noone, K.; Persson, A.; Chapin, F.S.; Lambin, E.F.; Lenton, T.M.; Scheffer, M.; Folke, C.; Schellnhuber, H.J.; Nykvist, B.; de Wit, C.A.; Hughes, T.; van der Leeuw, S.; Rodhe, H.; Sörlin, S.; Snyder, P.K.; Costanza, R.; Svedin, U.; Falkenmark, M.; Karlberg, L.; Corell, R.W.; Fabry, V.J.; Hansen, J.; Walker, B.; Liverman, D.; Richardson, K.; Crutzen, P.; Foley, J., 2009: "A Safe Operating Space for Humanity", in: *Nature*, 461: 472–475.

343. Rockström, J.; Steffen, W.; Noone, K.; Persson, A.; Chapin, F.S.; Lambin, E. F.; Lenton, T.M.; Scheffer, M.; Folke, C.; Schellnhuber, H.J.; Nykvist, B.; de Wit, C.A.; Hughes, T.; van der Leeuw, S.; Rodhe, H.; Sörlin, S.; Snyder, P. K.; Costanza, R.; Svedin, U.; Falkenmark, M.; Karlberg, L.; Corell, R.W.; Fabry, V.J.; Hansen, J.; Walker, B.; Liverman, D.; Richardson, K.; Crutzen, P.; Foley, J., 2009: "Planetary Boundaries: Exploring the Safe Operating Space for Humanity", in: *Ecology and Society*, 14,2, art. 32.

344. Williams, J.; Crutzen, P.J., 2010: "Nitrous Oxide from Aquaculture", in: *Nature Geoscience*, 3: 143.

345. Gleick, P.H.; et al., 2010: "Climate Change and the Integrity of Science", in: *Science Letters*, 328: 689–690.

346. Zalasiewicz, J.; Williams, M.; Steffen, W.; Crutzen, P., 2010: "The New World of the Anthropocene", in: *Environmental Science and Technology*, 44: 2228–2231.

347. Zalasiewicz J., Williams, M.; Steffen, W.; Crutzen, P., 2010: "Response to the Anthropocene Forces Us to Reconsider Adaptationist Models of Human-Environment Interactions", in: *Environmental Science and Technology*, 44,16: 6008, doi:10.1021/es102062w.

348. Pierazzo, E.; Garcia, R.R.; Kinnison, D.E.; Marsh, D.R.; Lee-Taylor, J.; Crutzen, P.J., 2010: "Ozone Perturbation from Medium-Size Asteroid Impacts in the Ocean", in: *Earth and Planetary Science Letters*, 299: 263–272.

349. Zander, R.; Duchatelet, P.; Mahieu, E.; Demoulin, P.; Roland, G.; Servais, C.; Auwera, J.V.; Perrin, A.; Rinsland, C.P.; Crutzen, P.J., 2010: "Formic Acid Above the Jungfraujoch During 1985–2007: Observed Variability, Seasonality, but no Long-Term Backround Evolution", in: *Atmospheric Chemistry and Physics*, 10: 10047–10065.

350. Brühl, C.; Lelieveld, J.; Crutzen, P.J.; Tost, H., 2012: "The Role of Carbonyl Sulphide as a Source of Stratospheric Sulphate Aerosol and Its Impact on Climate", in: *Atmospheric Chemistry and Physics*, 12: 1239–1253 (2012)

351. Crutzen, P.J.: "Sherry Rowland: Ozone and Advocacy", in: *Nature Geoscience*, 5: 311 (2012).

352. De Fries, R.S.; Ellis, E.C.; Stuart Chapin III, F.; Matson, P.A.; Turner II, B.L.; Agrawal, A.; Crutzen, P.J.; Field, C.; Gleick, P.; Kareiva, P.M.; Lambin, E.; Liverman, D.; Ostrom, E.; Sanchez, P.A.; Syvitski, J., 2012: "Planetary Opportunities: A Social Contract for Global Change Science to Contribute to a Sustainable Future", in: *BioScience*, 62: 603–606.

353. Smith, K.A.; Mosier, A.R.; Crutzen, P.J.; Winiwarter, W., 2012: "The Role of N_2O Derived from Crop-Based Biofuels, and from Agriculture in General, in Earth's Climate", in: *Philosophical Transactions of the Royal Society B*, 367: 1169–1174.

354. Reay, D.S.; Davidson, E.A.; Smith, K.A.; Smith, P.; Melillo, J.M.; Dentener, F.; Crutzen, P.J., 2012: "Global Agriculture and Nitrous Oxide Emissions", in: *Nature Climate Change*, 2, 410–416.

355. Williams, J.; Crutzen, P.J., 2013: "Perspectives on Our Planet in the Atmosphere", in: *Environmental Chemistry*, 10: 269–280, doi:http://dx.doi. org/10.1071/EN13061.

356. Elshorbany, Y.F.; Crutzen, P.; Steil, B.; Pozzer, A.; Tost, H.; Lelieveld, J., 2014: "Global and Regional Impacts of HONO on the Chemical Composition of Clouds and Aerosols", in: *Atmospheric Chemistry and Physics*, 14: 1167–1184.

357. Foley, S.F.; Gronenborn, D.; Andreae, M.O.; Kadereit, J.W.; Esper, J.; Scholz, D.; Pöschl, U.; Jacob, D.E.; Schöne, B.R.; Schreg, R.; Vött, A.; Jordan, D.; Lelieveld, J.; Weller, C.G.; Alt, K.W.; Gaudzinski-Windheuser, S.; Bruhn, K.-C.; Tost, H.; Sirocko, F.; Crutzen, P.J., 2013: "The Palaeoanthropocene—The Beginnings of Anthropogenic Environmental Change", in: *Anthropocene*, 3: 83–88.

358. Lenhart, K.; Weber, B.; Elbert, W.; Steinkamp, J.; Müller, C.; Clough, T.; Crutzen, P.; Pöschl, U.; Keppler, F., 2015: "Nitrous Oxide and Methane Emissions from Lichens and Mosses", in: *Global Change Biology*, doi:10. 1111/gcb.12995.

359. Weber, B.; Wu, D.; Tamm, A.; Ruckteschler, N.; Rodríguez-Caballero, E.; Steinkamp, J.; Meusel, H.; Elbert, W.; Behrend, T.; Sörgel, M.; Cheng, Y.; Crutzen, P.; Su, H.; Pöschl, U., 2015: "Biological Soil Crusts Accelerate the Nitrogen Cycle Through Large NO and HONO Emissions in Drylands", in: *PNAS*.

360. Crutzen, P.J., 2014: "The Anthropocene: When Humankind overrides Nature", in: Schmidt, F.; Nuttall, N. (Eds.): *Contributions towards a sustainable world—in dialogue with Klaus Töpfer* (München: Oekom Verlag): 21–27.

361. Zalasiewicz, J.; Water, C.N.; Williams, M.; Barnosky, A.D.; Cearreta, A.; Crutzen, P.; Ellis, E.; Ellis, M.A.; Fairchild, I.J.; Grinevald, J.; Haff, P.K.; Hajdas, I.; Leinfelder, R.; McNeill, J.; Odada, E.O.; Poirier, C.; Richter, D.; Steffen, W.; Summerhayes, C.; Syvitski, J.P.M.; Vidas, D.; Wagreich, M.; Wing, S.L.; Wolfe, A.P.; Zhisheng A.; Oreskes, N., 2015: "When did the Anthropocene begin? A mid-twentieth century boundary level is stratigraphically optimal", in: *Quaternary International*, 383: 196–203.

2.4 Other Publications (Unrefereed)

A1. Crutzen, P.J., 1969: "Koldioxiden och klimatet (Carbon Dioxide and Climate)", in: *Forskning och Framsteg*, 5: 7–9.

A2. Crutzen, P. J., 1971: "On Some Photochemical and Meteorological Factors Determining the Distribution of Ozone in the Stratosphere: Effects on Contamination by NO_x Emitted from Aircraft", Technical Report UDC 551.510.4, Institute of Meteorology, University of Stockholm.

A3. Crutzen, P.J., 1972: "Gas-Phase Nitrogen and Methane Chemistry in the Atmosphere", Report AP-10, Institute of Meteorology, University of Stockholm, 20 pp.

A4. Crutzen, P.J., 1972: "The Photochemistry of the Stratosphere with Special Attention Given to the Effects of NOx Emitted by Supersonic Aircraft", First Conference on CIAP, United States Department of Transportation, 80–88.

A5. Crutzen, P.J., 1972: "Liten risk för klimatändring (Small Risk for Climatic Change; in Swedish)", in: *Forskning och Framsteg*, 2: 27.

A6. Crutzen, P.J., 1974: "Artificial Increases of the Stratospheric Nitrogen Oxide Content and Possible Consequences for the Atmospheric Ozone", Technical Report UDC 551.510.4:546.2, Institute of Meteorology, University of Stockholm.

A7. Crutzen, P.J., 1974: "Väderforskning med matematik (Weather Research with Mathematics; in Swedish)", in: *Forskning och Framsteg*, 6: 22–23, 26.

A8. Crutzen, P.J., 1975: "Physical and Chemical Processes Which Control the Production, Destruction and Distribution of Ozone and Some Other Chemically Active Minor Constituents", in: *GARP Publications Series 16*, World Meterological Organization, Geneva, Switzerland.

A9. Crutzen, P.J., 1975: "A Two-Dimensional Photochemical Model of the Atmosphere Below 55 km", in: Hard, T.M.; Brodrick, A.J. (Eds.):

Estimates of Natural and Man-Caused Ozone Perturbations due to NOₓ. *Proceedings of 4th CIAP Conference.* DOT-TSC-OST-75-38 (Cambridge: U.S. Department of Transportation): 264–279.

A10. Crutzen, P.J., 1976: "Ozonhöljet tunnas ut: Begränsa spray-gaserna (The Ozone Shield is Thinning: Limit the Use of Aerosol Propellants; in Swedish)", in: *Forskning och Framsteg*, 5: 29–35.

A11. Heath, D.F.; Krueger, A.J.; Crutzen, P.J., 1976: *Influence of a Solar Proton Event on Stratospheric Ozone* (Greenbelt, Maryland: Goddard Space Flight Center).

A12. Crutzen, P.J., 1977: "The Stratosphere-Meosphere", in: White, O.R. (Ed.): *Solar Output and Its Variations* (Boulder, Colorado: Colorado Associated University Press): 13–16.

A13. Crutzen, P.J.; Fishman, J.; Gidel, L.T.; Chatfield, R.B., 1978: "Numerical Investigations of the Photochemical and Transport Processes Which Affect Halocarbons and Ozone in the Atmosphere", in: *Annual Summary of Research* (Fort Collins, CO: Dept. of Atmospheric Science, Colorado State University).

A14. Fishman, J.; Crutzen, P.J., 1978: "The Distribution of the Hydroxyl Radical in the Troposphere", Atmospheric Science Paper 284 (Fort. Collins, CO; Dept. of Atmos. Sci., Colorado State University).

A15. Crutzen, P.J.; Gidel, L.T.; Fishman, J., 1979: "Numerical Investigations of the Photochemical and Transport Processes Which Affect Ozone and Other Trace Constituents in the Atmosphere", in: *Annual Summary of Research* (Fort Collins, CO: Dept. of Atmospheric Science, Colorado State University).

A16. Fishman, J.; Crutzen, P.J., 1979: "A Preliminary Estimate of Stratospheric Ozone Depletion by the Release of Chlorocarbon Chemicals as Calculated by a Two-Dimensional Photochemical Model of the Atmosphere to 55 km", First Quarterly Progress Report EPA Grant R804921-03

A17. Crutzen, P.J., 1981: "Atmospheric Chemical Processes of the Oxides of Nitrogen Including Nitrous Oxide", in: Delwiche, C.C. (Ed.): *Denitrification, Nitrification and Atmospheric Nitrous Oxide* (New York: Wiley): 17–44.

A18. Birks, J.W.; Crutzen, P.J., 1983: "Atmospheric Effects of a Nuclear war", in: *Chemistry in Britain*, 19: 927–930.

A19. Galbally, I.E.; Crutzen, P.J.; Rodhe, H., 1983: "Some Changes in the Atmosphere over Australia That May Occur due to a Nuclear War", in: Denborough, M.A. (Ed.): *Australia and Nuclear War* (Canberrra, Australia: Croom Helm Ltd.): 161–185, 270 pp.

A20. Brühl, C.; Crutzen, P.J., 1984: "A Radiative Convective Model to Study the Sensitivity of Climate and Chemical Composition to a Variety of Human Activities", in: Ghazi, A. (Ed.): *Proceedings of a Working Party Meeting, Brussels, 18th May 1984*, CEC, pp. 84–94.

A21. Crutzen, P.J., 1985: "The Global Environment After Nuclear War", in: *Environment*, 27: 6–11.

A22. Crutzen, P.J., 1985: "Global Aspects of Atmospheric Chemistry: Natural and Anthropogenic Influences", Presented at the III. Bi-National Colloquium of the Alexander von Humboldt Foundation (Bonn, FRG) at Northern University, Evanston, Illinois, USA, September 17–20, 1985.

A23. Crutzen, P.J.; Andreae, M.O., 1985: "Atmospheric Chemistry", in: Malone, T.F.; Roederer, J.G. (Eds.): *Global Change* (Cambridge: Cambridge University Press): 75–113.

A24. Crutzen, P.J.; Galbally, I.E., 1985: "Atmospheric Conditions After a Nuclear War", in: Marini-Bettolo, G.B. (Ed.): *Chemical Events in the Atmosphere and Their Impact on the Environment* (Città del Vaticano: Pontificiae Academiae Scientiarum Scripta Varia): 457–502.

A25. Crutzen, P.J.; Hahn, J., 1985: "Atmosphärische Auswirkungen eines Atomkrieges", in: *Physik in unserer Zeit*, 16: 4–15.

A26. Klose, W.; Butin, H.; Crutzen, P.J.; Führ, F.; Greim, H.; Haber, W.; Hahlbrock, K.; Hüttermann, A.; Klein, W.; Klug, W.; Moosmayer, H.U.; Obländer, W.; Prinz, B.; Rehfuess, K.E.; Rentz, O., 1985: "Forschungsbeirat Waldschäden/Luftverunreinigungen der Bundesregierung und der Länder, Zwischenbericht Dezember 1984", in: *Bericht über den Stand der Erkenntnisse zur Ursache.*

A27. Crutzen, P.J., 1986: *Globale Aspekte der atmosphärischen Chemie: Natürliche und anthropogene Einflüsse, Vorträge Rheinisch-Westfälische Akademie der Wissenschaften* (Opladen: Westdeutscher Verlag GmbH): S. 41–72.

A28. Klose, W.; Butin, H.; Crutzen, P.J., 1986: u.a. Forschungsbeirat Waldschäden/Luftverunreinigungen der Bundesregierung und der Länder, 2. Bericht, 229 S.

A29. Crutzen, P.J., 1987: "Recent Depletions of Ozone with Emphasis on the Polar "Ozone Hole"", in: *Källa*, 28 (Stockholm; in Swedish).

A30. Crutzen, P.J., 1987: "Climatic Effects of Nuclear War, Annex 2", in: *Effects of Nuclear War on Health and Health Services*, Report A40/11 of the World Health Organization to the 40th World Health Assembly, 18 March 1987; WHO, Geneva.

A31. Crutzen, P.J., 1987: *Ozonloch und Spurengase - Menschliche Einflüsse auf Klima und Chemie der Atmosphäre* (München: Max-Planck-Gesellschaft, Jahrbuch 1987): S. 27–40.

A32. Darmstadter, J.; Ayres, L.W.; Ayres, R.U.; Clark, W.C.; Crosson, R.P.; Crutzen, P.J.; Graedel, T.E.; McGill, R.; Richards, J.F.; Torr, J.A., 1987: Impacts of World Development on Selected Characteristics of the Atmosphere: An Integrative Approach, Oak Ridge National Laboratory, 2 Volumes, ORNL/Sub/86-22033/1/V2, Oak Ridge, Tennessee 37931, USA.

A33. Crutzen, P.J., 1988: "Das Ozonloch – Menschliche Einflüsse auf die Chemie der globalen Atmosphäre", in: *Gewerkschaftliche Monatshefte 12'88 „Der blaue Planet in der Krise"*, 731–745.

A34. Brühl, C.; Crutzen, P.J., 1989: "The Potential Role of Odd Hydrogen in the Ozone Hole Photochemistry", in: Crutzen, P.J.; Gerard, J.-C.; Zander, R. (Eds.): *Our Changing Atmosphere* (Belgium: Université de Liège, Institut d'Astrophysique, B-4200 Cointe-Ougree): 171–177.

A35. Crutzen, P.J.; Hao, W.M.; Liu, M.H.; Lobert, J.M.; Scharffe, D., 1989: "Emissions of CO_2 and Other Trace Gases to the Atmosphere from Fires in the Tropics", in: Crutzen, P.J.; Gerard, J.-C.; Zander, R. (Eds.): *Our Changing Atmosphere*, Proceedings of the 28th Liège International Astrophysical Colloqium (Belgium: Université de Liège): 449–471.

A36. Graedel, T.E.; Crutzen, P.J., 1989: "The Changing Atmosphere", in: *Scientific American*, 160: 58–68 (in deutsch: Veränderungen der Atmosphäre. *Spektrum der Wissenschaften*, 11: 58–68).

A37. Lelieveld, J.; Crutzen, P.J.; Rodhe, H., 1989: "Zonal Average Cloud Characteristics for Global Atmospheric Chemistry Modelling", Report CM-76, UDC 551.510.4, Glomac 89/1. International Meteorological Institute in Stockholm, University of Stockholm, 54 pp.

A38. Crutzen, P.J., 1990: "Auswirkungen menschlicher Aktivitäten auf die Erdatmosphäre: Was zu forschen, was zu tun? *DLR-Nachrichten*", in: Heft, 59: 5–13.

A39. Crutzen, P.J., 1990: "Comments on George Reid's "Quo Vadimus" Contribution "Climate"", in: Garland, G.D.; Apel, John R.(Eds.): *Quo Vadimus. Geophysics for the Next Generation, Geophysical Monograph 60, IUGG Vol 10* (Washington, USA: American Geophysical Union): 47.

A40. Crutzen, P.J., 1990: "Global Changes in Tropospheric Chemistry", Proceedings of Summer School on Remote Sensing and the Earth's Environment, Alpbach, Austria, 26 July–4 August 1989, pp. 105–113.

A41. Crutzen, P.J.; Brühl, C., 1990: "The Potential Role of HO_x and ClO_x Interactions in the Ozone Hole Photochemistry", in: O'Neil, A. (Ed.): *Dynamics, Transport and Photochemistry in the Middle Atmosphere of the Southern Hemisphere* (Dordrecht: Kluwer): 203–212.

A42. Crutzen, P.J.; Brühl, C., 1990: "The Atmospheric Chemical Effects of Aircraft Operations", in: Schumann (Ed.): *Air Traffic and the Environment —Background, Tendencies and Potential Global Atmospheric Effects.* Proceedings of a DLR International Colloqium Bonn, Germany, November 15/16, 1990 (Heidelberg: Springer): 96–106.

A43. Horowitz, A.; von Helden, G.; Schneider, W.; Simon, F.G.; Crutzen, P.J.; Moortgat, G.K., 1990: "Oxygen photolysis at 214 nm and 25°C", in: Boikov, R.D.; Fabian, P. (Eds.): *Proceedings of the Quadrennial Ozone Symposium, Göttingen 8–13 August 1988* (Deepak Publ. Co.) 690–693.

A44. Brühl, C.; Crutzen, P.J.; Danielsen, E.F.; Graßl, H.; Hollweg, H.-D.; Kley, D., 1991: *Umweltverträglichkeitsstudie für das Raumtransportsystem*

SÄNGER, Teil 1 Unterstufe (Max-Planck-Institut für Meteorologie Hamburg): 142 pp.

A45. Crutzen, P.J., 1991: "Methane's Sinks and Sources", in: *Nature*, 350: 380–382.

A46. Lelieveld, J.; Crutzen, P.J., 1991: "Climate Discussion and Fossil Fuels", in: *Oil Gas—European Magazine*, 4: 11–15.

A47. Crutzen, P.J., 1992: "Ozone Depletion: Ultraviolet on the Increase", in: *Nature*, 356: 104–105.

A48. Crutzen, P.J., 1992: "Menschliche Einflüsse auf das Klima und die Chemie der globalen Atmosphäre", in: *Stadtwerke der Zukunft* - ASEW-Fachtagung Kassel, 1991 (Bochum: ASEW, Köln, Ponte Press): 7–27.

A49. Graedel, T.E.; Crutzen, P.J., 1992: "Ensemble Assessments of Atmospheric Emissions and Impacts", in: *Energy and the Environment in the 21st Century* (Cambridge: Energy Laboratory, Massachusetts Institute of Technology): 1–24.

A50. Sander, R.; Lelieveld, J.; Crutzen, P.J., 1992: "Model Calculations of the Nighttime Aqueous Phase Oxidation of S(IV) in an Orographic Cloud", in: Peeters, J.; Guyot, E. (Eds.): *Proceedings of Joint CEC/EUROTRAC Workshop and LACTOZ-HALIPP Working Group on Chemical Mechanisms Describing Tropospheric Processes, Leuven, Belgium, September 23–25, 1992, Air Pollution Research Report 45, SA, Brussels, 285–290.*

A51. Brühl, C.; Crutzen, P.J.; Graßl, H.; Kley, D., 1993: "The Impact of the Spacecraft System Sänger on the Composition of the Middle Atmosphere", in: *AIAA Fourth International Aerospace Planes Conference, Orlando/Florida, 1–4 December 1992* (Washington DC: American Institute of Aeronautics and Astronautics): 1–9.

A52. Crutzen, P.J., 1993: "Die Beobachtung atmosphärisch-chemischer Veränderungen: Ursachen und Folgen für Umwelt und Klima", in: *Klima: Vorträge im Wintersemester 1992/93,* (Heidelberger Verlagsanstalt: Sammelband der Vorträge des Studium Generale der Ruprecht-Karls-Universität Heidelberg): 31–48.

A53. Kanakidou, M.; Dentener, F.J.; Crutzen, P.J., 1993: "A Global Three-Dimensional Study of the Degradation of HCFC's and HFC-134a in the Troposphere", *Proceedings of STEP-HALOCSIDE/AFEAS Workshop on Kinetics and Mechanisms for the Reactions of Halogenated Organic Compounds in the Troposphere, Dublin, Ireland, March 23–25, 1993,* Campus Printing Unit, University College Dublin, 113–129.

A54. Grooß, J.U.; Peter, T.; Brühl, C.; Crutzen, P.J., 1994: "The Influence of High Flying Aircraft on Polar Heterogeneous Chemistry", in: Schumann, U.; Wurzel, D. (Eds.): *Proceedings of an International Scientific Colloquium on Impact of Emissions from Aircraft and Spacecraft upon the*

Atmosphere, Köln, Germany, April 18–20, 1994, DLR-Mitteilung 94-06, 229–234.

A55. Lelieveld, J.; Crutzen, P.J., 1994: "Emissionen klimawirksamer Spurengase durch die Nutzung von Öl und Erdgas", in: *Energiewirtschaftliche Tagesfragen*, 7: 435–440.

A56. Kanakidou, M.; Crutzen, P.J.; Zimmermann, P.H., 1994: "Estimates of the Changes in Tropospheric Chemistry Which Result from Human Activity and Their Dependence on NOx Emissions and Model Resolution", *Proceedings of the Quadrennial Ozone Symposium, June 4–13, 1992, Charlottsville, Virginia, U.S.*, NASA Conference Publication 3266, 66–69.

A57. Müller, R.; Crutzen, P.J., 1994: "On the Relevance of the Methane Oxidation Cycle to "Ozone Hole Chemistry"", *Proceedings of the Quadrennial Ozone Symposium, June 4–13, 1992, Charlottsville, Virginia, U.S.*, NASA Conference Publication 3266, 298–301.

A58. Peter, T.; Crutzen, P.J., 1994: "Das Ozonloch: Wie kam es dazu und was sollten wir daraus lernen?", in: Fröhlich, C. (Ed.): *Der Mensch im Strahlungsfeld der Sonne*. Konstanz und Wandel in Natur und Gesellschaft (Wissenschaftliches Studienzentrum, Davos: Forum Davos): 31–44.

A59. Steil, B.; Brühl, C.; Crutzen, P.J.; Dameris, M.; Ponater, M.; Sausen, R.; Roeckner, E.; Schlese, U.; Roelofs, G.J., 1994: "A Chemistry Model for Use in Comprehensive Climate Models", in: Schumann, U.; Wurzel, D. (Eds.): *Proceedings of an International Scientific Colloquium on Impact of Emissions from Aircraft and Spacecraft upon the Atmosphere, Köln, Germany, April 18–20, 1994, DLR-Mitteilung 94-06, 235–240.*

A60. Andreae, M.O.; Cofer III, W.R.; Crutzen, P.J.; Hobbs, P.V.; Hollander, J. M.; Kuhlbusch, T.; Novakov, R.; Penner, J.E., 1995: "Climate Impacts of Carbonaceous and Other Non-sulfate Aerosols: A Proposed Study", Lawrence Berkely Laboratory Document—PUB-5411.

A61. Crutzen, P.J., 1995: "On the Role of Ozone in Atmospheric Chemistry", in: Bandy, A.R. (Ed.): *The Chemistry of the Atmosphere. Oxidants and Oxidation in the Earth's Atmosphere*, Proceedings of the 7th BOC Priestley Conference, Lewisburg, Pennsylvania, U.S.A., June 25–27, 1994 (Cambridge, UK: The Royal Society of Chemistry): 3–22.

A62. Crowley, J.N.; Campuzano-Jost, P.; Carl, S.A.; Crutzen, P. J.; Finkbeiner, M.; Helleis, F.; Horie, O.; Moortgat, G.K.; Müller, R.; Roehl, C., 1996: "Laboratory Investigations of the Production and Loss of Hypochlorite in the Stratosphere", in: Pyle, J.A.; Harris, N.R.P.; Amanatidis, G.T. (Eds.): *Proceedings of the 3rd European Workshop on "Polar Stratospheric Ozone", Schliersee, Germany, September 18–22, 1995, Air Pollution Research Report 56, European Commission, Luxembourg, 1996, 679–683.*

A63. Crutzen, P.J., 1996: Das stratosphärische Ozonloch: Eine durch menschliche Aktivitäten erzeugte chemische Instabilität in der Atmosphäre. Festvortrag und Festansprache anlässlich der Verleihung des Internationalen Rheinlandpreises für Umweltschutz 1996 in Köln, TÜV Rheinland, 17–29.

A64. Vogt, R.; Crutzen, P.J., 1996: "Modelling of Halogen Chemistry in the Remote Marine Boundary Layer", in: Borrell, P.M.; Cvitas, T.; Kelly, K.; Seiler, W., (Eds.): *Proceedings of the EUROTRAC Symposium '96, Garmisch-Partenkirchen, Germany, 25–29 March 1996 on "Transport and Transformation of Pollutants in the Troposphere, Vol. 1, Clouds, Aerosols, Modelling and Photo-oxidants"* (Southampton: Computational Mechanics Publications): 445–449.

A65. Crutzen, P.J., 1997: "Entdeckung des Ozonlochs - Wissen und Vision", in: *GIT Labor-Fachzeitschrift*, 41: 110–112.

A66. Crutzen, P.J., 1997: "Mesospheric Mysteries", in: *Science*, 277: 1951–1952.

A67. Crutzen, P.J., 1997: "Problems in Global Atmospheric Chemistry", in: Larsen, B.; Versino, B.; Angeletti, G. (Eds.): *The Oxidizing Capacity of the Troposphere*, Proceedings of the 7th European Symposium on 'Physico-chemical Behaviour of Atmospheric Pollutants', Venice, Italy, 2–4 Ocotber 1996 (Brussels: European Commission): 1–13.

A68. Crutzen, P.J., 1997: "Die Beobachtung atmosphärisch-chemischer Veränderungen: Ursachen und Folgen für Umwelt und Klima. (Festvortrag anläßlich der Hauptversammlung der Max-Planck-Gesellschaft in Bremen am 6. Juni 1997)", in: *Max-Planck-Gesellschft. Jahrbuch 1997* (Göttingen: Generalverwaltung der Max-Planck-Gesellschaft München, Van den Hoek & Ruprecht): 51–71.

A69. Crutzen, P.J.; Braesicke, P., 1997: "Rule-of-Thumb: Converting Potential Temperature to Altitude in the Stratosphere", in: *EOS*, 78: 410.

A70. Crutzen, P.J.; Lawrence, M., 1997: "Ozone Clouds over the Atlantic", in: *Nature*, 388: 625–626.

A71. Lelieveld, J.; Crutzen, P.J.; Jacob, D.J.; Thompson, A.M., 1997: "Modeling of Biomass Burning Influences on Tropospheric Ozone", in: van Wilgen, B.W.; Andreae, M.O.; Goldammer, J.G.; Lindsay, J.A. (Eds.): *Fire in the Southern African Savannas* (Witwatersrand University Press): 217–238.

A72. Perner, D.; Klüpfel, T.; Hegels, E.; Crutzen, P.J.; Burrows, J.P., 1997: "First Results on Tropospheric Observations by the Global Ozone Monitoring Experiment, GOME, on ERS 2", Proceedings of the 3rd. ERS Symposium on "Space at the Service of Our Environment", Flerence, Italy, March 17–21, 1997, ESA SP-414, 3 Vols., 647–652.

A73. Steil, B.; Dameris, M.; Brühl, C.; Crutzen, P.J.; Grewe, V.; Ponater, M.; Sausen, R., 1997: Development of a Chemistry Module for GCMs: First Results of a Multi-annual Integration. DLR, Institut für Physik der Atmosphäre, Report No. **74**, U. Schumann (Ed.), DLR-Oberpfaffenhofen, 50 pp.

A74. Crutzen, P.J., 1998: "What is Happening to Our Precious Air? The Dramatic Role of Trace Components in Atmospheric Chemistry", in: *Science Spectra*, 14: 22–31.

A75. Crutzen, P.J., 1998: "ERASMUS Lecture: Changing Atmospheric Chemistry: Causes and Consequences for Environment and Climate", in: *European Review*, 6,1: 7–23.

A76. Crutzen, P.J., 1998: "How the Atmosphere Keeps Itself Clean and How this is Affected by Human Activities", in: *Pure and Applied Chemistry*, 70,7: 1319–1326 (IUPAC Symposium on "Degradation Processes in the Environment", 24–28 May 1998, Dubrovnik, Croatia).

A77. Hegels, E.; Harder, H.; Klüpfel, T.; Crutzen, P.J.; Perner, D., 1998: "On the Global Distribution of the Halogen Oxides from Observations by GOME (Global Ozone Monitoring Experiment) on ERS-2", Harris, N.R. P.; Kilbane-Dawe, I.; Amanatidis, G.T. (Eds.): *Proceedings of the 4th European Symposium "Polar Stratosperic Ozone 1997", Schliersee, Germany, September 22–26, 1997, Air Pollution Research Report 66, European Commission, Luxembourg, 1998, 343–346.*

A78. Hegels, E.; Crutzen, P.J.; Klüpfel, T.; Perner, D.; Burrows, J.P.; Ladstätter-Weißenmayer, A.; Eisinger, M.; Callies, J.; Hahne, A.; Chance, K.; Platt, U.; Balzer, W., 1998: "Satellite Measurements of Halogen Oxides by the Global Ozone Monitoring Experiment, GOME, on ERS2: Distribution of BrO and Comparison with Ground Based Observations", in: Bojkov, R.D.; Visconti, G. (Eds.): *Proceedings of the Quadrennial Ozone Symposium on "Atmospheric Ozone", l'Aquila, Italy, September 12–21, 1996* (S. Atto, Italy: Edigrafital S.p.A.): 293–296.

A79. Müller, R.; Grooß, J.-U.; McKenna, D.S.; Crutzen, P.J.; Brühl, C.; Russell III, J.M.; Tuck, A.F., 1998: "HALOE Observations of Ozone Depletion in the Arctic Vortex During Winter and Early Spring 1996–1997", in: Harris, N.R.P.; Kilbane-Dawe, I.; Amanatidis, G.T. (Eds.): *Proceedings of the 4th European Symposium "Polar Stratosperic Ozone 1997", Schliersee, Germany, September 22–26, 1997, Air Pollution Research Report 66, European Commission, Luxembourg, 1998, 301–304.*

A80. Müller, R.; Crutzen, P.J.; Grooß, J.-U.; Brühl, C.; Gernandt, H.; Russell III, J.M.; Tuck, A.F., 1998: "Chlorine Activation and Ozone Depletion in the Arctic Stratospheric Vortex During the First Five Winters of HALOE Observations on the UARS", in: Bojkov, R.D.; Visconti, G. (Eds.): *Proceedings of the Quadrennial Ozone Symposium on "Atmospheric Ozone", l'Aquila, Italy, September 12–21, 1996* (S. Atto, Italy: Edigrafital S.p.A.): 225–228.

A81. Waibel, A.; Fischer, H.; Welling, M.; Wienhold, F.G.; Peter, T.; Carslaw, K.S.; Brühl, C.; Grooß, J.-U.; Crutzen, P. J., 1998: "Nitrification and Denitrification of the Arctic Stratosphere During Winter 1994–1995 due to Ice Particle Sedimentation", in: Bojkov, R.D.; Visconti, G. (Eds.): *Proceedings of the Quadrennial Ozone Symposium on "Atmospheric Ozone", l'Aquila, Italy, September 12–21, 1996* (S. Atto, Italy: Edigrafital S.p.A.): 233–236.

A82. Crutzen, P.J., 1999: "Die Beobachtung atmosphärisch-chemischer Veränderungen: Ursachen und Folgen für Umwelt und Klima", in:

Leopoldina, 44: 351–368 (Jahrbuch 1998 der Deutschen Akademie der Naturforscher Leopoldina, Halle/Saale).

A83. Crutzen, P.J., 1999: "Verschmutzung und Selbstreinigung der Atmosphäre", in: *Naturw. Rdsch.*, 52: 1–5.

A84. Crutzen, P.J., 1999: "Das stratosphärische Ozonloch: Durch menschliche Aktivitäten erzeugte chemische Instabilität in der Atmosphäre", in: *Mainz wie es lebt und denkt. Ein Mainzer Mosaik, Vereinigung der Freunde des Lions Club Mainz-Schönborn e.V.* (Herausgeber):181–189.

A85. Crutzen, P.J., 1999: "Global Problems of Atmospheric Chemistry—The Story of Man's Impact on Atmospheric Ozone", in: Möller, D. (Ed.): *Atmospheric Environmental Research* (Heidelberg: Springer): 3–30.

A86. Crutzen, P.J., 1999: "An Essay in Atmospheric Chemistry and Global Change", in: Brasseur, G.P.; Orlando, J.J.; Tyndall, G.S. (Eds.): (New York–Oxford: Oxford University Press): 486.

A87. Ladstätter-Weißenmayer, A.; Burrows, J.P.; Crutzen, P.J.; Richter, A., 1999: "Biomass Burning and Its Influence on the Troposphere, in Atmospheric Measurements from Space", in: *ESA WPP*-161: 1369–1374.

A88. Lelieveld, J.; Ramanathan, V.; Crutzen, P.J., 1999: The Global Effects of Asian Haze. *IEEE Spectrum*, December 1999, 50–54.

A89. Crutzen, P.J., 2000: "Dowsing the Human Volcano", in: *Nature*, 407: 674–675.

A90. Crutzen, P.J., 2000: "Developments in Tropospheric Chemistry", in: Zerefos, C.S.; et al. (Eds.): *Chemistry and Radiation Changes in the Ozone Layer* (The Netherlands: Kluwer Academic Publishers): 1–12.

A91. Crutzen, P.J., 2000: "The Changing Chemistry of the Atmosphere", in: Parthier, B.; Simon, D. (Eds.): *Climate Impact Research: Why, How and When.* Joint International Symposium, Berlin, October 28–29, 1997 (Berlin: Akademie Verlag): 47–68.

A92. Crutzen, P.J.; Stoermer, E.F., 2000: "The "Anthropocene"", in: *IGBP Newsletter*, 41: 17–18.

A93. Steil, B.; Brühl, C.; Crutzen, P.J.; Manzini, E., 2000: "A MA-GCM with Interactive Chemistry (MAECHAM4-CHEM), Simulations for Present, Past and Future", *Proceedings Quadrennial Ozone Symposium*, Sapporo 2000, NASDA, 221–222.

A94. Crutzen, P.J., 2001: "The Role of Tropical Atmospheric Chemistry in Global Change Research: The Need for Research in the Tropics and Subtropics", *Proceedings of the Preparatory Session 12–14 November 1999 and the Jubilee Plenary Session 10–13 November 2000 on "Science and the Future of Mankind: Science for Man and Man for Science"*, Vatican City, Pontificia Academia Scientiarum, Vatican City, 2001, 110–114.

A95a. Crutzen, P.J., 2001: "Was ist Luft?", in: *Süddeutsche Zeitung Magazin*, 40: 20–23.

A95b. Crutzen, P.J., 2001: "Was ist Luft?", in: Stiekel, B. (Ed.): *Kinder fragen, Nobelpreisträger antworten* (München: Wilhelm Heyne Verlag): 129–136.

A95c. Crutzen, P.J., 2003: "What is Air?", in: Stiekel, B. (Ed.): *Nobel Book of Answers* (Atheneum Books for Young Readers): 166–191.

A96. Crutzen, P.J., 2001: "The Antarctic Ozone Hole, a Human-Caused Chemical Instability in the Stratosphere. What Should We Learn from It?", in: Bengtsson, L.O.; Hammer, C.U. (Eds.): *Geosphere—Biosphere Interactions and Climate* (Cambridge: Cambridge University Press): 1–11.

A97. Crutzen, P.J.; Sander, R.; Vogt, R., 2001: "The Influence of Aerosols on the Photochemistry of the Atmosphere", in: Jaenicke, R. (Ed.): *Dynamics and Chemistry of Hydrometers*. Final Report of the Collaborative Research Centre 233 "Dynamik und Chemie der Hydrometeore" (Weinheim: WYLEY-VCH Verlag GmbH): 130–147.

A98. Crutzen, P.J.; Oberlander, E.; Peters, W.; Römpp, A., 2001: "Overview of Atmospheric Chemistry", in: Moortgat, G. (Ed.): *Chemical, Physical and Biogenic Processes*. Notes from the 3rd COACh International School (Mainz: Max-Planck-Institut für Chemie): 7–17.

A99. Sander, R.; Crutzen, P.J., 2001: *Bodennahes Ozonloch in der Arktis. Spektrum der Wissenschaft*, Jan. 2001, 12–13.

A100. Keene, W.C.; Lobert, J.M.; Maben, J.R.; Scharffe, D.; Crutzen, P.J., 2001: "Emissions of Volatile Inorganic Halogens, Carboxylic Acids, Ammonia and Sulfur Dioxide from Experimental Burns of Southern African Biofuels", in: *Eos Transaction of AGU*, 82,47: F112.

A101. Crutzen, P.J., 2002: "Geology of Mankind—The Anthropocene", in: *Nature*, 415: 23.

A102. Crutzen, P.J., 2002: "The Importance of Tropical Atmospheric Chemistry in Global Change Research", in: Ginkel, H.; Barrett, B.; Court, J.; Velasquez, J. (Eds.): *Human Development and the Environment* (The United Nations University): 213–219.

A103. Crutzen, P.J., 2002: "Eine schillernde Hypothese…mit der sich's leben läßt und ohne die man als Forscher auskommt", in: *GAIA*, 1/2002: 19–21.

A104. Crutzen, P.J., 2002: "A Critical Analysis of the Gaia Hypothesis as a Model for Climate/Biosphere Interactions", in: *GAIA*, 2/2002: 96–103.

A105. Crutzen, P.J., 2002: "The Effects of Industrial and Agricultural Practices on Atmospheric Chemistry and Climate During the Anthropocene", in: *Journal of Environment Science and Health*, 37: 423–424.

A106. Crutzen, P.J., 2002: "Atmospheric Chemistry in the "Anthropocene"", in: Steffen, W.; Jäger, J.; Carson, D.J.; Bradshaw. C. (Eds.): *Challenges of a Changing Earth. Proceedings of the Global Change Open Science Conference, Amsterdam, The Netherlands, 10–13 July 2001* (Springer): 45–48.

A107. Crutzen, P.J., 2002: "The "Anthropocene"", in: Boutron, C. (Ed.): *ERCA, Vol. 5, From the Impacts of Human Activities on our Climate and Environment to the Mysteries of Titan* (EDP Sciences): 1–5

A108. Levin, Z.; Rudich, Y.; Crutzen, P.J.; Andreae, M.; Bott, A., 2002: "The Role of Cloud Processing and of Organic Matter on the Formation of

Soluble Layers on Mineral Dust Particles and the Impact on Clouds Characteristics", Final Report to GIF, April 2002.

A109. Oberlander, E.A.; Brenninkmeijer, C.A.M.; Crutzen, P.J.; Lelieveld, J.; Elansky, N.F., 2002: "Why Not Take the Train? Trans-Siberian Atmospheric Chemistry Observations across Central and East Asia", in: *EOS*, 83: 509/515/516.

A110. Crutzen, P.J., 2003: "Schutz der Ozonschicht – ein Beipspiel gelungener Umweltpolitik", in: Altner, G.; Leitschuh-Fecht, H.; Michelsen, G.; Simonis, U.E.; von Weizsäcker, E.U. (Eds.): *Jahrbuch Ökologie 2004* (Verlag C.H.Beck): 132–145.

A111. Crutzen, P.J., 2003: "High Flyer", in: *NewScientist*, 5 July 2003, 44–47.

A112. Crutzen, P.J., 2003: "Het antropoceen: op de drempel naar de toekomst", in: van Bekkum, D.W.; Priem, H.N.A.; van der Zwaan, G.J. (Eds.): *Systeem Aarde, cahiers bio-wetenschappen en maatschappij*, 60–64.

A113. Crutzen, P.J., 2004: "Anti-Gaia", in: *Global Change and the Earth System* (Berlin: Springer): 72.

A114. Crutzen, P.J., 2004: "The Ozone Hole", in: *Global Change and the Earth System* (Berlin: Springer): 236–237.

A115. Crutzen, P.J., 2004: "Keynote Address 3", in: Pachauri, R.K. (Ed.): *Partnerships for Sustainable Development Addressing the WEHAB Agenda. TERI—Delhi Sustainable Development Summit 2004* (New Delhi: TERI Press): 229–234.

A116. Wallström, M.; Bolin, B.; Crutzen, P.J.; Steffen, W., 2004: The Earth's Life-Support System is in Peril. *International Herald Tribune*, January 20, 2004.

A117. Crutzen, P.J., 2005: "Das stratosphärische Ozonloch: eine durch Menschen verursachte chemische Instabilität in der Atmosphäre. Was können wir daraus lernen?:, in: Acham K (Ed.): *Vermächtnis und Vision der Wissenschaft, Zeitdiagnosen 7* (Wien (Austria): Passagen Verlag): 31–38.

A118. Crutzen, P.J., 2006: "The "Anthropocene"", in: Ehlers, E.; Krafft, T. (Eds.): *Earth System Science in the Anthropocene—Emerging Issues and Problems* (Heidelberg: Springer): 13–18.

A119. Crutzen, P.J.: "Impact of China's Air Pollution", in: *Frontier in Ecology and the Environment*, 4: 340.

A120. Brenninkmeijer, C.; Slemr, F.; Schuck, T.; Scharffe, D.; Koeppel, C.; Pupek, M.; Jöckel, P.; Lelieveld, J.; Crutzen, P.; Rhee, T.S.; Hermann, M.; Weigelt, A.; Reichelt, M.; Heintzenberg, J.; Zahn, A.; Sprung, D.; Fischer, H.; Ziereis, H.; Schlager, H.; Schumann, U.; Dix, B.; Friess, U.; Platt, U.; Ebinghaus, R.; Martinsson, B.; Nguyen, H.N.; Oram, D.; O'Sullivan, D.; Penkett, S.; van Velthoven, P.; Röckmann, T.; Pieterse, G.; Assonov, S.; Ramonet, M.; Xueref-Remy, I.; Ciais, P.; Reimann, S.; Vollmer, M.; Leuenberger, M.; Valentino, F.L., 2007: "The CARIBIC Aircraft System for Detailed, Long-Term, Global-Scale Measurement of Trace Gases and Aerosol in a Changing Atmosphere", in: *IGACtivities*, 37: 2–9.

A121. Crutzen, P.J.: "Atmospheric Chemistry and Climate in the Anthropocene", in: Bindé, J. (Ed.): *Making Peace with the Earth—What Future for the Human Species and the Planet?* (Berghahn Books, UNESCO Publishing): 113–120.

A122. Winiwarter, W.; Crutzen, P.; Mosier, A.; Smith, K.: "N_2O in Treibhausgasbilanzen von Biotreibstoffen – eine globale Perspektive", in: *Verband Deutscher Landwirtschaftlicher Untersuchungs- und Forschungsanstalten, Kongressband 2008 Jena, Erhöhte Biomassenachfrage – eine neue Herausforderung für die Landwirtschaft. VDLUFA Schriftenreihe Bd. 64* (Darmstadt: VDLUFA-Verlag): 75–82.

A123. Crutzen, P.J.; Mosier, A.; Smith, K.; Winiwarter, W., 2009: "Atmospheric N2O Releases from Biofuel Production Systems: A Major Factor Against "CO_2 Emission Savings": A Global View", in: Zerefos, C.; Contopoulos, G.; Skalkeas, G. (Eds.): *Twenty Years of Ozone Decline—Proceedings of the Symposium for the 20th Anniversary of the Montreal Protocol* (Springer-Verlag): 67–70.

A124. Elansky, N.F.; Belikov, I.B.; Berezina, E.V.; Brenninkmeijer, C.A.M.; Buklikova, N.N.; Crutzen, P.J.; Elansky, S.N.; Elkins, J.V.; Elokhov, A.S.; Golitsyn, G.S.; Gorchakov, G.I.; Granberg, I.G.; Grisenko, A.M.; Holzinger, R.; Hurst, D.F.; Igaev, A.I.; Kozlova, A.A.; Kopeikin, V.M.; Kuokka, S.; Lavrova, O.V.; Lisitsyna, L.V.; Moeseenko, K.B.; Oberlander, E.A.; Obvintsev, Y.I.; Obvintseva, L.A.; Pankratova, N.V.; Postylyakov, O.V.; Putz, E.; Romashkin, P.A.; Safronov, A.N.; Shenfeld, K.P.; Skorokhod, A.I.; Shumsky, R.A.; Tarasova, O.A.; Turnbull, J.C.; Vartiainen, E.; Weissflog, L.; Zhernikov, K.V., 2009: *Atmospheric Composition Observations over the Northern Eurasia Using the Mobile Laboratory (Troica Experiments)* (Obukhov Institute of Atmospheric Physics, Russian Academy of Sciences, International Science and Technology Center, European Union): 1–75.

A125. Crutzen, P.J., 2010: *Cooling the Earth's Surface by Stratospheric Sulphur Injections*, Berlin, WZB.

A126a. Ramanathan, V.; Agrawal, M.; Akimoto, H.; Aufhammer, M.; Devotta, S.; Emberson, L.; Hasnain, S.I.; Iyngararasan, M.; Jayaraman, A.; Lawrance, M.; Nakajima, T.; Oki, T.; Rodhe, H.; Ruchirawat, M.; Tan, S.K.; Vincent, J.; Wang, J.Y.; Yang, D.; Zhang, Y.H.; Autrup, H.; Barregard, L.; Bonasoni, P.; Brauer, M.; Brunekreef, B.; Carmichael, G.; Chung, C.E.; Dahe, J.; Feng, Y.; Fuzzi, S.; Gordon, T.; Gosain, A.K.; Htun, N.; Kim, J.; Mourato, S.; Naeher, L.; Navasumrit, P.; Ostro, B.; Panwar, T.; Rahman, M.R.; Ramana, M. Rupakheti, M.V.; Settachan, D.; Singh, A.K.; Helen, G. S.; Tan, P.V.; Viet, P.H.; Yinlong, J.; Yoon, S.C.; Chang, W.-C.; Wang, X.; Zelikoff, J.; Zhu, A., 2008: *Atmospheric Brown Clouds*. Regional Assessment Report with focus on Asia, UNEP.

A126b. Ramanathan, V.; Agrawal, M.; Akimoto, H.; Aufhammer, M.; Devotta, S.; Emberson, L.; Hasnain, S.I.; Iyngararasan, M.; Jayaraman, A.; Lawrance, M.; Nakajima, T.; Oki, T.; Rodhe, H.; Ruchirawat, M.; Tan, S.K.; Vincent,

J.; Wang, J.Y.; Yang, D.; Zhang, Y.H.; Autrup, H.; Barregard, L.; Bonasoni, P.; Brauer, M.; Brunekreef, B.; Carmichael, G.; Chung, C.E.; Dahe, J.; Feng, Y.; Fuzzi, S.; Gordon, T.; Gosain, A.K.; Htun, N.; Kim, J.; Mourato, S.; Naeher, L.; Navasumrit, P.; Ostro, B.; Panwar, T.; Rahman, M.R.; Ramana, M.V.; Rupakheti, M.; Settachan, D.; Singh, A.K.; Helen, G.S.; Tan, P.V.; Viet, P.H.; Yinlong, J.; Yoon, S.C.; Chang, W.-C.; Wang, X.; Zelikoff, J.; Zhu, A., 2008: *Atmospheric Brown Clouds*. Regional Assessment Report with focus on Asia. Summary, UNEP.

A127. Crutzen, P.J., 2010: "Erdabkühlung durch Sulfatinjektionen in die Stratosphäre", in: Altner, G.; Leitschuh, H.; Michelsen, G.; Simonis, U.E.; von Weizsäcker, E.U. (Eds.): *Jahrbuch Ökologie 2010 (Die Klima-Manipulateure)* (Stuttgart: S. Hirzel Verlag): 33–39.

A128. Smith, K., Crutzen, P.J.; Mosier, A.; Winiwarter, W., 2010: "The Global N2O Budget: A Reassessment", in: Smith, K. (Ed.): *Nitrous Oxide and Climate Change,* Earthscan, May 2010.

A129. Crutzen, P.J., 2010: "Anthropocene Man", in: *Nature,* 467: S10.

A130. Ajai, L.; Bengtsson, L.; Breashears, D.; Crutzen, P.J.; Fuzzi, S.; Haeberli, W.; Immerzeel, W.W.; Kaser, G.; Kennel, C.; Kulkarni, A.; Pachauri, R.; Painter, T.H.; Rabassa, J.; Ramanathan, V.; Robock, A.; Rubbia, C.; Russell, L.; Sánchez Sorondo, M.; Schellnhuber, H.J.; Sorooshian, S.; Stocker, T.F.; Thompson, L.G.; Toon, O.B.; Zaelke, D.; Mittelstraß, J., 2011: "Fate of Mountain Glaciers in the Anthropocene", A Report by the Working Group on 2–4 April, Commissioned by the Pontifical Academy of Sciences, Vatican City, 15 pp.

A131. Crutzen, P.J.; Schwägerl, C.: *Living in the Anthropocene: Toward a New Global Ethos.* Yale Environment 360, Online-Publication of the Yale School of Forestry & Environmental Studies, Yale University, New Haven (Connecticut), 24 January 2011.

A132. Crutzen, P.J., 2011: "Die Geologie der Menschheit", in: Crutzen, Paul; Davis, Mike; Mastrandrea, Michael D. (Eds.): *Das Raumschiff Erde hat keinen Notausgang* (Suhrkamp Verlag) edition unseld.

A133. Crutzen, P.J., 2012: "Ozone and Advocacy (Sherry Rowland)", in: *Nature Geoscience,* 5: 311.

A134. Crutzen, P.J., 2012: "Climate, Atmospheric Chemistry and Biogenic Processes in the Anthropocene", in: Kant, H.; Reinhardt, C. (Eds.): *100 Jahre Kaiser-Wilhelm-/Max-Planck-Institut für Chemie (Otto-Hahn-Institut): Facetten seiner Geschichte* (Berlin: Veröffentlichungen aus dem Archiv der Max-Planck-Gesellschaft, Band 22): 241–249.

A135. Crutzen, P.; Lax, G.; Reinhardt, C., 2013: "Paul Crutzen on the Ozone Hole, Nitrogen Oxides, and the Nobel Prize", in: *Angewandte Chemie - International Edition,* 52: 48–50, doi:10.1002/anie.201208700.

This photo of Prof. Paul J. Crutzen was taken in July 2015 in the Max Planck Institute of Chemistry in Mainz by Carsten Costard

Chapter 3
The Influence of Nitrogen Oxides on Atmospheric Ozone Content

Paul J. Crutzen

Abstract The probable importance of NO and NO_2 in controlling the ozone concentrations and production rates in the stratosphere is pointed out. Observations on and determinations of nitric acid concentrations in the stratosphere by Murcray et al. (1968) and Rhine et al. (1969) support the high NO and NO_2 concentrations indicated by Bates/Hays (1967). Some processes which may lead to production of nitric acid are discussed. The importance of O (^{1}S), possibly produced in the ozone photolysis below 2340 Å, on the ozone photochemistry is mentioned. The author wishes to express his gratitude to the European Space Research Organization for a post-doctoral fellowship.

3.1 Introduction

It has long been assumed that the main reaction which balanced the production of odd oxygen particles by photodissociation of molecular oxygen was that between atomic oxygen and ozone. In recent years it has become clear, however, that this reaction is not sufficiently fast (Schiff 1969). In a search for other destruction mechanisms reactions between OH, HO_2 and O_3 have been proposed (Hunt 1966; Hampson 1965). It has, however, been indicated in a previous study (Crutzen 1969) that this hypothesis does not succeed in explaining the ozone observations between 30 and 35 km.

Bates/Hays (1967) have indicated that N_2O, possibly produced by microbiological action in the soil and diffusing upwards through the troposphere, may partly be converted to 'odd nitrogen' (NO and NO_2) by a photodissociation process in the stratosphere. As will be shown in this paper, on this hypothesis the NO and NO_2

This manuscript was first published as: "The influence of nitrogen oxides on the atmospheric ozone content" in: *Quaterly Journal of the Royal Meteorological Society*, Vol. 96, No. 408, April 1970: 320–325. Wiley Online Library grants its authors the free republication of their texts.

109

concentrations have a direct controlling effect on the ozone distributions in a large part of the stratosphere, and consequently on the atmospheric ozone production rates.

Recently measured nitric acid concentrations in the stratosphere (Murcray et al. 1968; Rhine et al. 1969) tend to support the suggestions by Bates/Hays (1967).

3.2 Reaction Scheme

$$NO + O_3 \xrightarrow{k_1} NO_2 + O_2 \tag{1}$$

$$NO_2 + hv \xrightarrow{j_2} NO + O \quad \lambda < 3975\text{Å} \tag{2}$$

$$NO_2 + O \xrightarrow{k_3} NO + O_2 \tag{3}$$

$$O_2 + hv \xrightarrow{j_4} 2O \quad \lambda < 2425\,\text{Å} \tag{4}$$

$$O + O_2 + M \xrightarrow{k_5} O_3 + M \tag{5}$$

$$O + O_3 \xrightarrow{k_6} 2O_2 \tag{6}$$

$$O_3 + hv \xrightarrow{j_7} O + O_2 \quad 3080\,\text{Å} < \lambda < 11,400\,\text{Å} \tag{7}$$

$$O_3 + hv \xrightarrow{j_8} O\left(^1D\right) + O_2 \quad \lambda < 3080\,\text{Å} \tag{8}$$

$$O\left(^1D\right) + M \xrightarrow{k_9} O + M \tag{9}$$

$$O\left(^1D\right) + H_2O \xrightarrow{k_{10}} 2OH \tag{10}$$

$$OH + O \xrightarrow{k_{11}} H + O_2 \tag{11}$$

$$H + O_2 + M \xrightarrow{k_{12}} HO_2 + M \tag{12}$$

$$HO_2 + O \xrightarrow{k_{13}} OH + O_2 \tag{13}$$

$$OH + O_3 \xrightarrow{k_{14}} HO_2 + O_2 \tag{14}$$

$$OH + OH \xrightarrow{k_{15}} H_2O + O \tag{15}$$

$$OH + HO_2 \xrightarrow{k_{16}} H_2O + O_2 \tag{16}$$

$$HO_2 + HO_2 \xrightarrow{k_{17}} H_2O_2 + O_2 \tag{17}$$

$$H_2O_2 + OH \xrightarrow{k_{18}} HO_2 + H_2O \tag{18}$$

$$H_2O_2 + h\nu \xrightarrow{j_{19}} 2OH \quad \lambda < 5650\,\text{Å} \tag{19}$$

The following rate coefficients (expressed in the centimeter molecule second system) have been applied:

$k_1 = 1 \cdot 7 \times 10^{-12} \exp(-1{,}310/T)$	Schofield (1967)
$j_2 = 5 \times 10^{-3}$	Nicolet (1965)
$k_3 = 3 \cdot 2 \times 10^{-11} \exp(-530/T)$	Schofield (1967)
$k_5 = 8 \times 10^{-35} \exp(445/T)$	Benson-Axworthy (1965)
$k_6 = 5 \cdot 6 \times 10^{-11} \exp(-2{,}850/T)$	Benson-Axworthy (1965)
$k_9 = 4 \times 10^{-11}$	Zipf (1969)
$k_{10} = 2 \times 10^{-10}$ (assumed)	
$k_{11} = 5 \times 10^{-11}$	Kaufman (1969)
$k_{12} = 4 \times 10^{-32}$	Schofield (1967)
$k_{13} = 2 \times 10^{-11}$	According to Kaufman (1969): $k_{13} \geq 10^{-11}$
$k_{14} = 10^{-13}$	According to Kaufman (1969): $k_{14} \leq 5 \times 10^{-13}$
$k_{15} = 1 \cdot 4 \times 10^{-11} \exp(-500/T)$	Schofield (1967)
$k_{16} = 10^{-11}$	According to Kaufman (1969): $k_{16} \geq 10^{-11}$
$k_{17} = 3 \times 10^{-12}$	Schofield (1967)
$k_{18} = 1 \cdot 6 \times 10^{-11} \exp(-900/T)$	Baulch et al. (1969)

The water vapour mixing ratio assumed in this study was 5×10^{-6}. Concentrations of nitrogen oxides are taken from Bates/Hays (1967) and are given in the table. Absorption cross-sections for ozone and molecular oxygen are from Vigroux (1953) and Brewer/Wilson (1965). Solar flux data have been taken from Brewer/Wilson (1965) and Johnson (1954).

It may be noticed that the very speculative reaction between HO_2 and O_3 is not used in the above scheme.

3.3   The Photochemical Equations

As shown by Nicolet (1965), NO and NO_2 are in mutual equilibrium through the reactions (1)–(3) and therefore

$$\frac{(NO)}{(NO_2)} = \frac{j_2 + k_3(O)}{k_1(O_3)} \tag{20}$$

where quantities within parentheses denote concentrations. This is even more exact for O and O_3 through reactions (5), (7) and (8).

$$\frac{(O)}{(O_3)} = \frac{j_7 + j_8}{k_5(O_2)(M)} \tag{21}$$

Making use of Eq. (20) we write for the time rate of change of odd oxygen and therefore ozone below 50 km

$$\frac{d}{dt}(O_3) = P - (D_1 + D_2 + D_3) \tag{22}$$

where

$$P = 2j_4(O_2) \tag{23a}$$

$$D_1 = 2k_3(O)(NO_2) \tag{23b}$$

$$D_2 = (k_{11}(O) + k_{14}(O_3))(OH) + k_{13}(O)(HO_2) \tag{23c}$$

$$D_3 = 2k_6(O)(O_3) \tag{23d}$$

3.4   Results

The daily production and destruction for odd oxygen below 50 km have been estimated for conditions at the Equator. The observed ozone distribution was taken from Dütsch (1964). The results of the computations are shown in Table 3.1.

It can be seen that the destruction of odd oxygen by the nitrogen oxides is of the same order of magnitude as the production by photodissociation of molecular oxygen. Reductions of odd oxygen by odd nitrogen is, according to these estimates, dominant between approximately 25 and 40 km. There are, however, indications in the Table that reduction by OH and HO_2 begins to be of larger importance around the stratopause.

Table 3.1 Number concentrations (cm^{-3}) and volume mixing ratios of some molecules and estimated production (P) and destruction rates (D$_1$, D$_2$ and D$_3$) over one day (8 (10) = 8 × 10^{10})

Altitude (km)	Concentrations at noon per cm^3						Volume mixing ratio NO.NO$_2$.HNO$_3$	Volume mixing ratio HNO$_3$	Volume mixing ratio			
	O$_3$	O	NO	NO$_2$	OH	HO$_2$			$\int P dt$	$\int D_1 dt$	$\int D_2 dt$	$\int D_3 dt$
50	8 (10)	6 (9)	2.6 (9)	9 (7)	2 (7)	3 (7)	1 (−7)	16 (− 9)	4 (11)	2 (11)	4.8 (11)	5 (10)
35	1.2 (12)	2.8 (8)	3.8 (9)	7.3 (9)	4.7 (6)	1 (8)	7 (−8)	6.3 (−9)	2.7 (11)	4.7 (11)	4.5 (10)	8 (9)
30	2.6 (12)	9 (7)	4.2 (9)	1.1 (10)	2.2 (6)	2 (8)	4 (−8)	8.7 (−9)	1.5 (11)	2.4 (11)	4 (10)	1.6 (10)
25	4.4 (12)	2.7 (7)	2.9 (9)	1.1 (10)	7.4 (3)	1.9 (8)	2.3 (−8)	7 (−9)	7.2 (10)	7.2 (10)	1.8 (10)	2.4 (9)
20	3.3 (12)	4.9 (6)	2 (9)	3 (9)	3.2 (3)	1.3 (8)	1 (−8)	3.7 (−9)	1.7 (10)	4 (9)	4 (9)	3 (8)
15	1 (12)	8.7 (5)	3 (9)	2.4 (9)	2.8 (5)	6.9 (7)	3 (−9)	2 (−9)	6 (8)	2 (8)	1 (9)	4.8 (6)

3.5   Nitric Acid

The detection of nitric acid (HNO_3) in the ozonosphere by Murcray et al. (1968) and the deduced number mixing ratios ($\sim 3 \times 10^{-9}$) provide us with very valuable information.

Nitric acid may be formed by the pair of reactions (see Leighton 1961; Nicolet 1965):

$$OH + NO_2 + M \rightarrow HNO_3 + M \qquad (24)$$

$$HO_2 + NO + M \rightarrow HNO_3 + M \qquad (25)$$

which are followed by

$$HNO_3 + hv \rightarrow OH + NO_2 \qquad (26)$$

These reactions occur mainly during daytime, because OH and NO are removed at night by reactions (1) and (14). Reaction rates k_{24} and k_{25} are not known, but to estimate expected maximum concentrations the value 10^{-31} has been assigned to them. The absorption of nitric acid has been measured by Dalmon (1943) between 2300 and 3100 Å. His data indicate a rise at shorter wavelengths, which is difficult to estimate. Leighton (1961) lists the value 5×10^{-6} for calculations in polluted air at ground level. The value applied in this study, 10^{-5}, is uncertain and may be too low. The derived maximum HNO_3 mixing ratios are listed in the Table and are of the same order of magnitude as the concentrations observed by Rhine et al. (1969). This is a clear indication that the nitrogen oxide concentrations are in fact of the order as those given by Bates/Hays (1967). It is interesting to note the possibility that in the lower stratosphere a large portion of the nitrogen oxides may appear as nitric acid. The actual NO and NO_2 concentrations in the stratosphere may be larger than those used here without leading to a surplus of the destruction over the production of odd oxygen. In the first place there is still uncertainty about the solar flux around 2100 Å, where Detwiler et al. (1961) give data which are almost three times larger than those used here. Furthermore the exact value of the rate coefficient k_5 is not well known and consequently the atomic oxygen concentrations may be lower than those used in this study.

It should also be pointed out that the water vapour mixing ratios may be different from those assumed here.

It can, of course, not be excluded that additional processes must be considered in the HNO_3 formation. A possibility for more OH and HO_2 are the processes

$$O_3 + hv \rightarrow O\,(^1S) + O_2\left(^3\sum_g^-\right), \quad \lambda \leq 2340\,\text{Å} \qquad (27)$$

followed by

$$O(^1S) + H_2O \rightarrow 2OH \qquad (28)$$

$k_{28} = 3 \times 10^{-10}$, Zipf (1969).

Reaction (27) is spin forbidden, but a quantum yield as low as 10^{-2} may be enough to make it of interest. The reason for this is that O (^1S) is much less rapidly deactivated than O (^1D):

$$O(^1S) + O_2 \rightarrow O + O_2 \qquad (29)$$

$$O(^1S) + N_2 \rightarrow O + N_2 \qquad (30)$$

$k_{29} \leq 5 \times 10^{-13}$, Zipf (1969)
$k_{30} \leq 1 \cdot 3 \times 10^{-15}$, Zipf (1969).

It cannot be dismissed that more OH will be produced by reaction (28) than by reaction (10) in parts of the stratosphere, in which case it may even be of direct importance for the ozone distribution (reactions (11)–(14)).

3.6 Conclusions

There is a distinct possibility that nitrogen oxides are of great importance in ozone photochemistry. In the first place we urgently need observations on their concentrations in the stratosphere. Investigations about the photodissociation products of N_2O and its origin (see Bates/Hays 1967) should be continued and extended in order to establish if N_2O is an important source for odd nitrogen in the upper atmosphere. If, however, most of the stratospheric NO and NO_2 is produced at very high levels by other processes some solar cycle influence on the ozone layer will be possible.

Another question which should be investigated is what products are formed in the photolysis of ozone below 2400 Å, in particular whether O (^1S) is formed.

The concentrations of nitric acid reported by Rhine et al. (1969) can be explained by considering reactions between OH, HO_2, NO and NO_2 (reactions 24 and 25), although other possibilities cannot be dismissed. Too little is known about reactions occurring in a nitrogen–oxygen–hydrogen atmosphere.

References

Bates, D.R.; Hays, P.B., 1967: "Atmospheric Nitrous Oxide", in: *Planetary and Space Science*, 150: 189–197.
Baulch, D.L.; Drysdale, D.D.; Lloyd, A.C., 1969: "High Temperature Reaction Rate Data", Report No. 3, (Leeds: Department of Physical Chemistry, The University).

Benson, S.W.; Axworthy, A.E., 1965: "Reconsiderations of the Rate Constants from the Thermal Decomposition of Ozone", in: *Journal of Chemical Physics,* 42,2: 614–2,615.

Brewer, A.W.; Wilson, A.W., 1965: "Measurements of Solar Ultraviolet Radiation in the Stratosphere", in: *Quarterly Journal of the Royal Meteorological Society,* 91: 452–461.

Crutzen, P.J., 1969: "Determination of Parameters Appearing in the 'Dry' and 'Wet' Photochemical Theories for Ozone in the Stratosphere", in: *Tellus,* 21: 368–388.

Dalmon, M.R., 1943: Recherches sur l'acide nitrique et ses solutions par les spectres d'absorption dans l'ultraviolet' Thèses présentées à la faculté des sciences de l'université de Paris.

Detwiler, C.R.; Garret, D.L.; Purcell, J.D.; Tousey, R., 1961: "The Intensity Distribution in the Ultraviolet Solar Spectrum", in: *Annals of Geophysics,* 17: 263–272.

Dütsch, H.U., 1964: "Uniform Evaluation of Umkehr Observations from the World Network", Part III, N.C.A.R., Boulder: Colorado.

Hampson, J., 1965: "Chemiluminescent Emission Observed in the Stratosphere and Mesosphere", Les problèmes météorologiques de la stratosphère et de la mésosphère, Presses Universitaires de France: 393–440.

Hunt, B.G., 1966: "Photochemistry of Ozone in a Moist Atmosphere", in: *Journal of Geophysical Research,* 71: 1385–1398.

Johnson, F.S., 1954: "The Solar Constant", in: *Journal of Meteorological,* 11: 431–439.

Kaufman, F., 1969: "Neutral Reactions Involving Hydrogen and Other Minor Constituents", in: *Canadian* Journal of *Chemistry,* 47: 1917–1927.

Leighton, P.A., 1961: *Photochemistry of Air Pollution* (New York: Academic Press).

Murcray, D.G.; Kyle, T.G.; Murcray, F.H.; Williams, W.J., 1968: "Nitric Acid and Nitric Oxide in the Lower Stratosphere", in: *Nature,* 218: 78–79.

Nicolet, M., 1965: "Nitrogen Oxides in the Chemosphere", in: *Journal of Geophysical Research,* 70: 679–689.

Rhine, P.E.; Tubbs, L.D.; Dudley, Williams, 1969: "Nitric Acid Vapour Above 19 km in the Earth's Atmosphere", in: *Application for Optics,* 8: 1500–1501.

Schiff, H.I., 1969: "Neutral Reactions Involving Oxygen and Nitrogen", in: *Canadian Journal of Chemistry,* 47: 1903–1916.

Schofield, K., 1967: "An Evaluation of Kinetic Rate Data for Reactions of Neutrals of Atmospheric Interest", in: *Planetary and Space Science,* 15: 643–670.

Vigroux, E., 1953: "Contribution à l'étude expérimental de l'absorption de l'ozone", in: *Annales de Physique,* 8: 709–763.

Zipf, E.C., 1969: "The Collisional Deactivation of Metastable Atoms and Molecules in the Upper Atmosphere", in: *Canadian Journal of Chemistry,* 47: 1863–1870.

Chapter 4
Biomass Burning as a Source of Atmospheric Gases CO, H$_2$, N$_2$O, NO, CH$_3$Cl and COS

Paul J. Crutzen, Leroy E. Heidt, Joseph P. Krasnec, Walter H. Pollock and Wolfgang Seiler

Biomass burning can contribute extensively to the budgets of several gases which are important in atmospheric chemistry. In several cases the emission is comparable to the technological source. Most burning takes place in the tropics in the dry season and is caused by man's activities.

The potential importance of deforestation and biomass burning for the atmospheric CO$_2$ cycle has received much attention and caused some controversy [1–3]. In this article we will show the probable importance of biomass burning as a trace gas source, which is caused by man's activities in the tropics. We used the results of our global biomass burning analysis [4] to derive some rough estimates of the sources of the important atmospheric trace gases CO, H$_2$, CH$_4$, N$_2$O, NO$_x$ (NO and NO$_2$), COS and CH$_3$Cl from the worldwide burning of biomass. Table 4.1 shows the results of our study for different activities and ecosystems.

Our approach has been to relate the emission quantities of these gases to those of CO$_2$ in fire plumes. We have determined such emission ratios during two major forest fires. The first was from a forest fire ~12.5 km south-west of Meeker, Colorado. The crown fire was fuelled by pinon and juniper timber, but most of the material burned consisted of annual and perennial grasses, shrub juniper, and sagebrush in the undergrowth. The dense plume rose to a height of ~4 km before flattening and moving horizontally. Samples were collected in stainless steel containers during flights through the smoke plume at different altitudes. Background samples were also collected, at plume altitude, 6 km before reaching the sampling area and again when leaving the fire area.

The second fire occurred in the Wild Basin area of the Rocky Mountain National Forest, not far from the Meeker site, in a mature spruce and fir forest in which there

This text was first published as: Paul J. Crutzen, Leroy E. Heidt, Joseph P. Krasnec, Walter H. Pollock and Wolfgang Seiler. "Biomass Burning as a Source of Atmospheric Gases CO$_2$, H$_2$, N$_2$O, NO, CH$_3$Cl and COS", in: *Nature*, Vol. 282, No. 5736, pp. 253–256. November 15, 1979. The permission to republish this artiucle was granted on 19 August 2015 by Ms. Claire Smith, Nature Publishing Group & Palgrave Macmillan, London, UK. The authors thank Drs. W. Lewis and C. McMahon for valuable comments. The National Center for Atmospheric Research is sponsored by the NSF.

Table 4.1 Summary of data for the annually burned area and biomass

Activity	Burned and/or cleared area	Total biomass cleared	Biomass exposed to fire	Annually burned biomass	Dead below ground biomass	Dead unburned above-ground biomass
Burning due to shifting agriculture	21–62 (41)	31–92 (62)	24–72 (48)	9–25 (17)	7–20 (14)	16–72 (44)
Deforestation due to population increase and colonisation	8.8–15.1 (12.0)	20–33 (26.5)	16–25 (20.5)	5.5–8.8 (7.2)	4.0–8.0 (6.0)	10.5–16.0 (13.3)
Burning of savanna and bushland	(600)	–	12.2–23.8 (18)	4.8–19 (11.9)	8–16 (12)	2.4–4.8 (3.6)
Wildfires in temperate forests	3.0–5.0 (4.1)	10.5–17.5 (14.0)	7.7–12.8 (10.3)	1.5–2.6 (2.1)	2.8–4.7 (3.8)	6.2–10.2 (8.2)
Prescribed fires in temperate forests	2.0–3.0 (2.5)	1.2–1.8 (1.5)	0.3–0.5 (0.4)	0.1–0.2 (0.2)	0.6–0.9 (0.8)	0.2–0.3 (0.3)
Wild fires in boreal forests	1.0–1.5 (1.3)	2.5–3.8 (3.2)	1.8–2.7 (2.3)	0.4–0.6 (0.5)	0.7–1.1 (0.9)	1.4–2.1 (1.8)
Burning of industrial wood and fuel wood	–	31–32 (31.5)	11–12 (11.5)	10–11 (10.5)	5.5	1[a]
Burning of agricultural wastes	–	–	19–23 (21)	17–21 (19)	27–31 (29)	1.9–2.3 (2.1)
Total	630–690 (660)	130–250 (180)	92–172 (132)	48–88 (68)	56–87 (72)	40–109 (74)

Units 100 Tg dry matter and 10^6 hectares; to convert dry matter to carbon multiply by 0.45. Data in parentheses represent average values
[a]Excluding wood used in long lasting structures

was much dead woody material on the ground and much less green vegetation than in the Meeker fire. In this case samples were taken on ground level. CO_2, CO, CH_4, H_2, COS and CH_3Cl were analysed by gas chromatography; N_2O was determined by mass spectrometry. To check for nonlinear gas-chromatographic responses at high concentrations in the samples, a volumetric measurement of CO_2 was also performed.

Some information on emissions from wood burning can also be derived from earlier field and laboratory experiments [5–9], the emission ratios are shown in Table 4.2 which includes only collections with CO_2 concentrations at least 10 % above background. At this stage we have not been able to measure gaseous emissions in tropical ecosystems. This is clearly a disadvantage, but it should not invalidate our main conclusions. Biomass burning has previously been considered unimportant as a global source of atmospheric trace gases [10]—our analysis shows that this is not the case.

Table 4.2 Compilation of product yields by volume relative to that of CO_2 from different studies

Excess of trace gas concentrations compared to excess CO_2 over background	H_2 (%)	CH_4 (%)	CO (%)	N_2O (%)	NO_x (%)	COS ($\times10^{-6}$)	CH_3Cl ($\times10^{-6}$)
Meeker forest fire (12 July 1977)							
Average (four observations)	3.3	2.1	12.4	0 38	–	–	–
Range	2.9–3.5	1.6–2.5	11.2–13.5	0.23–0.50	–	–	–
Wild Basin fire (20 September 1978)							
Average (seven observations)	–	2.2	19.9	0.06	–	15.8	23.4
Range	–	1.0–3.4	15.8–25.1	0.02–0.08	–	5.4–28.6	4.4–57.2
Agricultural wastes [5]	–	1.2	6.2	–	–	–	–
	–	0.15–4.6	3.3–15.8	–	–	–	–
Landscape refuse [6]	–	–	14.6	–	0.28	–	–
Grass, stubble and straw [7]							
Average	–	–	6.3	–	–	–	–
Range	–	–	2.9–14.9	–	–	–	–
Eight pine slash fires [8] (average)	–	–	19.7	–	0.65(NO + NO_2)	–	–
Wood burning in fire places [9]	–	–	18.0	–	–	–	–
Average ratios used for source estimates	3.3	1.6	14	0.22	0.47	15.8	23,4
Range	2.9–3.5	1.0–2.2	5.5–21.6	0.02–0.50	0.28–0.65	5.4–28.6	4.4–57.2

The forest fire plume data were obtained from wild fire observations. The remaining data were obtained from controlled fires or laboratory measurements

4.1   Importance of Trace Gases in the Atmosphere

Some of the trace gases such as CO_2, N_2O and CH_4 considered in this study contribute to the atmospheric greenhouse effect by their absorption of terrestrial thermal radiation [11].

Other trace gases such as carbon monoxide (CO) strongly affect the tropospheric concentrations of the highly reactive hydroxyl (OH) radical through the reaction

$$CO + OH \rightarrow CO_2 + H \tag{1}$$

It has, therefore, been proposed that a growth in tropospheric CO from industrial activities would lead to a decrease in tropospheric OH concentrations [12]. As the only significant tropospheric sink for many gases (especially hydrocarbons and chlorinated hydrocarbons) is reaction with OH, a rise in the tropospheric CO content would cause enhanced concentrations of these gases in the troposphere and a greater transfer to the stratosphere, with possible effects on stratospheric ozone. Other gases which affect the concentration of OH in the stratosphere include methane (CH_4) and molecular hydrogen (H_2).

Although the oxidation of methane (CH_4) is important, molecular hydrogen (H_2) can also add to the reservoir of stratospheric water vapour. This will affect the formation of stratospheric OH, with further consequences on the chemical and thermal balance of the stratosphere [13].

Little of the nitric oxide (NO) emitted at ground level can reach the stratosphere because of the efficient precipitative scavenging of its oxidation products NO_2 and HNO_3. The presence of NO in the troposphere affects the formation of ozone, through the oxidation of carbon monoxide, methane and other hydrocarbons [14]. It is therefore conceivable that increasing industrial inputs of CO and NO in the atmosphere will lead to increasing levels of tropospheric ozone [14].

As nitrous oxide (N_2O) does not react in the troposphere, it will have a long tropospheric residence time and reach the stratosphere where its oxidation leads to the formation of NO and NO_2. These gases appear in catalytic cycles of reactions leading to less ozone above 25 km and more ozone below this altitude [15].

Carbonyl sulphide (COS) and methyl chloride (CH_3Cl) have similar importance in stratospheric chemistry. During periods of little volcanic activity it is possible that the photodissociation of COS is the main source of sulphur in the stratospheric sulphate layer. Although little is known about the sources of COS there may be a growing anthropogenic input of it [16]. The decay of CH_3Cl in the stratosphere releases Cl and ClO, which are extremely efficient in limiting stratospheric ozone by catalytic reactions. An important source of methyl chloride is its emanation from the oceans, but its production by biomass burning has also been proposed [17].

4.2   Global Source Estimates

The data collected in Table 4.2 show an encouraging uniformity. Individual average emission ratios for several gases do not deviate more than a factor of 6 from each other. The largest variations in emission ratios are found for N_2O, COS and CH_3Cl. This is not surprising as one may expect these emissions to be particularly dependent on the type of material being burned. For example, the relative nutrient content in the green parts of the biomass is much larger than in stemwood [18]. We are not aware of any published emission data for the gases N_2O, H_2, CH_3Cl and COS. With the exception of the lower values obtained by Darley et al. [5] and Boubel et al. [7], the CO/CO_2 emission ratios have all been determined to be in the range of 12–20 %.

Because of the uniform CO/CO_2 ratios reported, we can give a reasonable indication of the magnitude of the CO emissions.

The same may apply to CH_4 and H_2. It is clear, however, that any estimations are much less reliable for NO_x, N_2O, COS and CH_3Cl. However, we can still indicate the potential importance of biomass burning as a source of these gases to the atmosphere.

If the data on total worldwide CO_2 emissions from wood burning of $2-4 \times 10^{15}$ g C year^{-1} from Table 4.1 are accepted, we can discuss its possible implications for the budgets of CO, H_2, CH_4, NO_x, N_2O, CH_3Cl and COS. The wood burning contribution can then be compared with previously derived estimates of the sources of these gases from other anthropogenic or natural processes. Note that our esti-mates on the global extent of biomass burning are substantially smaller than those of other workers such as Wong [2] and Woodwell et al. [3]. We cannot defend the data compiled in Table 6.1 here due to lack of space, however, acceptance of higher biomass burning rates will clearly lead to larger estimates of trace gas emission rates.

Adopting an average relative CO/CO_2 volume emission factor of 14 % from the data collected in Table 4.2, the average estimated input of CO would be 8.4×10^{14} g CO year^{-1}, with a range of about $2.4-16.6 \times 10^{14}$ g CO year^{-1}. Seiler [19] has estimated that the technological input of 6.4×10^{14} g CO year^{-1} is the dominant source of atmospheric CO, and a global CO emission of 2.9×10^{14} g year^{-1} from burning has been estimated by Bach [20]. Clearly, the CO input form fires is probably also important for the global CO budget and for the interpretation of the global CO distribution.

With a relative H_2/CO_2 emission ratio of 3.3 %, the production of H_2 would amount to 15×10^{12} g H_2 year^{-1}, with an estimated range of $9-21 \times 10^{12}$ g H_2 year^{-1}. Schmidt [21] estimated the total worldwide source of H_2 to be 36.5×10^{12} g year^{-1}, including an anthropogenic contribution of 25.5×10^{12} g year^{-1}. Although our analysis shows that fires are probably important no such contribution was considered in his budget.

Assuming an average CH_4/CO_2 emission ratio factor of 1.6 %, the global source of CH_4 can be calculated as 6×10^{13} g year^{-1} with a range of $2.5-11 \times 10^{13}$ g CH_4 year^{-1}. The combustion vegetation could contribute to the CH_4 budget of the atmosphere, which has an estimated input of $40-83 \times 10^{13}$ g CH_4 year^{-1} [22]. However, methane is not the only hydrocarbon gas emitted from the burning of vegetation: emissions of many reactive hydrocarbons, including acetylene (C_2H_2) have also been detected [5] and for some of these gases wood burning may be an important source.

As explained above, the derivation of global source estimates for N_2O, NO_x, COS and CH_3Cl is not trivial. The resulting fluxes which we will derive for these gases are, therefore, less reliable than those derived for CO, H_2 and CH_4. The simplest way of estimating the flux would be to apply the same procedure as for CO, H_2 and CH_4. With an approximate N_2O/CO_2 emission factor of 0.22 %, the estimated nitrous oxide source would be 20×10^{12} g N_2O year^{-1}. Similar calcu-lations would yield a source of 14×10^{12} g N year^{-1} for NO_x, 1.1×10^{11} g S year^{-1} for COS, and 1.9×10^{11} g Cl year^{-1} for CH_3Cl.

Although this approach indicates the potential importance of emission strengths, it is a dubious estimation. An upper limit to the NO_x emissions can be obtained by estimating the mass of bound nitrogen in the burned plant material and by assuming that almost all bound nitrogen is emitted as NO_x—an assumption consistent with other experimental studies [8, 23, 24]. The average of eight experimental slash fires gave relative volume emission ratios NO_x/CO_2 of 4 % for duff and 2.1 % for needles. These ratios are close to the values that can be expected if essentially all bound nitrogen in the plant material were volatilised as NO_x. No NO_x emissions were detected during the combustion of cellulose materials, presumably because burning temperatures were too low to allow appreciable NO_x production from atmospheric N_2 and O_2. The relative volume yield NO_x/CO_2 from an entire fire (in which about 55 % of all available fuel was consumed) was estimated to be 0.65 % [8]. In living plants the nutrients N and S are mainly concentrated in those parts which are most easily burned, such as leaves, small twigs and bark. Before burning, however, much plant tissue is dry and dead, and has actually lost nutrients. Using the information in our Table 6.1 and Tables 53 and 54 in the compilation by Rodin and Bazilevich [18], the average maximum global N/C ratio in the burned biomass can be roughly estimated to be between 1.5 and 2.5 %. Given this information we derive a maximum global average NO_x emission of 50×10^{12} g N year^{-1} from biomass burning, with a range of 20–100×10^{12} g N year^{-1} if most bound nitrogen is indeed emitted as NO_x. The mean estimate is about equal to that mentioned previously by Delwiche and Likens [25].

The assumption that all fixed nitrogen is released as NO_x is far from established. A study by Clements and McMahon [26] confirms that most of the NO_x released during burning arises from fuel nitrogen. They conclude that environmentally significant amounts of NO_x are formed by the burning of forest fuels. However, this study also indicated that the NO_x yield is not 100 % and that it actually increases with the nitrogen content of the fuel: a rough average of the NO_x yield observed in their studies is about 30 %. From a mass-balance perspective it is therefore conceivable that other trace gases besides NO_x are released by biomass burning. These could include N_2O, NH_3 and HCN. NH_3 is an important gas involved in the atmospheric nitrogen cycle. HCN has not so far been observed in the atmosphere, but it could be rather stable as it is not removed photochemically in the troposphere.

If we assume that our maximum observed N_2O/CO_2 emission ratio of ~ 0.4 % is produced from the burning of the nutrient-rich portions of the vegetation (such as the needles and leaves, which have an average N/C composition ratio of 4–6 %), we may guess that the N_2O volume emissions could be ~ 8 % of those of NO_x. Consequently we may estimate an average global N_2O source of $\sim 13 \times 10^{12}$ g N_2O year^{-1}. Finally, comparing the relative emission ratios of COS and CH_3Cl to those of N_2O obtained from the Wild Basin fire in Colorado, the global emission rate of COS can be roughly extrapolated to be 2.4×10^{11} g S year^{-1} (3×10^7 molcules cm^{-2}s^{-1} at the Earth's surface) and that of CH_3Cl to be 4.2×10^{11} g Cl year^{-1} ($\approx 4 \times 10^7$ molecules cm^{-2} s^{-1}). Judging from these estimates, biomass burning could produce much more COS than is destroyed in the stratosphere [16]. This may indicate the presence of another sink for COS in the troposphere or at the Earth's surface. On the other hand, the production of CH_3Cl from biomass burning seems to be an order of magnitude lower than its destruction in the troposphere by reaction with OH [27]. Therefore, from this limited analysis it seems that the oceanic source of CH_3Cl should be more important [17].

The indicated emissions of NO_x and N_2O both seem very important from a global perspective. A NO_x source strength of several times 10^{13} g N year^{-1} is comparable to or larger than the other two major sources of NO_x (industrial processes [28] and lightning [29]). The production of NO_x by the burning of vegetation during the dry season has been clearly demonstrated by Lewis and Weibezahn [30], who measured an increase in rainfall acidity by 10^4 from the beginning to the end of the dry season in the Aragua Valley in Venezuela. According to these authors the release of NO_x by seasonal vegetation burning and its conversion to HNO_3 were responsible for this phenomenon. The S/N ratios in plant material are too low for H_2SO_4 to be nearly as important as HNO_3 in the acidification of the precipitation. The observations by Lewis and Weibezahn [30] supported our method of extrapolation from mid-latitude to tropical conditions.

The indicated production rates of N_2O by burning are comparable to the removal rate in the stratosphere, which is still the only well-documented sink for nitrous oxide. The large oceanic source of N_2O once derived by Hahn [31] is now strongly disputed and more recent observations point to a much smaller source [32, 33]. It seems, therefore, that vegetation burning as a source of N_2O could be of a similar substantial importance.

Note that in our source calculations for the atmospheric trace gases, we did not consider the potential release from the heating of the topsoil organic matter or from the 40–80×10^{14} g C of matter which is exposed to fire but left behind as dead, unburned above-ground biomass (see Table 4.1). The topsoil organic matter is especially rich in nutrients and may make imporant contributions to the cycling of atmospheric trace gases and nutrient elements. The latter has already been pointed out by Lewis [34].

The analysis of emissions from the burning of vegetation has also revealed substantial yields of aldehydes, hydrogen peroxide (G. Kok, personal communication), and reactive and oxygenated hydrocarbons [7, 35, 36]. Because of the simultaneous production of NO, there may be a significant production of ozone in the fire-affected air. Such production of ozone has been observed [37].

Despite the limited amount of observations, we must conclude that biomass burning, especially in the tropics where the rates of biomass production and biomass burning are unparalleled, has the potential to contribute in an important way to the global budgets of several major atmospheric trace gases. We also need to consider that tropical emissions occur in a photochemically very active and dynamically important region, in which substantial transfer of tropospheric air to the stratosphere takes place.

Observations of biomass burning in the tropics are needed which will enable us to assess far more reliably the role of biomass burning in biogeochemical cycles and in atmospheric chemistry. This research is especially important in view of the increasing use of forest resources in the developing countries.

References

1. Wong, C.S., 1978: in: *Marine Pollution Bulletin*, 9: 257–264.
2. Woodwell, G.M.; et al., 1978: in: *Science*, 199: 141–146.
3. Bolin, B., 1977: in: *Science*, 196: 613–615.
4. Seiler, W.; Crutzen, P.J.: in: *Climatic Change* (in the press).
5. Darley, E.F.; Burleson, F.R.; Maleer, E.H.; Middleton, J.T., 1966: in: *Journal of the Air Pollution Control Association*, 11: 685–690.
6. Gerstle, R.W.; Kemnilz, D.A., 1967: in: *Journal of the Air Pollution Control Association*, 17: 324–327.
7. Boubel, R.W.; Darley, E.F.; Schuck, E.A., 1961: in: *Journal of the Air Pollution Control Association*, 19: 497–500.
8. Malte, P.C., 1975: in: *Washington State University Bulletin*, 339 (Washington: Pullman).
9. Short, J.E., 1974: *Air Quality Source Testing Rep.* No. 74–1 (Albuquerque: Department of Environmental Health).
10. Robinson, E.; Robbins, R.C., 1970: in: Singer, W.F. (Ed.): *Global Effects of Environmental Pollution* (Dordrecht: Reidel).
11. Wang, W.C.; Yung, Y.L.; Lacis, A.A.; Mo, T.; Hansen, J.E.; 1976: in: *Science*, 194: 685–6900.
12. Wofsy, S.C., 1976: in: *Annual Review of Earth and Planetary Sciences*, 4: 441–469.
13. Nicolet, M., 1975: in: *Reviews of Geophysics and Space Physics*, 13: 593.
14. Fishman, J.; Crutzen, P.J., 1978: in: *Nature*, 274: 855–858.
15. Duewer, W.H.; Wuebbles, D.J.; Ellsaesser, H.W.; Chang, J.S., 1977: in: *Journal of Geophysical Research*, 82: 935–942.
16. Crutzen, P.J., 1976:in: *Geophysical Research Letters*, 3: 73–76.
17. Lovelock, J.E., 1975: in: *Nature*, 256:193–197.
18. Rodin, L.Ye; Bazilevich, N.I., 1966: *Production and Mineral Cycling in Terrestrial Vegetation* (Edinburgh: Oliver and Boyd).
19. Seiler. W., 1974: in: *Tellus*, 26: 116.
20. Bach, W., 1976: in: *Bonner met. Abh.*, 24: 1–51.
21. Schmidt, H.U., 1974: in: *Tellus*, 26: 78–90.
22. Ehhalt, D.H., 1974: in: *Tellus*, 26: 58.
23. Isichei, A.O.; Sanford, W.W., 1978: SCOPE/UNEP *Workshop on Nitrogin Casting in West African Ecosystem* (IITA: Ibadan).
24. Nyc, P.H.; Greenland, D.J., 1960: *The Soil Under Shifting Cultivation* (Tech. Commun. No. 51, Harpenden: Commonwealth Bureau of Soils).
25. Delwiche, C.C.; Likens, G.E., 1977: in: *Global Chemical Cycles end Their Alterations by Man* (ed. Stumm, W.) 73–88 (Berlin: Dahlen Konferenzen).
26. Clements, H.B.; McMahon, C.K., *Thermochim Acta* (submitted).
27. Chang, J.S.; Penner, J.E., 1978: in: *Atmospheric Environment*, 12: 1867–1873.
28. Söderlund, R.; Svensson, B.H., 1976: in: *Ecological Bulletins, Stockholm*, 22: 23.
29. Clements, W.L.; Stedman, D.H.; Dickerson, R.; Rusch, D.W.; Cicerone, R.J., 1977: in: *Journal of the Atmospheric Sciences*, 34: 143.
30. Lewis, W.M. Jr.; Weibezahn, F.H., in: *Science* (submitted).
31. Hahn, J., 1974: In: *Tellus*, 26: 160.
32. Cohen Y.; Gordon, L.I., 1978: in: *Journal of Geophysical Research*, 84: 347–354.
33. Weiss, R.F., 1978: in: *EOS*, 59: 1101.
34. Lewis, W.M. Jr., 1975: in: *Mineral Cycling in Southeastern Ecosystems* (eds. Howell, F.G.; Gentry, J.B.; Smith, M.H.) 833–846 (US ERDA, 1975).
35. Feldstein, N.; Duckworth, S.; Wohlers, H.C.; Linsky, B., 1963: in: *Journal of the Air Pollution Control Association*, 13: 542–545.
36. Fritschen, L.; et al.,1970: USDA Forest Service Res. Pap. PNW-97 (1970).
37. Evans, L.F.; Weeks, I.A.; Eccleston, A.J.; Packham, D.R., 1977: in: *Environmental Science and Technology*, 11: 896–900.

Chapter 5
The Atmosphere After a Nuclear War: Twilight at Noon

Paul J. Crutzen and John W. Birks

As a result of a nuclear war vast areas of forests will go up in smoke—corresponding at least to the combined land mass of Denmark, Norway and Sweden. In addition to the tremendous fires that will burn for weeks in cities and industrial centers, fires will also rage across croplands and it is likely that at least 1.5 billion tons of stored fossil fuels (mostly oil and gas) will be destroyed. The fires will produce a thick smoke layer that will drastically reduce the amount of sunlight reaching the earth's surface. This darkness would persist for many weeks, rendering any agricultural activity in the Northern Hemisphere virtually impossible if the war takes place during the growing season.

5.1 Introduction

The immediate effects of a global nuclear war are so severe that any additional longterm effects might at first thought be regarded as insignificant in comparison. However, our investigation into the state of the atmosphere following a nuclear exchange suggests that other severely damaging effects to human life and the delicate ecosystems to which we belong will occur during the following weeks and months. Many of these effects have not been evaluated before.

Previous investigations of the atmospheric effects following a nuclear war have been concentrated primarily on the expected large depletions of ozone in the stratosphere [1, 2]. Reduction of the stratospheric ozone shield allows increased levels of harmful ultraviolet (uv) radiation to penetrate to the surface of the earth.

This text was first published as: Crutzen, P.J.; Birks, J.W. (1982). "The atmosphere after a nuclear war: Twilight at noon". *Ambio*, 11 (2/3): 114–125. The permission to republish this text was granted on 28 August 2015 by editor-in-chief of Ambio, Prof. Dr. Bo Söderström and by Prof. Dr. Per Hedenqvist, Executive Director of the Royal Swedish Academy of Sciences. The author is especially thankful to the following scientists for critically reading earlier versions of this article and providing us with advice in their various fields of expertise: Robert Charlson, Tony Delany, Jost Heintzenberg, Rupert Jaenicke, Harold Johnston, Chris Junge, Jeffrey Kiehl, Carl Kisslinger, James Lovelock, V. Ramanathan, V. Ramaswami, Henning Rodhe, Steven Schneider, Wolfgang Seiler, Robert Sievers, Darold Ward, Ellen Winchester, Jack Winchester and Pat Zimmerman.

P.J. Crutzen and H.G. Brauch (eds.), *Paul J. Crutzen: A Pioneer on Atmospheric Chemistry and Climate Change in the Anthropocene*, Nobel Laureates 50, DOI 10.1007/978-3-319-27460-7_5

Such ozone depletion results from the injection of oxides of nitrogen (NO_x) by large nuclear weapons having yields greater than one megaton. Should the nations having nuclear arsenals choose to use their large warheads in a nuclear war, then the earth's protective ozone layer would be much depleted, and the consequent adverse effects associated with the increased flux of ultraviolet radiation would occur. Our conclusions for such a scenario concur with those found in the 1975 report of the US National Academy of Sciences [1].

As assumed in Ambio's reference scenario, it is now believed that the most likely nuclear war is one in which few weapons having yields greater than 1 Mt are used, with preference given to the detonation of large numbers of smaller yield weapons. For such a nuclear war, very little NO_x would be injected above 15 km into the stratosphere by the nuclear bursts, and thus depletion of the ozone layer would not occur as a direct result of the explosions. Nonetheless, other profound effects on the atmosphere can be expected.

In discussing the state of the atmosphere following a nuclear exchange, we point especially to the effects of the many fires that would be ignited by the thousands of nuclear explosions in cities, forests, agricultural fields, and oil and gas fields. As a result of these fires, the loading of the atmosphere with strongly light absorbing particles in the submicron size range (1 micron = 10^{-6} m) would increase so much that at noon solar radiation at the ground would be reduced by at least a factor of two and possibly a factor of greater than one hundred. In addition, fires inject large quantities of oxides of nitrogen and reactive hydrocarbons, the ingredients of photochemical smog. This creates the potential for photochemical smog throughout much of the Northern Hemisphere which may persist for several months after the particulate matter has been deposited on the ground. Such effects have been largely overlooked or not carefully examined in previous considerations of this problem. They are, therefore, considered in some detail in this study.

5.2 Nuclear War Scenarios

The explosion of nuclear weapons produces oxides of nitrogen by heating air to temperatures well above 2000 K. When the major constituents of the air—nitrogen and oxygen—are heated to high temperature, nitric oxide (NO) is formed. The equilibrium between N_2, O_2 and NO is rapidly approached at the temperatures characteristic of the nuclear explosions:

$$N_2 + O_2 \leftrightarrow 2\,NO$$

As the temperature of the heated air falls, the reactions which maintain equilibrium become slow and NO cannot revert to the innocuous oxygen and nitrogen. Consequently, nuclear explosions produce NO in much the same way as it is formed as a pollutant in automobile and aircraft engines. A review of the mechanisms forming NO in nuclear explosions is provided in Appendix 1. The oxides of nitrogen are important trace atmospheric constituents and play a very important role in atmospheric photochemistry. They are key constituents in the formation of photochemical smog in the troposphere, and the catalytic reaction cycle leading to ozone destruction is the principal means by which ozone concentrations are regulated in the stratosphere. In Appendix 1 it is estimated that there are 1×10^{32} molecules of NO formed for each megaton of explosion yield. As will be discussed later, large amounts of nitric oxide would also be formed by the many fires that would be started during a nuclear war.

With regard to direct NO_x formation in nuclear explosions, we consider two nuclear war scenarios. Scenario I is Ambio's reference scenario [3]. In this scenario bombs having a total yield of 5750 Mt are detonated. The latitudinal and vertical distributions of the 5.7×10^{35} molecules of nitric oxide produced in these explosions are determined by the weapon sizes and targets projected for this scenario. Since most of the weapons have yields less than 1 Mt, most of the NO_x is deposited in the troposphere, and the effect on the chemistry of the stratosphere is much less than if the bomb debris were deposited mainly in the stratosphere. The assumed NO input pattern for the Scenario I war is provided in Table 5.1.

The Scenario II war is similar to those used in previous studies by investigators using one-dimensional models and is included here mostly for historical reasons. This scenario considers a total yield of 10,000 Mt uniformly distributed between 20° and 60° in the Northern Hemisphere. The vertical distribution of NO is calculated assuming equal yields of 1 and 10-Mt weapons, i.e. 5000 1-Mt weapons and 500 10-Mt weapons are detonated. For this scenario, equal quantities of NO_x are injected above and below 18 km, as seen in Table 5.2. Thus, the tropospheric effects for the Scenario II war are similar to those for the Scenario I war. However, the Scenario II war also results in an additional large perturbation of the stratospheric ozone layer.

5.3 Fires

From an atmospheric point of view, the most serious effects of a nuclear war would most likely result from the many fires which would start in the war and could not be extinguished because of nuclear contaminations and loss of water lines, fire equipment and expert personnel. The devastating effects of such fires in urban areas were indicated by Lewis [4]. Here we show that the atmospheric effects would be especially dramatic. Several types of fires may rage. Besides the fires in urban and industrial centers, vast forest fires would start, extensive grasslands and agricultural land would burn, and it is likely that many natural gas and oil wells would be ruptured as a result of the nuclear explosions, releasing huge quantities of oil and natural gas, much of which would catch fire. To give an estimate of the possible effects, we will consider as a working hypothesis that 10^6 km^2 of forests will burn (this corresponds roughly to the combined area of Denmark, Norway and Sweden) and that breaks in gas and oil production wells will release gaseous effluents from

Table 5.1 Distribution of NO_x produced by nuclear explosions for Scenario I ($\times 10^{32}$ molecules)

Alt. (km)	60–30°S	30°S-EQ	EQ-20°N	20–40°N	40–60°N	60°N-NP	Sum
30	–	–	–	0.7	–	–	0.7
29	–	–	–	0.7	1	–	1.7
28	–	–	–	2.3	1	–	3.3
27	–	–	–	2.3	3	–	5.3
26	–	–	–	2.3	3	–	5.3
25	–	–	–	2.3	3	–	5.3
24	–	–	–	3.7	3	–	6.7
23	–	–	–	3.7	5	–	8.7
22	–	–	–	3.7	5	–	8.7
21	–	–	–	3.7	5	–	8.7
20	–	–	–	2.1	5	–	7.1
19	–	–	–	2.1	2.8	–	4.9
18	–	0.3	1.1	0.1	2.8	–	4.3
17	–	1.1	3.5	10.4	0.2	–	15.2
16	0.7	3.5	10.8	30.7	24.5	–	70.2
15	2.3	8.9	27.5	30.7	72.9	–	142.3
14	2.3	8.9	27.5	116.8	72.9	1.1	229.5
13	3.7	13.0	39.7	247.7	121.5	3.5	429.1
12	8.5	12.1	36.7	225.1	276.6	3.5	562.5
11	16.6	6.6	20.4	329.4	533.5	11.9	918.4
10	14.6	0.5	1.5	327.3	470.2	26.4	840.5
9	24.4	–	–	183.2	775.8	25.0	1008.4
8	24.4	–	–	13.2	775.8	36.7	850.1
7	13.6	–	–	–	434.4	36.5	484.5
6	1.0	–	–	–	21.0	20.4	42.4
5	–	–	–	–	–	1.5	1.5
Sum	112.1	54.9	168.7	1544.2	3618.9	166.5	5665.3

the earth corresponding to the current rate of worldwide usage. In our opinion these are underestimates of the real extent of fires that would occur in a major nuclear war (see also Box 5.1).

5.3.1 Gaseous and Particulate Emissions from Forest Fires

In the US and especially in Canada and the USSR, vast forests are found close to important urban strategic centers, so that it may be expected that many wildfires would start burning during and after the nuclear exchange. Although it is hard to estimate how much forest area might burn, a total of 10^6 km^2, spread around in the Northern Hemisphere, is probably an underestimate, as it is only about 20 times

Table 5.2 Distribution of NO_x produced by nuclear explosions tor Scenario II ($\times 10^{32}$ molecules)

Alt. (km)	20–40°N	40–60°N	Sum
31	62	62	124
30	62	62	124
29	188	188	376
28	188	188	376
27	188	188	376
26	188	188	376
25	312	312	624
24	312	312	624
23	312	312	624
22	312	312	624
21	175	175	350
20	175	175	350
19	80	80	160
18	54	54	108
17	80	80	160
16	125	125	250
15	375	375	750
14	375	375	750
13	625	625	1250
12	625	625	1250
11	350	350	700
10	25	25	50
Sum	5000	5000	10,000

larger than what is now annually consumed by wildfires [5]. This amounts to 4 % of the temperate and boreal forest lands, and is not larger than that of the urban areas combined [6]. Furthermore, Ward et al. [7] have pointed out that effective fire control and prevention programs have reduced the loss of forests in the US (exclusive of Alaska) from 1.8×10^5 km^2 in the early 1930s to less than 1.6×10^4 km^2 by the mid 1970s. The US Forest Service is quoted as estimating that a nuclear attack on the US of ~ 1500 Mt would burn a land area of 0.4–6×10^6 km^2 in the US [8]. All this information indicates that our assumption of 10^6 km^2 of forest area that could be consumed by fire is not an overestimate.

An area of 10^6 km^2 of forest contains on the average about 2.2×10^{16} g dry matter or about 10^{16} g of carbon phytomass [6] and about 10^{14} g of fixed nitrogen, not counting the material which is contained in soil organic matter. Typically, during forest wildfires about 25 % of the available phytomass is burned [5], so that 2.5×10^{15} g of carbon would be released to the atmosphere. During wildfires about 75 kg of particulate matter is produced per ton of forest material burned or 450 kg of carbon [7], so that 4×10^{14} g of particulate matter is injected into the atmosphere by the forest fires. Independently, we can use the information by Ward et al. [7] to estimate the global biomass and suspended particulate matter expected to be produced by wildfires which would be started by the nuclear war. According to these authors the forest area now burned annually in the US, excluding Alaska, is about 1.8×10^4 km^2, which delivers 3.5×10^{12} g particulate matter to the atmosphere. Accordingly, a total area of 10^6 km^2 would inject 2×10^{14} g particulate matter into the atmosphere, which should come from 3×10^{15} g of burned forest material, or 1.3×10^{15} g C. This is a factor of two less than the earlier derived estimate, so we will use a range of $1.3–2.5 \times 10^{15}$ g of carbon as the global atmospheric gaseous release and $2–4 \times 10^{14}$ g as particulate matter.

In forest fires most of the carbon is released as CO_2 to the atmosphere. The forest fire contribution to the atmospheric CO_2 content, which totals 7×10^{17} g of carbon, is rather insignificant. The repercussions of the forest fires are, however, much more important for the contribution of other gases to the atmosphere, e.g. carbon monoxide (CO). With a relative release rate ratio $CO:CO_2$ of about 15 % [9], the production of CO would amount to $2–4 \times 10^{14}$ g C, which is roughly equal to or two times larger than the present atmospheric CO content [10]. Within a short period of time, average concentrations of CO at midlatitudes in the Northern Hemisphere would increase by up to a factor of four, and much larger CO increases may be expected on the continents, especially in regions downwind (generally east of the fires).

Box 5.1

The attenuations of sunlight at great distances from forest fires have been documented for many years Phenomena such as "dark days", "dry fog", "Indian summer" and "colored rain" are now attributed to smoke produced by fires in forests, prairies and peat bogs. The great forest fires during October 13–17, 1918 in Minnesota and adjacent sections of Wisconsin produced smoke that had strong optical effects and could even be smelled as far away as the eastern US coast. A report from Cincinnati, Ohio is particularly descriptive (H. Lyman [17]): "At 3 PM the smoke and haze became denser, but the sun's light and its disk could be seen until 3:35 PM, at which time the sun was entirely obscured. Objects at this time could not be seen at a distance of 300 ft." More than 100 forest fires in northwestern Alberta and northeastern British Columbia resulted in the "Great Smoke Pall" of September 24–30, 1950 with press reports carrying accounts of smoke being observed as far away as England, France, Portugal, Denmark and Sweden (H. Wexler [17]). Most of Canada and the eastern one-third of the continental US were particularly affected. In the eastern US the smoke was confined to the altitude range of about 2.5–4.5 km, so that there was no reduced visibility at the ground. However, the sun was so obscured that it was visible to the naked eye without discomfort and had what was typically described as a violet or lavender color. Measurements in Washington, DC indicated that the solar intensity was reduced by a factor of two on September 25–26 in the absence of clouds.

Accompanying those emissions there will also be significant inputs of tens of Teragrams (1 Teragram = 1 Tg = 10^{12} g) of reactive hydrocarbons to the atmosphere, mostly ethylene (C_2H_4) and propylene (C_3H_6), which are important ingredients in urban, photochemical smog formation. More important, phytomass consists roughly of about 1 % fixed nitrogen, which is mainly contained in the smaller-sized material such as leaves, bark, twigs and small branches, which are preferentially burned during fires. As a rough estimate, because of the forest fires we may expect an input of 15–30 Tg of nitrogen into the atmosphere [7]. Such an emission of NO would be larger than the production in the nuclear fireballs and comparable to the entire annual input of NO_x by industrial processes. Considering the critical role of NO in the production of tropospheric ozone, it is conceivable that a large accumulation of ozone in the troposphere, leading to global photochemical smog conditions, may take place. An increase of ozone due to photochemical processes in forest fire plumes has indeed been observed by several investigators [11, 12].

5.3.2 Particulate Matter from Forest Fires and Screening of Sunlight

The total production of 2–4 × 10^{14} g of particulate matter from the burning of 10^6 km^2 of forests is comparable on a volume basis to the total global production of particulate matter with diameter less than 3 microns (μm) over an entire year (or 200–400 million tons, 13). The physical and chemical nature of this material has been reviewed [14].

The bulk of the mass (>90 %) of the particulate matter from forest fires consists of particles with diameters of less than 1 μm and a maximum particle number density at a diameter of 0.1 μm. The material has a very high organic matter content (40–75 %) and much of it is formed from gaseous organic precursors. Its composition is on the average: 55 % tar, 25 % soot and 20 % ash. These particles strongly absorb sunlight and infrared radiation. The light extinction coefficient, b_s (m), is related to the smoke density, d (g/m^3), by the relationship $b_s = ad$, where a is approximately 4–9 m^2/g [14–17]. With most smoke particles in the submicron size range, their average residence time in the atmosphere is about 5–10 days [13]. If we assume that the forest fires will last for 2 months [18], a spread of 2–4 × 10^{14} g of aerosol over half of the Northern Hemisphere will cause an average particle loading such that the integrated vertical column of particles is equal to 0.1–0.5 g/m^2. As a result, the average sunlight penetration to the ground will be reduced by a factor between 2 and 150 at noontime in the summer. This would imply that much of the Northern Hemisphere would be darkened in the daytime for an extended period of time following the nuclear exchange. The large-scale atmospheric effects of massive forest fires have been documented in a number of papers [18–22]. Big forest fires in arctic regions are commonly accompanied by huge fires in peat bogs, which may burn over two meters in depth without any possibility of being extinguished [18]. The production of aerosol by such fires has not been included in the above estimates.

5.3.3 Gas, Oil and Urban Fires

In addition to the above mentioned fires there are also the effects of fires in cities and industrial centers, where huge quantities of combustible materials and chemicals are stored. As an example, if the European 95-day energy stockpile is roughly representative for the world [23], about 1.5×10^{15} g C fossil fuel (around 1.5 thousand million tons) is stored globally. Much of this would be destroyed in the event of a nuclear war. Therefore, if the relative emission yields of particulate matter by oil and gas fires are about equal to those of forest fires, similar rates of production of atmospheric aerosol would result. Although it may be enormously important, in this study we will not consider the global environmental impacts of the burning and release of chemicals from urban and industrial fires, as we do not yet have enough information available to discuss this matter in a quantitative manner.

Even more serious atmospheric consequences are possible, due to the many fires which would start when oil and gas production wells are destroyed, being among the principal targets included in the main scenario provided for this study [5]. Large quantities of oil and gas which are now contained under high pressure would then flow up to the earth's surface or escape into the atmosphere, accompanied by huge fires. Of course, it is not possible for the nuclear powers to target all of the more than 600,000 gas and oil wells of the world. However, certain regions of the world where production is both large and concentrated in small areas are likely to be prime targets in a nuclear war. Furthermore, the blowout of a natural gas well results in the release of gas at a much greater rate than is allowed when under control and in a production network. For example, one of the more famous blowouts, "The Devil's Cigarette Lighter", occurred at Gassi Touil in the Sahara. This well released 15×10^6 m^3 of gas per day until the 200-m high flame was finally extinguished by explosives and the well capped [24]. Fewer than 300 such blowouts would be required to release natural gas (partly burned) to the atmosphere at a rate equal to present consumption. Descriptions of other blowouts such as the Ekofisk Bravo oil platform in the North Sea [25], a sour gas well (27 % H$_2$S) in the province of Alberta, Canada [26] and the Ixtoc I oil well in the Gulf of Mexico [27] may be found in the literature.

As an example of how very few weapons could be used to release large quantities of natural gas, consider the gas fields of the Netherlands. The 1980 production of 7.9×10^{10} m^3 of natural gas in Groningen amounted to 38 % of that for all of Western Europe and 5 % of that for the entire world [24]. Most of the gas production in the Netherlands is concentrated in a field of about 700 km^2 area. It seems likely that a 300-kt nuclear burst would uncap every gas well within a radius of 1 km either by melting the metal pipes and valves, by snapping the pipes off at the ground by the shock wave, or by breaking the well casings via shock waves propagated in the earth. This is in consideration of the following facts [28]: (1) the fireball radius is 0.9 km, (2) for a surface burst the crater formed is approximately 50 m deep and 270 m in diameter, (3) the maximum overpressure at 1 km is 3.1 atmospheres (atm), (4) the maximum dynamic pressure at 1 km is 3.4 atm, and (5) the maximum wind speed at 1 km is 1700 km/h. Considering then that a 300-kt bomb has a cross-section of greater than 3 km^2 for opening gas wells, fewer than 230 such weapons are required to cover the entire 700 km^2 Groningen field of the Netherlands. This amounts to less than 69 Mt of the 5750 Mt available for the Scenario I nuclear war.

Offshore oil and gas platforms might also be targets of a nuclear war. For example, in 1980 the United Kingdom and Norway produced 2.1 × 10^6 barrels of oil per day from a total of 390 wells (about 40 platforms) in the North Sea [24]. Considering that a 100-kt weapon would be more than sufficient to destroy an offshore platform, only 4 Mt of explosive yield need be used to uncap these wells, which produce 3.5 % of the world's petroleum.

One can point out many other regions of the world where gas and oil production is particularly concentrated. Production in the US is considerably more dispersed than in other countries, however. For comparison, in 1980 the US produced an average of 8.6 × 10^6 barrels of oil per day from about 530,000 wells whereas the USSR production was 12.1 × 10^6 barrels per day from only 80,000 wells [24]. The oil and gas fields of the Soviet Union, particularly the oil producing Volga-Ural region and the gas and oil fields of the Ob region, are highly localized and particularly vulnerable to nuclear attack.

Much of the gas and oil released as a result of nuclear attacks will burn. This is another source of copious amounts of particulate matter in the atmosphere. However, it is also likely that a fraction of the gas would escape unburned to the atmosphere where it would be gradually broken down by photochemical reactions. Much of the escaping oil may likewise burn, but an appreciable portion of it may volatilize as in the Ixtoc I blowout in the Gulf of Mexico, which resulted in the world's largest oilspill. In this case it is estimated that only 1 % of the oil burned, while 50–70 % evaporated [27]. We next consider the influence of these emissions on the gaseous composition of the atmosphere.

Natural gas consists usually of a mixture of 80–95 % (by volume) methane (CH_4) and the remaining 5–20 % heavier hydrocarbons, mainly ethane (C_2H_6) and propane (C_3H_8), and varying amounts of carbon dioxide and nitrogen. Current global consumption of natural gas amounts to about 10^{15} g of carbon per year, which is 20 % of the total fossil fuel consumption rate [29]. The current atmospheric content of ethane is equal to about 6 × 10^{12} g of carbon, based on observations indicating amounts of 1 ppbv (1 ppbv = 10^{-9} by volume) in the Southern and 2 ppbv in the Northern Hemisphere [30, 31]. Consequently the rapid release of C_2H_6 by blow-outs during a nuclear war can increase by many-fold the atmospheric concentrations of this gas, which has an atmospheric residence time of about 2 months. Similar conclusions can be drawn with regard to the higher hydrocarbons. Although relative increases of methane in the atmosphere will take place at a relatively slower pace—as its present atmospheric abundance is much larger, 3×10^{15} g of carbon—even here the atmospheric concentrations may multiply if a sufficiently large percentage of the gas wells are being destroyed. Once destroyed, it seems unlikely that quick repair can be possible in a chaotic world in which little expert personnel and equipment will be available, while the fields will furthermore be heavily contaminated with radioactivity.

Box 5.2

Reaction Cycle C1. In the presence of sufficient NO the oxidation of CO to CO_2 results in the formation of ozone as follows:	Reaction Cycle C2. The oxidation of methane in the atmosphere leads to ozone formation as follows:	Reaction Cycle C3. In the absence of sufficient NO in the atmosphere the oxidation of CO leads to ozone destruction as follows:
R1 $CO + OH \rightarrow H + CO_2$	R6 $CH_4 + OH \rightarrow CH_3 + H_2O$	R1 $CO + OH \rightarrow CO_2 + H$
R2 $H + O_2 + M \rightarrow HO_2 + M$	R7 $CH_3 + O_2 + M \rightarrow CH_3O_2 + M$	R2 $H + O_2 + M \rightarrow HO_2 + M$
R3 $HO_2 + NO \rightarrow OH + NO_2$	R8 $CH_3O_2 + NO \rightarrow CH_3O + NO_2$	R11 $H_2O + O_3 \rightarrow OH + 2O_2$
R4 $NO_2 + hv \rightarrow NO + O$	R9 $CH_3O + O_2 \rightarrow CH_2O + HO_2$	$C_3 CO + O_3 \rightarrow CO_2 + O_2$
R5 $O + O_2 + M \rightarrow O_3 + M$	R3 $HO_2 + NO \rightarrow OH + NO_2$	
C1 $CO + 2 O_2 \rightarrow CO_2 + O_3$	R4 $NO_2 + hv \rightarrow NO + O$ (Twice)	
	R5 $O + O_2 + M \rightarrow O_3 + M$ (Twice)	
	R10 $CH_2O + hv \rightarrow CO + H_2$	
	$C_2 CH_4 + 4 O_2 \rightarrow CO + H_2 + H_2O + 2 O_3$	

Of course it is impossible to guess how many oil and gas well destructions would result from a nuclear war, how much gas will burn and how much will escape unburned to the atmosphere. As an example to indicate the atmospheric effects, let us assume that quantities of oil and gas will continue to burn corresponding to present usage rates, with 25 % of the present production gas escaping unburned into the atmosphere. We do not know whether the latter assumption is realistic. If not, the chosen conditions may represent a gross underestimate of the atmospheric emissions which could take place during and after a nuclear war. This is, of course, especially the case when the world's oil and gas production fields are targeted as foreseen in the main scenario of this study. We simulate NO_x emissions from oil and gas field fires with those provided by current industrial rates. This adds 20 Tg of nitrogen to the NO_x source from forest fires.

5.4 Tropospheric Photochemistry

For the Scenario I nuclear war most of the bomb cloud remains in the troposphere. The sudden input of a large quantity of nitric oxide of 5.7×10^{35} molecules (12 Tg nitrogen) by nuclear explosions and the more gradual input of NO_x from forest fires and gas and oil well fires, mainly in the Northern Hemisphere, will cause important changes in the course of the photochemical reactions taking place. Of course, these reactions should occur only in regions where sufficient sunlight would still penetrate. Alternatively, these reactions begin to occur after an appreciable fraction of the aerosol loading of the atmosphere has diminished because of removal of the particulate matter by rain or dry deposition. The following discussion is, therefore, mainly aimed at illustrating the sort of photochemical effects that may take place. The presence of NO in the troposphere favors chemical processes leading to the production of ozone, e.g. during the oxidation of carbon monoxide (CO) and methane (CH_4), which are present at part per million levels as normal constituents of the troposphere. The production of ozone in these cases takes place with OH, HO_2, NO and NO_2 as catalysts via the cycles of reaction C1 and C2 shown in Box 5.2. Under present non-war conditions, it appears that a large fraction of the troposphere does not contain enough NO for ozone production to take place. For such conditions the oxidation of CO occurs instead via the reaction cycle C3 of Box 5.2. In contrast to reaction cycle C1, cycle C3 leads to ozone destruction. From a comparison of reaction cycles C1 and C3, it follows that ozone production takes place as long as the atmospheric concentration of NO exceeds 1/4000 that of O_3, which is the ratio of rate coefficients for the reactions R11 and R3 [32, 33]. If enough NO were present everywhere in the troposphere for all atmospheric oxidation of CO and CH_4 to occur via reaction cycles C1 and C2, the globally averaged, vertical column integrated photochemical production of ozone in the troposphere would be much larger ($\sim 5 \times 10^{11}$ molecules/cm^2/s) than can be balanced by destruction at the earth's surface ($\sim 6 \times 10^{10}$ molecules/cm^2/s) and by photochemical removal via the reactions

$$\text{R12}\ O_3 + h\nu \rightarrow O(^1D) + O_2$$
$$\text{R13}\ O(^1D) + H_2O \rightarrow 2\,OH$$

which is estimated at 8×10^{10} molecules/cm^2/s [34, 35]. Reactions R12 and R13 constitute the main pathway for the production of hydroxyl radicals (OH), which initiate many oxidation processes in the atmosphere.

Box 5.3

Reaction Cycle C4. Atmospheric oxidation of ethane forms ozone as follows. The carbon monoxide (CO) produced may also be oxidized to form additional ozone via cycle C1.

$$R14\ C_2H_6 + OH \rightarrow C_2H_5 + H_2O$$
$$R15\ C_2H_5 + O_2 + M \rightarrow C_2H_5O_2 + M$$
$$R16\ C_2H_5O_2 + NO \rightarrow C_2H_5O + NO_2$$
$$R17\ C_2H_5O + O_2 \rightarrow CH_3CHO + HO_2$$
$$R18\ CH_3CHO + OH \rightarrow CH_3(C{=}O) + H_2O$$
$$R19\ CH_3(C{=}O) + O_2 + M \rightarrow CH_3(C{=}O)O_2 + M$$
$$R20\ CH_3(C{=}O)O_2 + NO_2 + M \rightarrow CH_3(C{=}O)O_2NO_2 + M$$
$$R21\ CH_3(C{=}O)O_2 + NO \rightarrow CH_3 + CO_2 + NO_2$$
$$R7\ CH_3 + O_2 + M \rightarrow CH_3O_2 + M$$
$$R8\ CH_3O_2 + NO \rightarrow CH_3O + NO_2$$
$$R9\ CH_3O + O_2 \rightarrow CH_2O + HO_2$$
$$R3\ HO_2 + NO \rightarrow OH + NO_2 (2\ times)$$
$$R4\ NO_2 + hv \rightarrow NO + O (5\ times)$$
$$R5\ O + O_2 + M \rightarrow O_3 + M (5\ times)$$
$$R10\ CH_2O + hv \rightarrow CO + H_2$$
$$\overline{}$$
$$C5\ C_2H_6 + 10\ O_2 \rightarrow 2\ H_2O + H_2 + CO_2 + CO + 5O_3$$

The photochemistry of the ethane and higher hydrocarbon oxidation in the atmosphere follows similar reaction paths as for methane, although reactions occur faster because of the higher reactivity of these molecules [33, 36]. In the case of ethane, there can be a net production of five ozone molecules per ethane molecule consumed, if sufficient NO is present in the atmosphere. The cycle of reactions, cycle C4, that produces ozone from ethane is shown in Box 5.3. The compound peroxyacetyl nitrate, $CH_3(C{=}O)O_2NO_2$, which appears in C4 is a strong phyto-toxicant and air pollutant, better known by the acronym PAN [37]. The compound, CH_2O, is formaldehyde and CH_3CHO is acetaldehyde.

Few observations of NO in the background atmosphere have been made, mainly due to the extreme difficulties which are involved in its measurement at low concentrations [38, 39]. The hypothesis that ozone production may take place only in a relatively small fraction of the troposphere is in accordance with present estimations of the sources and sinks of tropospheric NO_x [40]. According to this compilation, the tropospheric sources of NO_x are dominated by industrial activities. This could imply that the current concentrations of tropospheric ozone in the Northern

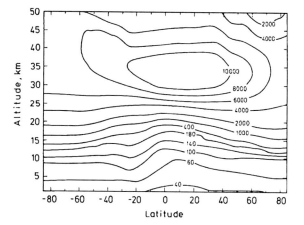

Fig. 5.1 Ozone mixing ratios (ppbv, 1 ppbv = 10^{-9}) in the present atmosphere as calculated by the 2-D model for August 1

Hemisphere are substantially larger than those which prevailed during pre-industrial times.

We have modeled the atmospheric photochemistry following a Scenario I nuclear war under the illustrative assumptions listed above. A description of the computer model used in this work is provided in Appendix 2. The mixing ratios of ozone in the present atmosphere as calculated by the unperturbed model for August 1 are provided in Fig. 5.1, and these are in good agreement with the observations [41]. The calculated ozone concentrations on August 1, 50 days after the start of the war, are shown in Fig. 6.2. We notice the possibility of severe world-wide smog conditions resulting in high concentrations of ozone. With time, at midlatitudes in the Northern Hemisphere there may be large accumulations of ethane (50–100 ppbv) and PAN (1–10 ppbv).

5.5 Effects of Tropospheric Composition Changes

For Ambio's Scenario I type of war the most significant effects in the atmosphere will occur as a result of the wide variety of large fires, which affect especially military, urban and industrial centers, agricultural fields, oil and gas production areas, and forests. In the preceding section, we have considered a scenario of events which, in our opinion, represents probably the minimum of what may occur: wildfires in 10^6 km^2 of forests, and the burning and escape of oil and natural gas at rates comparable to present industrial usage. The estimated atmospheric effects are very large. The fires would create sufficient quantities of airborne particulate matter in the atmosphere to screen out a large fraction of the solar radiation for many weeks, strongly reducing or even eliminating the possibility of growing agricultural crops over large areas of the Northern Hemisphere. Dark aerosol deposits on the vegetation would likewise severely limit plant productivity. In addition, if the war should start during the summer months, as envisaged in the war scenario of this study, much cropland would be destroyed directly by fast-moving fires. Also of special concern are the heavy deposits of air pollutants from the atmosphere which would take place in the months during and following the war. If an appreciable fraction of the NO_x formed in the nuclear explosions and in the resulting fires were to be deposited in rain, the rainwater would be highly acidic with an average pH of less than 4.

If the production of aerosol by fires is large enough to cause reductions in the penetration of sunlight to ground level by a factor of a hundred, which would be quite possible in the event of an all-out nuclear war, most of the phytoplankton and herbivorous zooplankton in more than half of the Northern Hemisphere oceans would die [42]. This effect is due to the fast consumption rate of phytoplankton by zooplankton in the oceans. The effects of a darkening of such a magnitude have been discussed recently in connection with the probable occurrence of such an event as a result of the impact of a large extraterrestrial body with the earth [43]. This event is believed by many to have caused the widespread and massive extinctions which took place at the Cretacious-Tertiary boundary about 65 million years ago.

For several weeks following the war the physical properties of the Northern Hemispheric troposphere would be fundamentally altered, with most solar energy input being absorbed in the atmosphere instead of at the ground. The normal dynamic and temperature structure of the atmosphere would therefore change considerably over a large fraction of the Northern Hemisphere, which will probably lead to important changes in land surface temperatures and wind systems. The thick, dark aerosol layer would likely give rise to very stable conditions in the troposphere (below 10 km) which would restrict the removal of the many fire-produced and unhealthy pollutants from the atmosphere. Furthermore, fires also produce as many as 6×10^{10} cloud condensation nuclei per gram of wood consumed. The effect of many condensation nuclei is to narrow the cloud droplet size distribution and suppress formation of rain droplets by coalescence, probably leading to a decrease in the efficiency with which clouds can produce rain [44]. The influence of large-scale vegetation fires on weather has been recognized by researchers for many years (e.g. [39]). After the settling of most of the particulate matter, ozone concentrations over much of the Northern Hemisphere could approach 160 ppbv for some months following the war. With time, substantial increases in other pollutants such as PAN to several ppbv may also occur. These species are important air pollutants which are normally present in the atmosphere at much lower concentrations (~ 30 ppbv for ozone and less than 0.1 ppbv for PAN) [39, 46, 47].

The effects of ozone on public health and plant growth have been studied for several decades, especially in the US in connection with the Los Angeles basin photochemical smog problem. The effects on agricultural plants may be particularly severe. A major EPA report [31], listed several examples of decreases in yields of agricultural crops. For instance: "A 30 % reduction in the yield of wheat occurred when wheat at antheses (blooming) was exposed to ozone at 200 ppbv, 4 ha day for 7 days... Chronic exposures to ozone at 50–150 ppbv for 4–6 ha day reduced yields

Fig. 5.2 Ozone mixing ratios (ppbv) on August 1, 50 days after the beginning of the Scenario I nuclear war. Inputs from forest fires and oil and gas well fires as described in the text

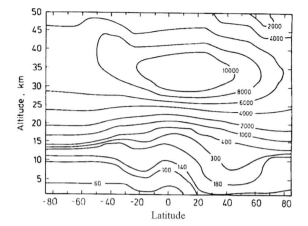

in soybeans and corn grown under field conditions. The threshold for measurable effects for ozone appear to be between 50 and 100 ppbv for sensitive plant cultivers... An ozone concentration of 50–70 ppbv for 4–6 h per day for 15–133 days can significantly inhibit plant growth and yield of certain species." (Figure 5.2)

As a result of the nuclear holocaust we have indicated the possibility of an increase of average ground level ozone concentration to 160 ppbv with higher values to be expected in areas in the wake of the mix of forest and gas and oil well fires assumed in this study. It follows, therefore, that agricultural crops may become subjected to severe photochemical pollutant stress in addition to the even greater damaging effects due to the large load of aerosol particles in the atmosphere.

We conclude, therefore, that the atmospheric effects of the many fires started by the nuclear war would be severe. For the war scenario adopted in this study, it appears highly unlikely that agricultural crop yield would be sufficient to feed more than a small part of the remaining population, so many of the survivors of the initial effects of the nuclear war would probably die of starvation during the first post-war years. This analysis does not address the additional complicating adverse effects of radioactivity or synergism due to concomitant use of chemical and biological warfare weapons.

The described impacts will be different if a nuclear war starts in the winter months. Forest areas burned may be half as large [7], photochemical reactions would be slower because of less solar radiation and lower temperatures. However, in wintertime, because of the low sun, the darkness caused by the fire-produced aerosol would be much worse.

In this work little discussion could be devoted to the health effects of fire-produced pollutants. They too, no doubt, will be more serious in winter than in summer.

5.6   Stratospheric Ozone Depletion

In the stratosphere, molecular oxygen, O_2, absorbs solar radiation of wavelengths shorter than 242 nm and dissociates into two oxygen atoms. These oxygen atoms combine with two oxygen molecules to form two ozone molecules as follows:

$$R14\ O_2 + hv \rightarrow O + O$$
$$R5\ O + O_2 + M \rightarrow O_3 + M\ (Twice)$$

This formation mechanism is quite different from that described previously for the troposphere and summarized in cycles C1 and C2 of Box 5.2. Whereas oxides of nitrogen promote ozone formation in the troposphere, in the stratosphere, where the chemical composition and light spectrum are quite different, the effect of oxides of nitrogen is to catalyze ozone destruction via the reactions:

$$R15\ NO + O_3 \rightarrow NO_2 + O_2$$
$$R16\ O + NO_2 \rightarrow NO + O_2$$
$$\underline{R17\ O_3 + hv \rightarrow O_2 + O}$$
$$Net:\quad 2O_3 \rightarrow 3O_2$$

It is now recognized that this cycle is the principal means by which ozone is limited in the natural stratosphere [48]. Also, whereas ozone is an undesirable pollutant in the troposphere, in the stratosphere ozone performs the necessary function of shielding the earth's surface from biologically damaging ultraviolet radiation.

Our model does not predict significant stratospheric ozone depletion for Ambio's reference Scenario I since as seen in Table 5.1, very little NO_x is deposited in the stratosphere for this scenario. However, for Scenario II (based on previous studies)—which considers the detonation of numerous weapons of large yield—the model predicts very large depletions. For this scenario the quantity of NO_x in the stratosphere of the Northern Hemisphere is increased by a factor of approximately twenty above the natural level [26]. The resulting large ozone depletions would begin in the Northern Hemisphere and eventually spread to the Southern Hemisphere. For purposes of illustration, the Scenario II nuclear war begins on June 11. The resulting ozone depletions on November 1 of the same year are shown in Fig. 6.3. These large ozone depletions are consistent with the one-dimensional model results of Whitten, Borucki and Turco [2] and with the result of Chang as reported by the US National Academy of Sciences [1].

Whitten et al. [2] considered total bomb yields in the range of 5000–10,000 Mt. They distributed the weapon yields either equally between 1 and 5-Mt weapons or equally between 1 and 3-Mt weapons. They also considered that the NO_x was either uniformly distributed throughout the Northern Hemisphere or spread uniformly between 30° and 70°N. Maximum depletion of the ozone column occurred two to three months following the NO_x injection and ranged from 35–70 %. The 35 %

depletion occurred for the 5000 Mt total yield distributed equally between 1 and 3-Mt bombs and spread uniformly over the entire Northern Hemisphere. The maximum of 70 % depletion occurred for a total bomb yield of 10,000 Mt distributed equally between 1 and 5-Mt explosions and confined to the region 30°–70°N. The time constant (e-folding time) for ozone recovery was approximately three years.

The NAS report [1] reaches similar conclusions. A 10,000 Mt war, confined to the Northern Hemisphere, is projected to result in a 30–70 % ozone column reduction in the Northern Hemisphere and a 20–40 % reduction in the Southern Hemisphere. Again, the characteristic recovery time was found to be approximately 3 years. Within 10 years the ozone column depletions were estimated to have decreased to 1–2 %.

Our two-dimensional model predicts a rather uniform 65 % depletion of the ozone column spread from 45°N to the North Pole by the 50th day following the war. The depletions become less toward the equator and beyond, being 57, 42, 26, 12 and 1 % at 35°N, 25°N, 15°N, 5°N and 5°S, respectively. As time progresses, the ozone depletions become less in the Northern Hemisphere, but NO_x is transported to the Southern Hemisphere and causes significant depletion there. Two years following the war in the Northern Hemisphere the ozone column depletions vary uniformly from 15 % at 5°N to 56 % at 85°N, with a 39 % depletion of the ozone column at 45°N. At the same time ozone column depletions range from 12 % at 5°S to 18 % at 85°S in the Southern Hemisphere (Fig. 5.3).

An important uncertainty in the model calculations for the stratosphere stems from the perturbations in the heating rates that accompany the large ozone depletions. Reduction of ozone causes a cooling of the stratosphere. By absorbing ultraviolet sunlight, ozone heats the atmosphere and causes the temperature inversion that is responsible for the high degree of resistance to vertical mixing. To a large extent the NO_x is partitioned into NO_2 in the stratosphere, and the absorption of solar radiation by this species also heats the stratosphere. We find that

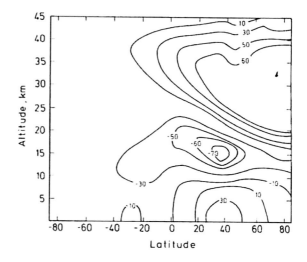

Fig. 5.3 Atmospheric ozone depletion (%) on August 1 of the same year as the Scenario II nuclear war. Negative values indicate ozone increases and show the opposite effects of NO_x injections on ozone in the upper and lower regions of the atmosphere

the net effect at midlatitudes in the perturbed stratosphere is heating below about 22 km and cooling above. The net heating below 22 km is due both to greater penetration of solar uv as a result of the reduced ozone column and the added heating in this region due to NO_2. This will undoubtedly affect the dynamics of the stratosphere and the temperature profile in the stratosphere in complex ways which we cannot predict. We can be confident, however, that the perturbation in the ozone column would be quite large for a Scenario II nuclear war.

Finally, we may point out that there is a possibility that even a nuclear war according to Scenario I, in which most NO_x is deposited in the troposphere, may cause ozone depletions in the stratosphere, if the hot fires in the oil and gas production regions become so powerful that the fire plumes penetrate into the stratosphere. Another means of upward transport may occur when the heavy, dark aerosol layer, initially located in the troposphere, is heated by solar radiation and starts to set up convection and wind systems which will transport an appreciable fraction of the fire effluents into the stratosphere. These speculative thoughts may be pursued further with currently available general circulation models of the atmosphere.

5.6.1 Past Nuclear Weapons Tests

In light of this discussion, one might naturally ask whether past nuclear weapon testing in the atmosphere resulted in significant ozone depletion. This topic has been the subject of considerable debate [49–58]. That nuclear explosions produce copious quantities of nitric oxide and that multi-megaton bursts deposit this NO in the stratosphere was first recognized by Foley and Ruderman [50]. The problem was presented as a possible test of whether NO_x from SST airplane exhaust would actually damage the ozone layer as suggested by Johnston [59] and Crutzen [60]. The approximately 300 Mt of total bomb yield in a number of atmospheric tests by the US and USSR in 1961 and 1962 introduced about 3×10^{34} additional molecules of NO to the stratosphere. Using a onedimensional model, Chang, Duewer and Wuebbles [55] estimated that nuclear weapon testing resulted in a maximum ozone depletion in the Northern Hemisphere of about 4 % in 1963. Analysis of the ground ozone observational data for the Northern Hemisphere by Johnston, Whitten and Birks [51] revealed a decrease of 2.2 % for 1960–1962 followed by an increase of 4.4 % in 1963–1970. These data are consistent with the magnitude of ozone depletion expected, but by no means is a cause-and-effect relationship established. Angell and Korshover attribute these ozone column changes to meteorological factors [53, 54]. The ozone increase began before most of the large weapons had been detonated and persisted for too long a period to be totally attributed to recovery from bomb-induced ozone depletion. Considering the large scatter in ozone measurements and our lack of understanding of all of the natural causes of ozone fluctuations, we cannot draw definite conclusions based on ground observations of ozone following the nuclear weapons tests of the late 1950s and early 1960s.

5.6.2 Solar Proton Events

From the previous discussion it is clear that we have no direct experimental evidence for stratospheric ozone depletion as a result of nuclear explosions. However, at least for altitudes above 30 km the sudden input of significant amounts of NO_x has clearly been shown to lead to large ozone destructions. In August 1972 a major solar proton event deposited large amounts of nitrogen oxides in the stratosphere, leading to ozone depletions poleward of about 60°N. The estimated ozone depletions calculated with a photochemical model were confirmed by satellite observations of stratospheric ozone [61].

5.7 Effects of Increased UV-B Radiation

Ozone in the stratosphere serves as a protective shield against the harmful effects of solar radiation in the wavelength region 240–320 nm (10^{-9} m). The flux of radiation in the wavelength region 290–320 nm ('uv-B' radiation) is particularly sensitive to very small changes in the ozone column [1]. This biologically active radiation is also absorbed by the proteins and nucleic acids within living cells, resulting in a large variety of photoreactions and consequent cell damage [62–64].

The expected adverse effects of increased levels of uv-B radiation include increased incidence of skin cancer in fairskinned races, decreased crop yields and a variety of stresses on terrestrial and aquatic ecosystems. Such effects have been considered in the past in connection with possible reduction of the ozone shield by the operation of fleets of SST airplanes [65] and by the continued release of chlorofluoromethanes used as refrigerants and as propellants in aerosol spray cans [66]. The information available is insufficient to allow quantification of most of these effects. Epidemiological data were used in the NAS study [1] to estimate that a 50 % ozone shield reduction lasting 3 years would lead to an increase of skin carcinoma and melanoma of 3–30 % at midlatitudes, with a geometric mean of about 10 %, that will persist for 40 years. This may be compared with the estimate made in the same study that during the first generation a 10,000 Mt war would increase the spontaneous cancer death rate by about 2 % as a result of exposure to low levels of ionizing radiation from radioactive fallout.

Effects of increased uv-B radiation on food crops are extremely difficult to predict. The sensitivity of plants to supplemented uv-B has been found to be highly variable from one species to another. For example, whereas peas and onions are sensitive, more important food crops such as soybeans and corn appear to have a higher tolerance [1]. Possible climatic changes following a nuclear war further complicate the picture for food crops. Crops are particularly sensitive to temperature, length of growing season and amount of precipitation. The coupling of significant changes in one or all of these factors with a change in the spectrum and intensity of light reaching the earth's surface could be particularly detrimental.

Reduction in stratospheric ozone and the concomitant increase in uv-B radiation would also stress natural ecosystems. As in agriculture, individual species of plants and animals differ considerably in their sensitivities to uv-B radiation. However, in natural ecosystems a direct effect on only one species may be propagated to a large number of species because of complex interdependences. For example, the food chain of the oceans is based on photosynthesis by phytoplankton, and these microscopic, green plants have been demonstrated to be quite sensitive to uv radiation [66]. It was estimated from uv-B irradiation experiments that a 16 % ozone reduction (the degree of ozone depletion projected by the NAS study for continued release of chlorofluoromethanes) could kill up to 50 % of the anchovies in the top 10 m of the clearest ocean water or else require them to substantially deepen their usual water depth [66, 67]. Avoidance could provide protection for many animals, but it is thought that few species can sense uv-B light.

The "effective" increases in uv-B radiation may be determined by integrating the product of the uv-B radiation flux and the appropriate "action spectrum" over wavelength. We have computed these integrals using the action spectrum for erythema (sunburn). This action spectrum is very similar to the absorption spectrum of DNA, as are most uv-B action spectra, and thus the results apply rather generally to cell damage of all types [68]. The relative increases in effective uv-B radiation are shown in Fig. 6.4 for several latitudes as a function of time following the nuclear war. As noted earlier, the uv-B increases are extremely large and persist for several years. The Scenario II nuclear war initially would result in increases in uv-B radiation by a factor greater than 5 throughout most of the Northern Hemisphere and greater than 10 between 55°N and the North Pole. These large increases in uv-B radiation are expected to persist long after the attenuation of light by atmospheric aerosol produced by the nuclear blasts and by the many fires is no longer significant. By comparison, the projected increase in effective uv-B radiation for continued release of chlorofluoromethanes at 1977 levels is 44 % [66] (Fig. 5.4).

Fig. 5.4 Relative increases in effective uv-B radiation based on the erythema action spectrum for the Scenario II nuclear war

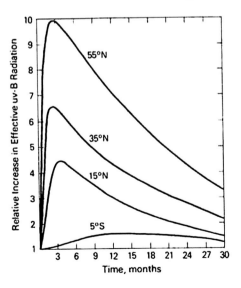

5.8   Long-Term Effects

Regarding possible climatic effects, little can be said with confidence. The increase in tropospheric ozone, methane and possibly other pollutant gases may lead to increased temperatures at the earth's surface [69, 70], while the dark aerosol produced by the fires will change the heat and radiative balance and dynamics of the earth and the atmosphere for awhile. Longer lasting effects may be caused by the changes in the reflective properties of the land surfaces because of many fires. In a recent study Hansen et al. [71] have been able to trace observed mean global temperatures over the past 100 years with a simple climate model by introducing changes in the atmospheric CO_2 content, volcanic activity and solar variability as the main driving forces. In their model the climate sensitivity was also tested for various global radiation perturbations which are relevant for this study: stratospheric aerosol, tropospheric aerosol (divided into opposite sulfate and soot effects), and atmospheric trace gas content (carbon dioxide, ozone, methane and nitrous oxide). From this study it is conceivable that climate could be sensitive over the short term to the tropospheric and stratospheric aerosol loading. It may be possible to test the impact of a nuclear war on climate with this and similar models, when these are supplied with reasonable estimates of the trace gas and aerosol composition of the earth's atmosphere. Whether the induced perturbation in the climate system could lead to longer lasting climatic changes will, however, be difficult to predict. In fact, it may seem unlikely that it will take place. The Krakatoa volcanic eruption of 1883 injected quantities of aerosol into the atmosphere comparable to those which would be caused by a nuclear war, and global mean temperatures were affected for only a few years [1]. Still, we must be cautious with a prediction as the physical characteristics of the aerosol produced by volcanos and fires are different, and much is still unknown about the fundamentals of climatic changes. For instance, we may ask questions such as whether the earth's albedo would be substantially altered after a nuclear war and thus affect the radiation balance or whether the deposition of soot aerosol on arctic snow and ice and on the glaciers of the Northern Hemisphere might not lead to such heavy snow and ice melting as to cause an irreversible change in one or more important climatic parameters.

5.9   Conclusions

In this study we have shown that the atmosphere would most likely be highly perturbed by a nuclear war. We especially draw attention to the effects of the large quantities of highly sunlight-absorbing, dark particulate matter which would be produced and spread in the troposphere by the many fires that would start burning in urban and industrial areas, oil and gas producing fields, agricultural lands, and forests. For extended periods of time, maybe months, such fires would strongly restrict the penetration of sunlight to the earth's surface and change the physical properties of the earth's atmosphere. The marine ecosystems are probably particularly sensitive to prolonged periods of darkness. Under such conditions it is likely that agricultural production in the Northern Hemisphere would be almost totally eliminated, so that no food would be available for the survivors of the initial effects of the war. It is also quite possible that severe, worldwide photochemical smog conditions would develop with high levels of tropospheric ozone that would likewise interfere severely with plant productivity. Survival becomes even more difficult if stratospheric ozone depletions also take place. It is, therefore, difficult to see how much more than a small fraction of the initial survivors of a nuclear war in the middle and high latitude regions of the Northern Hemisphere could escape famine and disease during the following year.

In this paper we have attempted to identify the most important changes that would occur in the atmosphere as a result of a nuclear war. The atmospheric effects that we have identified are quite complex and difficult to model. It is hoped, however, that this study will provide an introduction to a more thorough analysis of this important problem.

Appendix 1: Production and Spatial Distribution of Nitric Oxide from Nuclear Explosions

There have been numerous estimates [49–52, 72] of the yield of nitric oxide per megaton (Mt) of explosion energy, and these have been reviewed by Gilmore [72]. Nitric oxide is produced by heating and subsequent cooling of air in the interior of the fireball and in the shock wave.

The spherical shock wave produces nitric oxide by heating air to temperatures above 2200 K. This air is subsequently cooled by rapid expansion and radiative emission, while the shock front moves out to heat more air. At a particular temperature the cooling rate becomes faster than the characteristic time constant for maintaining equilibrium between NO and air. For cooling times of seconds to milliseconds the NO concentration 'freezes' at temperatures between 1700 and 2500 K, corresponding to NO concentrations of 0.3–2 %. Gilmore [72] estimates a yield of 0.8×10^{32} NO molecules per Mt for this mechanism.

The shock wave calculation of NO production does not take into account the fact that air within the fireball center contains approximately one-sixth of the initial explosion energy, having been heated by the radiative growth mechanism described earlier. This air cools on a time scale of several seconds by further radiative emission, entrainment of cold air, and by expansion as it rises to higher altitudes. These mechanisms are sufficiently complex that one can only estimate upper and lower limits to the quantity of NO finally produced.

A lower limit to total amount of NO finally produced may be obtained by assuming that all of the shock-heated air is entrained into the fireball and again heated to a high enough temperature to reach equilibrium. This is possible since the thickness of the shell of shock- heated air containing NO is smaller than the radius of the fireball. To minimize the cooling rate, and thus the temperature at which equilibrium is not re-established rapidly, it is assumed that this air mass cools only by adiabatic expansion as the fireball rises and by using a minimum rise velocity. The resulting lower limit to total NO production is 0.4×10^{32} molecules per Mt [72].

Since the interior of the fireball is much hotter than the surrounding, shock-heated air, it will rise much faster and possibly pierce through the shell of shock-heated air to mix with cold, undisturbed air above it. Thus, an upper limit to NO production may be obtained by assuming that none of the 0.8×10^{32} NO molecules per Mt produced in the shock wave are entrained by the hot fireball interior. Instead, one assumes that the interior is cooled totally by entrainment of cold, undisturbed air to produce additional NO. The upper limit to total NO production is then estimated to be 1.5×10^{32} molecules per Mt [72]. Thus, the range of uncertainty for total NO_x formation is $0.4–1.5 \times 10^{32}$ molecules per Mt.

For the purposes of this study we assume a nitric oxide yield of 1.0×10^{32} molecules per Mt. One can make strong arguments against either of the extreme values. This estimate of NO production applies only to detonations in the lower atmosphere.

In a nuclear war some bombs may be exploded at very high altitudes for the purpose of disrupting radio and radar signals. The ionization of air by gamma rays, X-rays and charged particles creates a phenomenon known as the "electromagnetic pulse" or "EMP" [73]. The partitioning of energy between the locally heated fireball, shock wave, and escaping thermal radiation changes dramatically as the altitude of the explosion increases above 30 km. As the altitude increases, the X-rays are able to penetrate to greater distances in the low density air and thus create very large visible fireballs. For explosions above about 80 km, the interaction of the highly ionized weapon debris becomes the dominant mechanism for producing a fireball, and for such explosions the earth's magnetic field will influence the distribution of the late-time fireball. Explosions above 100 km produce no local fireball at all. Because of the very low air density, one-half of the X-rays are lost to space, and the one-half directed toward the earth deposits its energy in the so-called "X-ray pancake" region as they are absorbed by air of increasing density. The X-ray pancake is more like the frustum of a cone pointing upward, with a thickness of about 10 km and a mean altitude of 80 km. The mean vertical position is essentially independent of the explosion altitude for bursts well above 80 km [73].

The absorption of X-rays by air results in the formation of pairs of electrons and positively charged ions. One ion pair is formed for each 35 eV of energy absorbed [74], and in the subsequent reactions approximately 1.3 molecules of NO are produced for each ion pair [75]. A 1-Mt explosion corresponds to 2.6×10^{34} eV of total energy. Thus, considering that only half of the X-rays enter the earth's atmosphere, the yield of NO is calculated to be 4.6×10^{32} molecules per Mt (i.e. this mechanism is about five times more effective at producing NO than the thermal mechanism described above).

In the course of a nuclear war up to one hundred 1-Mt bombs might be detonated in the upper atmosphere for the purpose of creating radio wave disturbances. The injection of NO would therefore be 4.8×10^{34} molecules or 1.1 Tg of nitrogen. Natural production of NO in the thermosphere due to the absorption of EUV radiation depends on solar activity and is in the range 200–400 Tg of nitrogen per year [40]. Thus the amount of NO injected by such high altitude explosions is about equal to the amount of NO produced naturally in 1 day and falls within the daily variability. In addition, the X-ray pancake is positioned at an altitude where nitrogen and oxygen species are maintained in photochemical equilibrium. Excess nitric oxide is rapidly destroyed by a sequence of reactions involving nitrogen and oxygen atoms as follows:

$$R22 \: NO + hv \rightarrow N + O$$
$$R23 \: N + NO \rightarrow N_2 + O$$
$$\overline{Net: \quad 2NO \rightarrow N_2 + O + O \rightarrow N_2 + O_2}$$

For these reasons, we expect that high altitude explosions of such magnitudes will have no significant global effect on the chemistry of the stratosphere and below.

Results of past tests of nuclear explosions show that nuclear clouds rise in the atmosphere and finally stabilize at altitudes that scale approximately as the 0.2 power of bomb yield. An empirical fit to observed cloud geometries at midlatitudes gives the following expressions for the heights of the cloud tops and cloud bottoms, respectively [50]:

$$H_T = 22 \: Y^{0.2}$$
$$H_B = 13 \: Y^{0.2}$$

where H is in kilometers and Y has units of megatons. Thus, bomb clouds from weapons having yields greater than about 1 Mt completely penetrate the tropopause at midlatitudes. For such explosions all of the NO_x produced in the fireball, and perhaps a significant fraction of that produced in the shock wave but not entrained by the bomb cloud, is deposited in the stratosphere. Oxides of nitrogen formed in nuclear explosions having yields less than 1 Mt have little effect on stratospheric ozone since: (1) only a minor fraction of the NO_x formed is deposited above the tropopause, (2) the residence time in the stratosphere increases with altitude of injection, and (3) the NO_x-catalytic cycle for ozone destruction is most effective at higher altitudes. In fact, below about 20 km NO_x additions to the atmosphere tend to result in ozone concentration increases [76, 77].

The stabilized nuclear bomb clouds have diameters ranging from 50–500 km depending on bomb yield. They are sheared by horizontal winds at constant latitude, and within a few weeks may be uniformly distributed around the earth at a constant latitude [78].

Appendix 2: Model Description

The computer model used in this study is a two-dimensional model of coupled photochemistry and dynamics. It treats transport in both the vertical and latitudinal directions by parameterization of these motions by means of eddy diffusion coefficients and mean motions. The model covers altitudes between the ground and 55 km and latitudes between the South Pole and North Pole, and it attempts to simulate the longitudinally averaged, meridional distributions of trace gases. Therefore, the main assumption is that composition variations in the zonal (East–West) directions are much smaller than those in the vertical and latitudinal directions. Although the 2-D model is a step forward from 1-D models, which take into account only variations in the vertical direction, the neglect of longitudinal variations in air composition will clearly introduce substantial deviations from reality, especially at lower altitudes, where the influence of chemical and biological processes at the earth's surface are large. One should keep these limitations of the 2-D model in mind especially when interpreting the results obtained for the troposphere.

The model photochemistry considers the occurrence of nearly one hundred reactions, which are now thought to be important in global air chemistry. It takes into account the reactions of ozone and atomic oxygen, and the reactive oxides of nitrogen, hydrogen and chlorine, which are derived from the oxidation of nitrous oxide (N_2O), water vapor (H_2O), methane (CH_4) and organic chlorine compounds. In the troposphere, the photochemistry of simple reactions leading to ozone formation in the presence of NO_x, carbon monoxide (CO), methane and ethane (C_2H_6) are taken into account. The influence of industrial processes is an important consideration of the model. A more detailed description of the model may be found elsewhere [77, 78]. Detailed descriptions of atmospheric photochemistry are given in a number of review articles [40, 79–81].

References

1. National Academy of Sciences, 1975: *Long Term Worldwide Effects of Multiple Nuclear-Weapon Detonations* (Washington, DC).
2. Whitten, R.C.; Borucki, W.J.; Turco, R.P., 1975: in: *Nature*, 257: 38.
3. Advisors, *Ambio*, this issue (see Ambio's Reference Scenario).
4. Lewis, K.N., 1979: in: *Scientific American*, 241: 35.
5. Seiler, W.; Crutzen, P.J., 1980: in: *Climate Change*, 2: 207.
6. 1979: in: Bolin, B.; Degens, E.T.; Kempe, S.; Ketner, P. (Eds.): *The Global Carbon Cycle*, SCOPE 13 (New York: Wiley): 491.
7. Ward, D.E.; McMahon, E.K.; Johansen, R.W., 1976: Paper 76-2.2, 69th Annual Meeting of the Air Pollution Control Association, Portland, Oregon, June 27–July 1, 15 p.
8. 1981: "FAS, Effects of Nuclear War", in: *Journal of the Federation of American Scientists*, 34: 3.
9. Crutzen, P.J.; Heidt, L.E.; Krasnec, J.P.; Pollock, W.H.; Seiler, W., 1979: in: *Nature*, 282: 253.
10. Seiler, W., 1974: in: *Tellus*, 26: 116.

11. Evans, L.F.; Weeks, I.A.; Eccleston, A.J.; Packham, D.R., 1977: in: *Environmental Science and Technology*, 11: 896.
12. Radke, L.F., Stith, J.L.; Hegg, D.A.; Hobbs, P.V., 1978: in: *Journal of the Air Pollution Control Association*, 28: 30.
13. Jaenicke, R., 1981: in: Berger, A. (Ed.): *Climatic Variations and Variability: Facts and Theories* (Dordrecht, Holland: D Reidel): 577–597.
14. McMahon, C.K.; Ryan, P.W., 1976: Paper 76-2.3, 69th Annual Meeting of the Air Pollution Control Association, Portland, Oregon, June 27–July 1, 21 p.
15. Vines, R.G.; Gibson, L.; Hatch, A.B.; King, N.K.; McArthur, D.A.; Packham, D.R.; Taylor, R.J., 1971: "CSIRO Division of Applied Chemistry", Technical Paper No 1, 32 p.
16. Chylek, P.; Ramaswamy, V.; Cheng, R.; Pinnick, R.G., 1981: in: *Applied Optics*, 20: 2980.
17. Waggoner, A.D.; Weiss, R.E.; Ahlquist, N.C.; Covert, D.S.; Will, S.; Charlson, R.J., 1981: in: *Atmospheric Environment*, 15: 1891.
18. Shostakovitch, V.B., 1925: in: *Journal of Forestry*, 23: 365.
19. Plummer, F.G., 1912: "Forest Fires", in: *US Department of Agriculture Forest Service Bulletin*, 117: 15–22.
20. Lyman, H., 1919: in: *Monthly Weather Review*, 46: 506.
21. Wexler, H., 1950: in: *Weatherwise*, December 3.
22. Smith, C.D., Jr., 1950: in: *Monthly Weather Review*, 77: 180.
23. 1981: Kommission der Europäischen Gemeinschaften, 26 p.
24. 1981: *International Petroleum Encyclopedia*, Vol. 14 (Tulsa, Oklahoma: PennWell Publishing Co.).
25. Gotaas, Y., 1980: in: *Journal of the Air Pollution Control Association*, 30: 789.
26. Leahey, D.M.; Brown, G.L.; Findlay, R.L., 1980: in: *Journal of the Air Pollution Control Association*, 30: 787.
27. Jernelöv, A.; Linden, O., 1981: in: *Ambio*, 10: 299.
28. 1977: *Nuclear Bomb Effects Computer*, Revised Edition (Lovelace Biomedical and Environmental Research Institute, Inc.).
29. International Institute for Applied Systems Analysis, 1981: *Energy in a Finite World: Paths to a Sustainable Future*.
30. Singh, H.B.; Hanst, P.L., 1981: in: *Geophysical Research Letters*, 8: 941.
31. Rudolph, J.; Ehhalt, D.H., 1981: in: *Journal of Geophysical Research*, 81: 11959.
32. Howard, C.J.; Evenson, K.M., 1977: in: *Geophysical Research Letters*, 4: 437.
33. Zahniser, M.; Howard, C.J., 1979: in: *Journal of Chemical Physics*, 73: 1620.
34. Fishman, J.; Crutzen, P.J., 1978: in: *Nature*, 274: 855.
35. Fabian, P.; Junge, C.E., 1970: in: *Archiv für Meteorologie, Geophysik und Bioklimatologie*, A19: 161.
36. Demerjian, K.L.; Kerr, J.A.; Calvert, J.G., 1974: in: *Advances in Environmental Science and Technology*, 4: 1.
37. 1978: *Air Quality Criteria for Ozone and Other Photochemical Oxidants*, EPA-600/8-78-004 (Washington, DC: US Environmental Protection Agency).
38. Noxon, J.F., 1978: in: *Journal of Geophysical Research*, 83: 3051.
39. McFarland, M.; Kley, D.; Drummond, J.W.; Schmeltekopf, A.L.; Winkler, R.H., 1979: in: *Geophysical Research Letters*, 6: 605.
40. Crutzen, P.J., 1979: in: *Annual Review of Earth and Planetary Science*, 7: 443.
41. Dütsch, H.V., 1974: in: *Canadian Journal of Chemistry*, 52: 1491.
42. Milne, D.H.; McKay, C.P., in press: "Response of Marine Plankton Communities to Global Atmospheric Darkening", Proceedings of the Conference on Large Body Impacts, Special Paper, Geological Society of America.
43. Alvarez, L.; Alvarez, W.; Asaro, F.; Michel, H., 1980: in: *Science*, 208: 1095.
44. Eagan, R.C.; Hobbs, P.V.; Radke, L.F., 1974: in: *Journal of Applied Meteorology*, 13: 553.
45. Knoche, W., 1937: in: *Meteorologische Zeitschrift*, 54: 243.

46. Lonneman, W.A.; Bufalini, J.J.; Seila, R.L., 1976: in: *Environmental Science and Technology,* 10: 374.
47. Nieboer, H.; van Ham, J., 1976: in: *Atmospheric Environment,* 10: 115.
48. Crutzen, P.J., 1970: in: *Quarterly Journal of the Royal Meteorological Society,* 96: 320.
49. Zel-dovich, Y.; Raizer, Y., 1967: in: *Physics of Shock Waves and High Temperature Phenomena* (New York: Academic Press): 565.
50. Foley, H.M.; Ruderman, M.A., 1973: in: *Journal of Geophysical Research,* 78: 4441.
51. Johnston, H.S.; Whitten, G.; Birks, J.W., 1973: in: *Journal of Geophysical Research,* 78: 6107.
52. Goldsmith, P.; Tuck, A.F.; Foot, J.S.; Simmons, E.L.; Newson, R.L., 1973: in: *Nature,* 244: 545.
53. Angell, J.K.; Korshover, J., 1973: in: *Monthly Weather Review,* 101: 426.
54. Angell, J.K.; Korshover, J., 1973: in: *Monthly Weather Review,* 104: 63.
55. Chang, J.S.; Duewer, W.H.; Wuebbles, D.J., 1979: in: *Journal of Geophysical Research,* 84: 1755.
56. Christie, A.D., 1976: in: *Journal of Geophysical Research,* 81: 2583.
57. Johnston, H.S., 1977: in: *Journal of Geophysical Research,* 82: 3119.
58. Bauer, E.; Gilmore, F.R., 1975: in: *Reviews of Geophysics and Space Physics,* 13: 451.
59. Johnston, H.S., 1971: in: *Science,* 173: 517.
60. Crutzen, P.J., 1971: in: *Journal of Geophysical Research,* 76: 7311.
61. Heath, D.F.; Krueger, A.J.; Crutzen, P.J., 1977: in: *Science,* 197: 886.
62. Wacker, A., 1963: in: *Progress in Nucleic Acid Research,* 1: 369.
63. Smith, K.C., 1964: in: Giese, A.C. (Ed.): *Photophysiology,* Vol. 2 (New York: Academic Press): 329.
64. Setlow, J.K., 1966: in: Ebert, M.; Howard, A. (Eds.): *Current Topics in Radiation Research,* Vol. 2 (Amsterdam: North-Holland): 195.
65. Nachtway, D.F.; Caldwell, M.M.; Biggs, R.H., (Eds.), 1975: in: *CIAP Monograph 5, Impacts of Climatic Change on the Biosphere, Part 1, Ultraviolet Radiation Effects,* DOT-TST-75-55 (Washington, DC: US Department of Transportation).
66. National Academy of Sciences, 1975: *Protection Against Depletion of Stratospheric Ozone by Chlorofluorocarbons* (Washington, DC).
67. Hunter, J.R.; Taylor, J.H.; Moser, H.G., 1979: in: *Photochemistry and Photobiology,* 29: 325.
68. Benger, D.; Robertson, D.F.; Davies, R.E., 1975: in: Nachtway, D.F.; Caldwell, M.M.; Biggs, R.H. (Eds.): *CIAP Monograph 5, Impacts of Climatic Change on the Biosphere, Part 1, Ultraviolet Radiation Effects,* DOT-TST-75-55 (Washington, DC: US Department of Transportation): 2-235–2-264.
69. Wang, W.C.; Yung, T.L.; Lacis, A.A.; Mo, T.; Hansen, J.E., 1976: in: *Science,* 194: 685.
70. Fishman, J.; Ramanathan, V.; Crutzen, P.J.; Liu, S.C., 1979: in: *Nature,* 282: 818.
71. Hansen, J.; Johnson, D.; Lacis, A.; Lebedeff, S.; Lee, P.; Rind, D.; Russell, G., 1981: in: *Science,* 213: 957.
72. Gilmore, F.R., 1975: in: *Journal of Geophysical Research,* 80: 4553.
73. Glasstone, S.; Dolan, P.J., 1977: *The Effects of Nuclear Weapons,* 3rd edn (Washington, DC: US Government Printing Office).
74. Banks, P.M.; Kockarts, G., 1973: *Aeronomy,* Part A (New York: Academic Press).
75. Rusch, D.W.; Gerárd, J.-C.; Solomon, S.; Crutzen, P.J.; Reid, R.C., 1981: in: *Planetary and Space Science,* 7: 767.
76. Duewer, W.H.; Wuebbles, D.J.; Ellsaesser, H.W.; Chang, J.S., 1977: in: *Journal of Geophysical Research,* 82: 935.
77. Hidalgo, H.; Crutzen, P.J., 1977: in: *Journal of Geophysical Research,* 82: 5833.
78. Crutzen, P.J., 1975: in: *Fourth Conference on CIAP* (Cambridge: US Department of Transportation): 276.

79. Levy II, H., 1974: in: *Advances in Photochemistry,* 9: 5325.
80. Crutzen, P.J., 1973: in: *Pure and Applied Geophysics,* 106108: 1385.
81. Logan, J.A.; Prather, M.J.; Wofsy, S.C.; McElroy, M.B., 1978: in: *Philosophical Transactions of the Royal Society of London,* 290: 187.

Chapter 6
Nitric-Acid Cloud Formation in the Cold Antarctic Stratosphere—A Major Cause for the Springtime Ozone Hole

Paul J. Crutzen and Frank Arnoldt

Abstract Large depletions in stratospheric ozone were first reported by Farman et al. [1] at Halley Bay (76 °S), and confirmed by satellite observations [2]. Chubachi [3] gives a detailed account of ozone decreases and temperatures in the lower stratosphere during the spring of 1982 at 69 °S. There is now evidence [2] for annual declines in total ozone by ~ 6 and 3 % in regions of total ozone minima and maxima, respectively, from September to mid-October since the late 1970s. We propose here a chemical mechanism for the formation of the ozone hole. It involves removal of gaseous odd nitrogen by ion- and/or aerosol-catalysed conversion of N_2O_5 and $ClONO_2$ to HNO_3 vapour, followed by heteromolecular HNO_3–H_2O condensation, leading to HNO_3–H_2O aerosols. At an altitude of 17 km, these processes start at temperatures below 205 ± 5 K, well above the condensation temperature of pure water vapour. We propose that the absence of gaseous odd nitrogen and catalytic methane oxidation reactions driven by sunlight in early spring lead to large OH concentrations which rapidly convert HCl to ClOX. Catalytic reactions of ClOX and BrOX cause drastic ozone destructions and can account for the springtime 'ozone hole' first observed by Farman et al. [1]. By our model the depletion would be mainly due to emissions of industrial organic chlorine compounds. Arctic regions may also become affected. The depletion lasts while HNO_3, but not HCl, is incorporated in the particles in the temperature range 205 ± 5 to 192 K.

Farman et al. [1] connected the ozone decrease with the extremely cold temperatures and stable dynamics of the lower stratosphere and also proposed that the depletion was due to the increasing transfer of industrial chlorine compounds to the stratosphere. This idea was later developed by McElroy et al. [4] and Solomon et al. [5]. McElroy et al. [4] consider reactions between ClO and BrO. Their scheme

This text was first published as: Paul J. Crutzen and Frank Arnoldt: "Letters to Nature" entitled "Nitric acid cloud formation in the cold Antarctic stratosphere: a major cause for the springtime 'ozone hole'", in: *Nature,* vol. 324/18, 25 December 1986: 651–655. The permission to republish this artiucle was granted on 19 August 2015 by Ms. Claire Smith, Nature Publishing Group & Palgrave Macmillan, London, UK.

P.J. Crutzen and H.G. Brauch (eds.), *Paul J. Crutzen: A Pioneer on Atmospheric Chemistry and Climate Change in the Anthropocene,* Nobel Laureates 50, DOI 10.1007/978-3-319-27460-7_6 requires the absence of NO and NO_2 from the stratosphere, brought about by reactions of N_2O_5 and $ClONO_2$ on the surfaces of the stratospheric cloud particles observed [6, 7] in polar regions:

$$N_2O_5 + H_2O \rightarrow 2HNO_3 \qquad\qquad (1)$$

$$ClONO_2 + H_2O \rightarrow ClOH + HNO_3 \qquad\qquad (2)$$

and furthermore the conversion of most HBr to BrO_x, and most HCl to ClO_x. For the latter assumption, however, they do not present a mechanism. Solomon et al. [5] assume that efficient reactions of $ClONO_2$ and HCl occur in polar clouds (Figs. 6.1 and 6.2):

$$ClONO_2 + HCl \rightarrow Cl_2 + HNO_3 \qquad\qquad (3)$$

leading to the catalytic reaction chain

$$OH + O_3 \rightarrow HO_2 + O_2 \qquad\qquad (4)$$

$$Cl + O_3 \rightarrow ClO + O_2 \qquad\qquad (5)$$

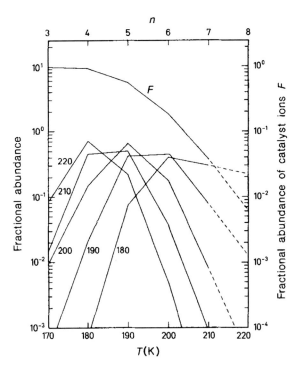

Fig. 6.1 Size distributions of $H^+(H_2O)_n$ cluster ions (upper abscissa and left ordinate) for an altitude of 17 km, an H_2O-vapour volume mixing ratio of 3×10^{-6} and temperatures of 180, 190, 200, 210 and 220 K (thermodynamical data of Lau et al. [17]). Also shown is the fraction F of $H^+(H_2O)_n$ ions (lower abscissa and right ordinate) with $n \geq 6$ that may serve as catalysts

$$ClO + HO_2 \rightarrow ClOH + O_2 \tag{6}$$

$$ClOH + hv \rightarrow Cl + OH \tag{7}$$

$$\text{net}: \quad 2O_3 \rightarrow 3O_2 \tag{D1}$$

The efficiency of the heterogeneous reactions (1)–(3) for stratospheric conditions has not been measured, so the above studies are speculative.

To explain the observations we present a chemical mechanism, involving the following steps: (i) conversion of NO_x to N_2O_5 during polar night:

$$NO + O_3 \rightarrow NO_2 + O_2 \tag{8}$$

$$NO_2 + O_3 \rightarrow NO_3 + O_2 \tag{9}$$

$$NO_3 + NO_2(+M) \rightarrow N_2O_5(+M) \tag{10}$$

(ii) formation of HNO_3 by ion- and/or aerosol-catalysed conversion of N_2O_5 and $ClONO_2$ at temperatures ≤ 215–210 K; (iii) heterogeneous, heteromolecular condensation of HNO_3 vapour with H_2O giving rise to stratospheric cloud particles and depletion of gaseous HNO_3 at temperatures $\leq 205 \pm 5$ K; (iv) efficient hydroxyl radical formation by cosmic ray ionization and ozone photolysis, amplified by photochemical methane oxidation reactions after sunlight returns to the polar regions; (v) conversion of HCl to ClOX, and HBr to BrOX, by reaction with OH; (vi) ozone destruction through ClOX and BrOX catalysis.

Conversion of N_2O_5 and $ClONO_2$ to HNO_3 and the depletion of gaseous NO_x and HNO_3, which are possible only at low temperatures, are key elements in our mechanism as they prevent OH removal by the otherwise most important catalytic reaction cycle:

$$OH + HNO_3 \rightarrow H_2O + NO_3 \tag{11}$$

$$NO_3 + hv \rightarrow NO_2 + O \tag{12}$$

$$NO_2 + OH(+M) \rightarrow HNO_3(+M) \tag{13}$$

$$\text{net}: \quad 2OH \rightarrow H_2O + O \tag{H1}$$

Conversion of N_2O_5 and $ClONO_2$ to HNO_3 via (1) and (2) may be catalysed by ions and/or aerosols. The ion-catalysed mechanism involves an H_2O molecule as a reaction partner, contained in a cluster ion [8–10]. Atmospheric ambient ions present at ~ 17 km, the centre of the height region of interest, are $H^+(H_2O)_n$, $H^+(CH_3CN)_m(H_2O)_m$ and $NO_3^-(HNO_3)_m(H_2O)_n$ as was found from recent balloon-borne mass-spectrometer measurements [11–13]. For $H^+(H_2O)_n$ ions, for example, N_2O_5 conversion may proceed via

$$N_2O_5 + H^+(H_2O)_n \rightarrow HNO_3 + H^+(H_2O)_{n-1}HNO_3 \qquad (1a)$$

$$H^+(H_2O)_{n-1}HNO_3 + H_2O \rightarrow HNO_3 + H^+(H_2O)_n \qquad (1b)$$

becoming exothermic only for $n \geq 6$. Ions of the type $NO_3^-(HNO_3)_m(H_2O)_n$ may be even more efficient catalysts as extra energy is gained due to stronger clustering of HNO_3 formed by the N_2O_5 reaction [10]. However, to serve as catalysts both positive and negative ions must be hydrated enough. This is possible only at low atmospheric temperatures. Figure 5.1 shows calculated relative abundances of the members of the $H^+(H_2O)_n$ family and also the fraction F of $H^+(H_2O)_n$ ions with $n \geq 6$. Taking the calculated F values, lower limits to the N_2O_5 lifetime, t_i, against ion-catalysed conversion can be estimated. For a total ion concentration of $\sim 10^4$ cm^{-3} and a typical rate coefficient for an ion-molecule reaction of 2×10^{-9} cm^3 s^{-1} one obtains $t_i \geq 5 \times 10^4/Fs$. Hence, at temperatures below about 215 K, t_i may become shorter than one month. It should be mentioned, however, that exothermicity does not necessarily guarantee a large reaction coefficient and t_i may therefore be larger than this. For a more detailed discussion of ion processes see Ref. [10].

Conversion of N_2O_5 and $ClONO_2$ on aerosols is rather uncertain as the reaction probability per collision, $\gamma\gamma$, is not well known. The lifetime t_{AC} of a gaseous molecule against collision with aerosols in the background atmosphere at ~ 17 km is $\sim 10^4$ s and thus the reaction time t_A becomes t_{AC}/γ. To make t_A one month, $\gamma = 2 \times 10^{-3}$ is required, which is much larger than γ measured [14] for concentrated sulphuric acid at 300 K.

In the cold Antarctic stratosphere, however, the aerosol content may become much larger than in the background stratosphere as found from solar extinction measurements [6, 7]. This excess aerosol, termed polar stratospheric clouds (PSCs) has previously [7] been explained in terms of H_2O condensation alone.

Here we shall argue that heteromolecular condensation of HNO_3 and H_2O occurs and leads to a great increase of the aerosol mass and surface area, which, in turn, may accelerate aerosol-catalysed N_2O_5 and $ClONO_2$ conversion, and more importantly leads to a depletion of HNO_3 vapour. Figure 5.1 shows extrapolations of measured [15] equilibrium saturation vapour pressures e_1 (H_2O) and e_2 (HNO_3) over an HNO_3–H_2O mixture with an HNO_3-mass fraction $f_2 = 0.5$. The extrapolations contain considerable uncertainties, especially for the solid state [10]. The case $f_2 = 0.5$ was chosen as this mixture gives the highest condensation temperature. Also shown in Fig. 5.2 are partial pressures for a_1 (H_2O) and a_2 (HNO_3) as found in the background atmosphere around 17 km altitude [29]. The procedure to derive such conditions is illustrated in Fig. 6.3 for liquid mixtures and a temperature of 195 K. The condition for heteromolecular condensation ($e_1 \leq a_1$ and $e_2 \leq a_2$) for $T = 185$ K is met at f_2 near 40 %. For this mixture $e_1 = a_1$, but $a_2 \gg e_2$, so that gaseous HNO_3 condenses out. This lowers both e_1 and the atmospheric partial

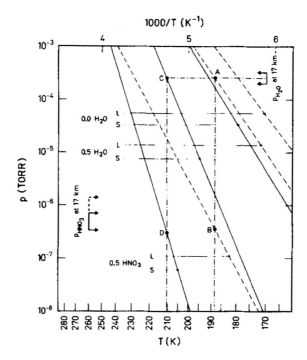

Fig. 6.2 Equilibrium saturation vapour pressures of H_2O and HNO_3 over solid (*solid curves*) and liquid (*broken curves*) H_2O–HNO_3 mixtures with HNO_3 mass fractions of 0.5 and 0.0 (pure H_2O). Also shown are partial pressures for H_2O and HNO_3 as observed in the background atmosphere around 17 km altitude [29]. Condensation temperatures for liquid mixture (points A and B) and solid (points C and D) mixtures are about 188 and 211 K respectively. For liquid and solid H_2O ($f_2 = 0.0$) condensation temperatures are about 179 and 191 K respectively

pressure of HNO_3 and increases e_2, so that about 1.5 times more H_2O than HNO_3 molecules condense out. Because $a_1 \gg a_2$, the resulting atmospheric pressures of HNO_3 will be much reduced, while those of H_2O are hardly affected.

An inspection of Fig. 5.2 reveals that the condensation temperature T_c for a liquid mixture with $f_2 = 0.5$ (points A and B) is about 188 K. For solid mixture one obtains $T_c = 211$ K (points C and D). A further cooling below T_c by only 3–4 K would be sufficient to deplete 50 % of the HNO_3 vapour, giving rise to an increase of the aerosol mass and surface area by factors of about 100 and 20 respectively. Condensation of pure H_2O vapour, which becomes possible only at temperatures below about 191 K (solid) to 179 K (liquid), gives rise to an additional strong increase of the aerosol mass. The heteromolecular HNO_3–H_2O condensation mechanism proposed here offers an explanation for recent extinction measurements [6, 7] showing a strong increase of the stratospheric total extinction at 1 μm wavelength over Antarctica as temperatures fall below about 205 K, which is close enough to 211 K considering the uncertainty of vapour pressure extrapolations and height distribution of the aerosol. Below about 191 K also other trace gases,

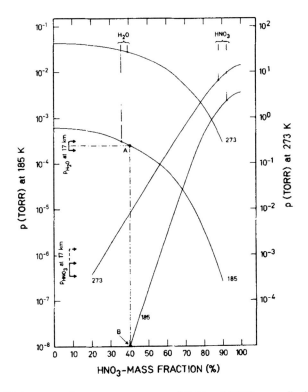

Fig. 6.3 Equilibrium saturation vapour pressures of H_2O and HNO_3 over liquid H_2O–HNO_3 mixtures against the HNO_3-mass fraction f_2 for temperatures of 273 and 185 K. Also shown are partial pressures of H_2O and HNO_3 as observed in the background atmosphere around 17 km altitude. Both sets of curves show the lowering of equilibrium vapour pressures of H_2O and HNO_3 over those of the one-component systems. At 185 K and $f_2 = 40\%$ large supersaturation of gaseous HNO_3 occurs and co-condensation of HNO_3 and H_2O takes place

particularly HCl, may become incorporated in aerosols [10]. With the increased aerosol surface, aerosol-catalysed N_2O_5 and $ClONO_2$ conversion may become important, depending on γ. However, ion-catalysed processes may be more efficient. It is possible that they are the missing source of HNO_3 in the polar winter stratosphere required in recent studies [10, 18].

Another ion process which has a role in polar-night stratospheric chemistry is hydroxyl radical formation by ion chemistry driven by galactic cosmic rays. The rate of this ion-assisted OH formation is one to two OH molecules per ionization event or ~ 40–80 OH molecules cm^{-3} s^{-1} at high geomagnetic latitudes. Because the galactic cosmic ray ionization rate varies with solar activity, an 11-year modulation will be induced. A more detailed discussion of ion and aerosol processes will be given in Ref. [10].

Modelling of stratospheric photochemical processes is complex and all details cannot be discussed here. Our model uses a standard set of chemical reactions [20, 21]. To emphasize the effects of chemistry alone, we adopted a very small eddy diffusion coefficient, because the south polar vortex seems to be such a stable dynamic system [22].

When gas-phase NO_x and HNO_3 are removed from the lower stratosphere, the OH radial reacts mainly with CH_4, HCl and HBr.

$$CH_4 + OH(+O_2) \rightarrow CH_3O_2 + H_2O \qquad (14)$$

$$HCl + OH \rightarrow Cl + H_2O \qquad (15)$$

$$HBr + OH \rightarrow Br + H_2O \qquad (16)$$

The results of the model calculations shown in Fig. 5.4, for altitudes of 17 and 22 km for 1979 and 1985, at 70 °S, are mainly intended to demonstrate the sensitivity of ozone to odd-nitrogen removal during springtime. They are not close simulations of the exact conditions in the Antarctic. The calculations were started with concentrations obtained with our two-dimensional model. Ozone and temperature data were taken from the measurements [3] made at Syowa (60 °S) in 1983. The small jumps in the curves occurring around day 147 indicate the date at which we have assumed that HNO_3 becomes incorporated in the cloud particles. The large changes in concentrations which occur near day 200, about 20 July, mark the return of daytime conditions (Sun above the horizon) for some period during the day. Results are shown for noon. In these calculations ClOX denotes the sum of Cl, ClO, ClOH, Cl_2O_2 and OClO. The last three compounds all dissociate rapidly in daylight, producing ClO_x (=Cl + ClO). BrOX is defined similarly. The stratospheric concentrations of total inorganic bromine were assumed to be 1 % of those of inorganic chlorine, which seems to be roughly in agreement with the observations, although the uncertainty is large [23, 24].

Besides reaction cycle D1, there are two other catalytic reaction chains that are important in the breakdown of stratospheric ozone during early spring when sunlight has returned to the Antarctic:

$$ClO + BrO \rightarrow Cl + Br + O_2 \qquad (17)$$

$$Cl + O_3 \rightarrow ClO + O_2 \qquad (5)$$

$$Br + O_3 \rightarrow BrO + O_2 \qquad (18)$$

$$net: \quad 2O_3 \rightarrow 3O_2 \qquad (D2)$$

as considered by McElroy et al. [4], and

$$ClO + ClO(+M) \rightarrow Cl_2O_2(+M) \qquad (19)$$

$$Cl_2O_2 + hv \rightarrow Cl + ClOO, \; \lambda < 350 \text{ nm} \qquad (20)$$

$$ClOO + M \rightarrow Cl + O_2 + M \tag{21}$$

$$Cl + O_3 \rightarrow ClO + O_2(2\times) \tag{5}$$

$$net: \quad 2O_3 \rightarrow 3O_2 \tag{D3}$$

as recently introduced by Molina and Molina [25] and R.A. Cox (personal communication). According to Molina and Molina, the photolysis rate of Cl_2O_2 is roughly that of ClOH. The product yield of (20) is still only based on indirect evidence [25]. For (19) we chose the rate expression $k = 6 \times 10^{-33} (300/T)^{21}$ cm^6 molecule^{-2} s^{-1} by R.A. Cox (personal communication).

We note from the results in Fig. 6.4 that during polar night much HBr can be converted to BrOX by (16). The conversion would be much enhanced if HBr reacted rapidly with NO_3, which seems quite likely. The conversion of HCl to ClOX is less efficient, because (15) is $\sim 1/25$ as fast as (16). During polar night, a large fraction of the OH radicals produced by the galactic cosmic rays is lost by reaction with CH_4. The situation changes drastically after sunlight returns, when,

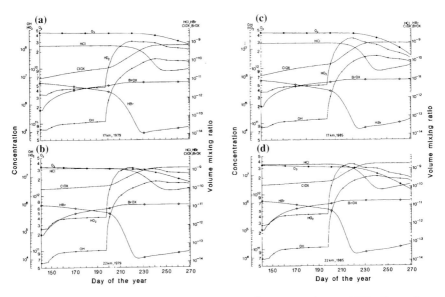

Fig. 6.4 **a**, **b**, calculated concentrations (molecules cm^{-3}) at noon, of OH and HO$_2$ (first scale on the left) and of O$_3$ (second scale on the left), and volume mixing ratios of HCl, ClOX, HBr and BrOX (right scale). Polar night conditions at 70 °S, here defined as Sun below horizon for 24 h period, lasted from 22 May (day 142) to 18 July (day 198). It was assumed that from day 147 nitric acid became incorporated in the polar stratospheric clouds. Calculations were performed for two altitudes, 17 km (**a**) and 22 km (**b**), and estimated conditions for the year 1979. **c**, **d**, Similar calculations as **a**, **b**, for 1985

starting from low odd hydrogen concentrations, reactions with CH_4 lead to rapid net production of odd hydrogen. The important reactions are:

$$CH_4 + OH(+O_2) \rightarrow CH_3O_2 + H_2O, \tag{14}$$

$$CH_3O_2 + HO_2 \rightarrow CH_3O_2H + O_2 \tag{22}$$

$$CH_3O_2H + hv \rightarrow CH_3O + OH, \; \lambda \leq 350 \text{ nm} \tag{23}$$

$$CH_3O + O_2 \rightarrow CH_2O + HO_2 \tag{24}$$

$$\text{net}: \quad CH_4 + O_2 \rightarrow CH_2O + H_2O$$

followed with about 30 % probability by

$$CH_2O + hv \rightarrow H + CHO, \; \lambda \leq 330 \text{ nm}; \tag{25}$$

$$H + O_2 + M \rightarrow HO_2 + M \tag{26}$$

$$CHO + O_2 \rightarrow CO + HO_2 \tag{27}$$

The net result of the oxidation of one methane molecule by OH is therefore an average net gain of 0.6 odd hydrogen radicals, providing for a rapid buildup of OH radicals until these are also significantly lost by

$$CH_3O_2H + OH \rightarrow CH_3O_2 + H_2O \tag{28a}$$

$$CH_3O_2H + OH \rightarrow CH_2O + H_2O + OH \tag{28b}$$

For the first weeks, the main primary source of OH is production by galactic cosmic rays. Later, the reactions

$$O_3 + hv \rightarrow O(^1D) + O_2, \; \lambda \leq 310 \text{ nm}; \tag{29}$$

$$O(^1D) + H_2O \rightarrow 2OH \tag{30}$$

become more important, further enhancing the OH concentrations. ClOX is efficiently produced from HCl after the polar night, leading to efficient destruction of ozone mainly through reaction cycle Dl and D3, with some contributions also from the BrO_x/ClO_x cycle D2. The calculations for two altitudes 17 and 22 km, shown in Fig. 5.4, demonstrate the sensitivity of stratospheric ozone to the perturbations that characterize the springtime Antarctic lower stratosphere, so that the observed trends in ozone removal can be explained by the chemical mechanisms presented above. The ozone decay continues as long as most HNO_3 is bound in the stratospheric polar clouds, that is, until the Antarctic stratospheric warming occurs in late September or early October [3, 22]. Then the evaporation of the polar clouds will reintroduce the various NOX gases into the lower stratosphere through known photochemical processes and most ClOX will be converted back to HCl. The decreasing ozone concentrations will delay the springtime solar warming of the lower stratosphere and this could gradually increase the period of unstable photochemistry. The ozone depletion will last as long as HNO_3, but not HCl is incorporated in PSCs. At an altitude of 17 km this occurs between 205 ± 5 and 192 K. Although less frequent than in the Antarctic, such conditions occur also in the Arctic stratosphere.

In this paper we could not present a detailed analysis of all features of the Antarctic ozone hole, because this requires consideration of and access to extensive meteorological, chemical and optical data sets [26]. We have, however, clearly shown how in the Antarctic lower stratosphere during winter and springtime unstable photochemical conditions can develop that lead to drastic destruction of ozone. Several other complications that are likewise not treated in this paper must be considered in future studies, such as the effects of the major eruptions of volcano El Chichon in 1982, the possibility of gravitational settling of the particles [26], and the reactions of odd hydrogen and other chemical compounds on the particles.

To test our chemical mechanism, measurements of odd nitrogen and chlorine gases as well as ambient ions would be very informative. No measurements of HNO_3 and HCl have so far been reported in the Antarctic for September-October. Available total column NO_2 observations [27, 28] agree with the mechanism proposed, but do not provide proof. Also laboratory studies of the binary H_2O–HNO_3 system at low temperatures and ion-catalysed reactions are needed.

References

1. Farman, J.C.; Gardiner, B.G.; Shanklin, J.D., 1985: in: *Nature*, 315: 207–210.
2. Stolarski, R.S.; et al.: in: *Nature*, 322: 808–811.
3. Chubachi, S., 1985: in: Kato, S. (Ed.): *Handbook for MAP, vol 18* (Urbana: SCOSTEP, Univ. Illinois): 453–457.
4. McElroy, M.B.; Salawitch, R.J.; Wofsy, S.C.; Logan, J.A., 1986: in: *Nature*, 321: 759–762.
5. Solomon, S.; Garcia, R.R.; Rowland, F.S.; Wuebbles, D.J., 1986: in: *Nature*, 321: 755–758.
6. McCormick, M.P.; Steele, H.M.; Hamill, P.; Chu, W.P.; Swissler, T.J., 1982: in: *Journal of the Atmospheric Sciences*, 39: 1387–1397.
7. Steele, H.M.; Hamill, P.; McCormick, M.P.; Swissler, T.J., 1983: in: *Journal of the Atmospheric Sciences*, 40: 2055–2067.
8. Ferguson, E.E.; Fehsenfeld, F.C.; Albritton, D.L., 1979: in: *Gas Phase Ion Chemistry*, 1: 45–81 (Academic Press).
9. Bohringer, H.; Fahey, D.W.; Fehsenfeld, F.C.; Ferguson, E.E., 1983: in: *Planetary and Space Science*, 31: 185–191.
10. Arnold, F.; Crutzen, P.J., (in preparation).
11. Arnold, F., 1982: in: *Proceedings of the DAHLEM Workshop on Atmospheric Chemistry* (Ed. Goldberg, E. D.): 273.
12. Arnold, F.; Bührke, T., 1983: in: *Nature*, 301: 293.
13. Ferguson, E.E.; Arnold, F., 1981: in: *Accounts of Chemical Research*, 14: 327.
14. Baldwin, A.C.; Golden, D.M., 1979: in: *Science*, 206: 562–563.

15. Clavelin, J.L.; Mirabel, P., 1979: in: *Annales de Chimie et de Physique*, 76: 533–537.

16. Forsythe, W.R.; Giauque, F., 1942: in: *Journal of the American Chemical Society*, 64: 48–61.

17. Lau, Y.K.; Ikuta, S.; Kebarle, P., 1982: in: *Journal of the American Chemical Society*, 104: 1462–1469.

18. Austin, J.A.; Garcis, R.R.; Russell, J.M.; Solomon, S.; Tuck, A.F., 1986: in: *Journal of Geophysical Research*, 90: 5477–5485.

19. Evans, W.J.F.; McElroy, C.T.; Galbally, J.E., 1985: in: *Geophysical Research Letters*, 12: 825–828.

20. DeMore, W.B.; et al., 1985: in: *Chemical Kinetics and Photochemical Data for Use in Stratospheric Modeling* (California: JPL Publication, NASA Jet Propulsion Laboratory): 85–37.

21. Crutzen, P.J.; Schmailzl, U., 1983: in: *Manet. Space Science,* 31: 1009–1032.

22. Labitzke, X., 1981: in: *Journal of Geophysical Research*, 86: 9665–9678.

23. Berg, W.W.; Crutzen, P.J.; Grahek, F.E.; Gitlin, S.N.; Sedlacek, W.A., 1980: in: *Geophysical Research Letters*, 7: 937–940.

24. Berg, W.W.; Heidt, L.E.; Pollock, W.; Sperry, P.D.; Cicerone, R.J., 1984: in: *Geophysical Research Letters*, 11: 429–432.

25. Molina, L.T.; Molina, M.: in: *The Journal of Physical Chemistry* (submitted).

26. McCormick, M.P.; Trepte, C.R.: in: *Journal of Geophysical Research* (submitted).

27. McKenzie, R.L.; Johnston, P.V., 1984: in: *Geophysical Research Letters*, 11: 73–75.

28. Shibasaki, K., 1985: in: Kato, S. (Ed.): *Handbook for MAP* 18: 506–509 (SCOSTER, Univ. Illinois, 1985).

29. World Meteorological Organization, 1985: *Atmospheric Ozone 1985*, WMO Rep. no. 16 (Geneva: WMO).

Chapter 7
Biomass Burning in the Tropics: Impact on Atmospheric Chemistry and Biogeochemical Cycles

Paul J. Crutzen and Meinrat O. Andreae

Biomass burning is widespread, especially in die tropics. It serves to clear land for shifting cultivation, to convert forests to agricultural and pastoral lands, and to remove dry vegetation in order to promote agricultural productivity and the growth of higher yield grasses.[2] Further-more, much agricultural waste and fuel wood is being combusted, particularly in developing countries. Biomass containing 2–5 petagrams of carbon is burned annually (1 petagram = 10^{15} grams), producing large amounts of trace gases and aerosol particles that play important roles in atmospheric chemistry and climate. Emissions of carbon monoxide and methane by biomass burning affect the oxidation efficiency of the atmosphere by reacting with hydroxyl radicals, and emissions of nitric oxide and hydrocarbons lead to high ozone concentrations in the tropics during the dry season. Large quantities of smoke particles are produced as well, and these can serve as cloud condensation nuclei. These particles may thus substantially influence cloud microphysical and optical properties, an effect that could have repercussions for the radiation budget and the hydrological cycle in the tropics. Widespread burning may also disturb biogeochemical cycles, especially that of nitrogen. About 50 % of the nitrogen in the biomass fuel can be released as molecular nitrogen. This pyrodenitrification process causes a sizable loss of fixed nitrogen in tropical ecosystems, in the range of 10–20 teragrams per year (1 teragram = 10^{12} grams).

7.1 Introduction

The use of fire as a tool to manipulate the environment has been instrumental in the human conquest of Earth, the first evidence of the use of fires by early hominids dating back to 1–1.5 million years ago [1]. Even today, most human-ignited vegetation fires take place on the African continent, and its widespread, frequently burned savannas bear ample witness to this. Although natural fires can occur even

This text was first published as: Crutzen, P.J.; Andreae, M.O. (1990). "Biomass Burning in the Tropics: Impact on Atmospheric Chemistry and Biogeochemical Cycles". *Science* **250** (4988): 1669–1678. The copyright belongs to the authors. We thank S. Brown, E.F. Bruenig, J. Clark, P. Feamside, I.Y. Fung, D.W. Griffith, J. Goldammer, C.S. Hall, D.O. Hall, A.L. Hammond, W.M. Hao, R. Herrera, R.A. Houghton, M. Keller, V.W.H. Kirchhoff, J.P. Lanly, J. Levine, J. Lobert, J.M. Logan, P. Matson, J.C. Menaut, N. Myers, Ph. Robertson, C.F. Rogers, B.J. Stocks, E. Sanhueza, and D. Schimel for comments.

© The Author(s) 2016
P.J. Crutzen and H.G. Brauch (eds.), *Paul J. Crutzen: A Pioneer on Atmospheric Chemistry and Climate Change in the Anthropocene*,
Nobel Laureates 50, DOI 10.1007/978-3-319-27460-7_7

165

in tropical forest regions [2, 3], the extent of fires has greatly expanded on all continents with the arrival of *Homo sapiens*. Measurements of charcoal in dated sediment cores have shown clear correlations between the rate of burning and human settlement [4]. Pollen records show a shift with human settlement from pyrophobic vegetation to pyrotolerant and pyrophilic species, testimony to the large ecological impact of human-induced fires.

Natural fires have occurred since the evolution of land plants some 350–400 million years ago and must have exerted ecological influences [5]. In fact, high concentrations of black carbon in the Cretaceous-Tertiary boundary sediments suggest that the end of the age of the reptiles some 65 million years ago was associated with global fires that injected enormous quantities of soot particles into the atmosphere [6].

Today, the environmental impact of the burning of fossil fuels and biomass is felt throughout the world, and concerns about its consequences are prominent in the public's mind. Although the quantities of fossil fuels burned have been well documented, most biomass burning takes place in developing countries and is done by farmers, pioneer settlers, and housewives, for whom keeping records of amounts burned is not an issue. Biomass burning serves a variety of purposes, such as clearing of forest and brushland for agricultural use; control of pests, insects, and weeds; prevention of brush and litter accumulation to preserve pasturelands; nutrient mobilization; game hunting; production of charcoal for industrial use; energy production for cooking and heating; communication and transport; and various religious and aesthetic reasons. Studies on the environmental effects of biomass burning have been much neglected until rather recently but are now attracting increased attention [7]. This urgent need has been recognized and will form an important element in the International Geosphere-Biosphere Programme [8].

In this article, we update quantitative estimates of the amounts of biomass burning that is taking place around the world and the resulting gaseous and particulate emissions and then discuss their atmospheric-chemical, climatic, and ecological consequences. Distinction should be made between net and prompt releases of CO_2. Net release occurs when land use changes take place by which the standing stock of biomass is reduced, for example, through deforestation. Biomass burning causes a prompt release of CO_2 but does not necessarily imply a net release of CO_2 to the atmosphere, as the C that is lost to the atmosphere may be returned by subsequent regrowth of vegetation. In either case, there is a net transfer of particulate matter and trace gases other than CO_2 from the biosphere to the atmosphere. Many of these emissions play a large role in atmospheric chemistry, climate, and terrestrial ecology.

7.2   Estimates of Worldwide Biomass Burning

In this section we derive some rough estimates of the quantities of biomass that are burned in the tropics through various activities, such as forest clearing for permanent use for agriculture and ranching, shifting cultivation, removal of dry savanna vegetation and firewood, and agricultural waste burning. In all cases, the available data are extremely scanty, allowing only a very uncertain quantitative assessment.

7.2.1 Clearing of Forests for Agricultural Use

Two types of forest clearing are practiced in the tropics: shifting agriculture, where for a few years the land is used and then allowed to return to forest vegetation during a fallow period, and permanent conversion of forests to grazing or crop lands. In both cases, during the dry season, undergrowth is cut and trees are felled and left to dry for some time in order to obtain good burning efficiency. The material is then set on fire. The efficiency of the first burn is variable. Observations in forest clearings in Amazonia gave a burning efficiency of about 28 % [9], similar to the value used by Seiler and Crutzen [10]. This relatively low efficiency is due to the large fraction of the biomass that resides in tree trunks, only a small portion of which is consumed in the first burn. The remaining material may be left to rot or dry but is often collected and set on fire again. Adequate statistics are not available on how much of the original above-ground biomass is finally burned. Taking reburn into account, we assume that in primary forests some 40 % is combusted [9]. For secondary forests, which have been affected by human activities and contain smaller sized material, we assume that 50 % is burned.

According to Seiler and Crutzen [10], shifting agriculture (also called slash-and-burn agriculture, field-forest rotation, or bush fallowing) was practiced by some 200 million people worldwide in the 1960s on some 300–500 million ha, with an annual clearing of some 20–60 million ha and a burning of 900–2500 Tg dm, that is, 400–1100 Tg of C; of this, 75 % takes place in tropical secondary forests and the remainder in humid savannas (dm = dry matter; 1 g dm $\approx$ 0.45 g of C). Originally, shifting cultivators typically practiced crop and fallow periods of 2–3 and 10–50 years, respectively. Because of growing populations and lack of forest areas, fallow periods in many regions have shortened so much that the land cannot recover to the required productivity, which causes shifting agriculture to decline [11]. On the other hand, in other regions it may still be expanding. According to Lanly [12], some 240 million ha were under traditional shifting agriculture by the end of the 1970s. On the basis of these statistics, Hao et al. [13] estimated that $\sim$24 million ha are cleared annually for shifting cultivation in secondary forests. This clearing exposes $\sim$2400–3000 Tg dm, that is, 1000–2400 Tg of C, to fire and thus leads to the release of 500–700 Tg of C. We combine the two ranges into an annual C release rate from shifting cultivation of between 500 and 1000 Tg. In traditional shifting agriculture, no net release of CO_2 to the atmosphere takes place because the forest is allowed to return to its original biomass density during the fallow period. The estimated rates, therefore, mainly represent prompt CO_2 release. However, because of overly frequent burning, the affected ecosystems often cannot recover to their original biomass, so that a net release of C to the atmosphere does result.

Permanent removal of tropical forests is currently progressing at a rapid rate. This process is driven by expanding human populations which require additional land, by large-scale resettlement programs, and by land speculation. The global rate of deforestation is subject to much uncertainty. The tropical forest survey of the Food and Agricultural Organization (FAO) of the United Nations for the latter part of the 1970s [12] has been the basis of several studies on net CO_2 release to the atmosphere. It now appears that the FAO statistics significantly underestimated deforestation rates, which, furthermore, may almost have doubled over the past decade [11, 14]. As the earlier work on tropical deforestation was clearly based on questionable information, we feel that there is little point in reviewing it. Instead, we will estimate the consequences of deforestation activities for trace gas emissions, using the statistics assembled by Myers [11] and Houghton [14]. This is the only available database that may be up to date and has also been adopted by the Intergovernmental Panel on Climate Change [15]. Although newer data have now been assembled by the FAO, unfortunately they have not been released in time to be included in the present review. The net CO_2 release to the atmosphere due to deforestation from these sources still allows for the wide range of 1.1–3.6 Pg of C per year [14] (this range is given because of uncertainty regarding the areal extent of deforestation [11] and the original and successional biomass loadings [16]). As about 60 % of the total biomass is located below ground, including soil organic matter, this net release of CO_2 implies that 0.5–1.4 Pg of C per year of biomass are exposed to fire. As only 40–50 % of the CO_2 release is through combustion (the rest is by microbial decomposition of organic matter), the resulting prompt release of CO_2 to the atmosphere would be in the range of 0.2–0.7 Pg of C per year.

Tropical savannas and brushland, typically consisting of a more or less continuous layer of grass interspersed with trees and shrubs, cover an area of about 1900 million ha [17]. Savannas are burned every 1–4 years during the dry season with the highest frequency in the humid savannas [18]. The extent of burning is increasing as a result of growing population pressures and more intensive use of rangeland. Although lightning may start some fires in savannas, most investigators are convinced that almost all are set by humans [4]. Only dried grass, litter, weeds, and shrubs are burned; the larger trees of fire-resistant species suffer little damage.

Menaut [18] has estimated that in the West African savannas 45–240 Tg of C per year are burned. The total area of this savanna region is 227 million ha, including 53 million ha of Sahel semidesert. No similarly detailed analysis on biomass burning has yet been attempted for other savanna regions. If we extrapolate to include all the savanna regions of the world (1900 million ha), we estimate that between 400 and 2400 Tg of C burn annually, and that most emissions are from the African continent. As, especially on the African continent, shifting cultivation also takes place in savanna regions, some double accounting could occur. The analysis by Seiler and Crutzen [10] indicates that a 30 % correction may have to be applied to the above range, reducing it to 300–1600 Tg of burned C per year.

7.2.2 *Fuel Wood, Charcoal, and Agricultural Waste*

In the developing countries, fuel wood and agricultural waste are the dominant energy sources for cooking, domestic heating, and some industrial activities. It is difficult to estimate the amount of wood burned each year. The number given by FAO [19] for 1987, 1050 Tg dm, is certainly an underestimate because it includes only wood that is marketed. Scurlock and Hall [20] estimate that the annual per capita biofuel need (firewood, crop residues, dungcakes) is about 500 kg in urban and 1000 kg in rural regions, and that perhaps two-thirds of the rural energy use in China comes from agricultural wastes. Altogether, they estimate that 14 % of the global energy and 35 % of the energy in developing countries is derived from biomass fuels, equivalent to 2700 Tg dm or 1200 Tg of C per year. Because of rapidly increasing populations in the developing world, this energy need is growing by several percent per year. An analysis of the situation in India [21], however, indicates a biofuel consumption of only 350 kg per capita per year in rural and 160 kg per capita per year in urban areas, adding up to a total consumption rate of 230 Tg per year for the Indian population of 760 million. About half of the biomass burned was firewood, the other half was mostly dung and crop residues.

It is clearly very difficult to extrapolate from this information. If the partitioning of biofuel between fuel wood and agricultural waste products derived for India were representative for the rest of the developing world, more than 1050 Tg dm of firewood and roughly an equal amount of agricultural waste products would be burned worldwide, that is, together at least about 950 Tg of C per year, 20 % less than the 1200 Tg of C per year estimated by Scurlock and Hall [20]. On the other hand, if the estimate of 230 Tg dm per year of biofuel burning for the Indian population is extrapolated to the total population in the developing world, then the amount is only about 600 Tg of C per year. Altogether we will assume a range of biofuel burning of 600–1200 Tg of C per year, with about equal contributions from firewood and agricultural waste products.

Burning of agricultural wastes in the fields, for example, sugar cane and rice straw, and stalks from grain crops, is another important type of biomass burning. The amount of residue produced equals about 1700 Tg dm per year in the developing world and a similar amount in the developed world [22]. It is difficult to estimate what fraction of this waste is burned. Rice straw makes up 31 % of the agricultural waste in the developing world, and, at least in Southeast Asia, burning of rice straw in the fields is the preferred method of waste disposal [23]. Sugar cane residues account for about 11 % of agricultural waste and are mostly disposed of by burning. We very tentatively guess that at least 25 % of the agricultural waste, about 200 Tg of C per year, are burned in the fields. Summarizing from the uncertain information that is available, we estimate that yearly some 300–600 Tg of C of firewood and 500–800 Tg of C of agricultural wastes are burned in the developing world. In the industrial world the corresponding figures are about one-tenth as large.

7.2.3 Prescribed Burning and Forest Wildfires

It is interesting to compare the quantities of tropical biomass burned with those due to fires in temperate and boreal forests. Although individual wildfires may be large, because of fire-fighting efforts, the area burned per year is relatively small. Stocks [24] estimates that about 8 million ha of temperate and boreal forests are subject to wildfires each year.

Prescribed burning is commonly used for forest management. It serves mainly to reduce the accumulation of dry, combustible plant debris in order to prevent destructive wildfires. Because it is limited to North America and Australia and the area involved is only 2–3 million ha per year [10], it has little impact on a global scale. Together some 150–300 Tg of C per year are burned by prescribed burning and wildfires, much less than through fires in the tropics.

7.3 Emissions to the Atmosphere

Table 7.1 summarizes the quantitative estimates of biomass burning in the tropics. We estimate that a total of 2700–6800 Tg of C are annually exposed to fires, of which 1800–4700 Tg of C are burned. The average chemical composition of dry plant biomass corresponds closely to the formula CH_2O. The nutrient element content varies with seasonal growth conditions; on a mass basis it is relatively low: about 0.3–3.8 % N, 0.1–0.9 % S, 0.01–0.3 % P, and 0.5–3.4 % K [25]. Consequently, although the emissions from biomass combustion are dominated by CO_2, many products of incomplete combustion that play important roles in atmospheric chemistry and climate are emitted as well, for example, CO, H_2, CH_4, other hydrocarbons, aldehydes, ketones, alcohols, and organic acids, and compounds

Table 7.1 Summary of the biomass exposed to fires, the total carbon released, the percentage of N to C in the fuel, and the total mass of N compounds released to the atmosphere by fires in the tropics

Source or activity	Carbon exposed (Tg C/year)	Carbon released (Tg C/year)	N/C ratio (% by weight)	Nitrogen released (Tg N/year)
Shifting agriculture	1000–2000	500–1000	1	5–10
Permanent deforestation	500–1400	200–700	1	2–7
Savanna fires	400–2000	300–1600	0.6	2–10
Firewood	300–600	300–600	0.5	1.5–3
Agricultural wastes	500–800	500–800	1–2	5–16
Total	2700–6800	1800–4700	–	15–46

containing the nutrient elements N and S, for example, NO, NH_3, HCN, and CH_3CN, SO_2, and COS. The smoke also contains particulate matter (aerosol) consisting of organic matter, black (soot) carbon, and inorganic materials, for example, K_2CO_3 and SiO_2. In Table 7.2, we combine our estimates of global amounts of biomass burning with the emission ratios for various important trace species and derived global rates of pyrogenic emissions.

In spite of the large uncertainties, it is quite evident from Table 7.2 that biomass burning results in globally important contributions to the atmospheric budget of several of the gases listed [26]. Because much of the burning is concentrated in limited regions and occurs mainly during the dry season (July to September in the Southern Hemisphere and January to March in the Northern Hemisphere), it is not surprising that the emissions result in levels of atmospheric pollution that rival those in the industrialized regions of the developed nations. This comparison applies especially to a group of gases that are the main actors in atmospheric photochemistry: hydrocarbons (for example CH_4), CO, and nitrogen oxides (NO_x). These

Table 7.2 Estimates of emissions in teragrams of C, H_2, CH_3Cl, N, S, or aerosol mass per year (TPM, total particulate matter; POC, particulate organic carbon; EC, elemental carbon; K, potassium)

Element or compound	Emission ratio	Emission from biomass burning	Total emissions from all sources
All C (from Table 7.1)		1800–4700	–
CO_2	≈90 %	1600–4100	–
CO	10 ± 5 %	120–510	600–1300
CH_4	1 ± 0.6 %	11–53	400–600
H_2	2.7 ± 0.8 %	5–16	36
CH_3Cl	$1.6 ± 1.5 × 10^{-4}$ %	0.5–2	2
All N (from Table 7.1)		15–46	–
NO_x	12.1 ± 5.3 %	2.1–5.5	25–60
RCN	3.4 ± 2.5 %	0.5–1.7	>0.4
NH_3	3.8 ± 3.2 %	0.5–2.0	20–60
N_2O	0.7 ± 0.3 %	0.1–0.3	12–14
N_2	≤50 %	≤11–19	100–170
SO_2	0.3 ± 0.15 %	1.0–4.0	70–170
COS	0.01 ± 0.005 %	0.04–0.20	0.6–1.5
TPM	30 ± 15 g/kg C	36–154	≈1500
POC	20 ± 10 g/kg C	24–102	≈180
EC	5.4 ± 2.7 g/kg C	6.4–28	20–30
K	0.4 ± 0.2 g/kg C	0.5–21	–

Emission ratios for C and S compounds are in moles relative to CO_2; those for N compounds are expressed as the ratios of emission relative to the N content of the fuel; the emissions of TPM, POC, EC, and K are in grams per kilogram of fuel C. The emission ratios have been derived from information in (26, 29–31, 36, and 46). In calculating the ranges of total emissions, we used only half the ranges of total C emissions (2500–3900 Tg of C per year) and the emission ratios (for instance, 7.5–12.5 % for CO). A similar procedure was followed for the N compounds gases have a strong influence on the chemistry of O_3, and OH, and thus on the oxidative state of the atmosphere. We will next discuss the most important emissions.

7.3.1 Carbon Dioxide

Our estimates of the amount of biomass exposed to fire worldwide (2.7–6.8 Pg of C per year; Table 7.1) and the resulting prompt CO_2 release to the atmosphere (1.8–4.7 Pg of C per year) are larger than earlier estimates [10, 13]. They are 30–80 % of the fossil fuel burning rate of 5.7 Pg of C per year [16].

We caution again that the prompt release of CO_2 to the atmosphere is not the same as the net CO_2 release from deforestation. The latter is estimated at 1.1–3.6 Pg of C per year [15]. However, these figures need to be reduced somewhat, as a fraction of the burned biomass is converted into elemental C (charcoal), which is not subject to destruction by microbial activity [5, 10]. There is hardly any information available on charcoal formation in fires. Fearnside and co-workers determined that in two forest clearings in Amazonia 3.6 % of the biomass C exposed to the fires remained in the partially burned vegetation as elemental C [9]. To this must be added the elemental C that is released in the smoke [27], so that the total elemental C yield may be about 4 % of the C exposed to fire, or alternatively 14 % of the C burned. From observations on a prescribed burn in a Florida pine forest [28], an elemental C yield of 5.4 % (3.6–7.4 %) of the C exposed or 9 % (6–16 %) of the C burned can be derived. From this limited data set we adopt charcoal yields of 5 and 10 % of the C exposed or burned, respectively. When these elemental C yields are applied to the estimate of biomass burning given above, a range of elemental C production of 0.2–0.6 Pg of C per year can be deduced, which thus may reduce the range of net CO_2 emissions of 0.5–3.4 Pg of C per year. This correction is extremely tentative because of the paucity of measurements on elemental C production from forest fires and the total absence of data on yields from savanna fires or agricultural waste burning.

7.3.2 CO, CH$_4$ and Other Hydrocarbons, H$_2$, CH$_3$Cl

Figure 7.1 shows the sequence of the emission of CO_2 (maximum in the flaming stage), CO (maximum in the smoldering stage), and of various other gaseous products from experimental fires conducted in our laboratory. The fraction of CO emitted depends on the fire characteristics: hot flaming fires with good O_2 supply produce only a few percent, whereas smoldering fires may yield up to 20 % CO [29–31]. Therefore, CO may serve as a marker of the extent of smoldering combustion, so that emissions of gases from smoldering combustion can be better estimated on the basis of emission ratios relative to CO rather than relative to CO_2.

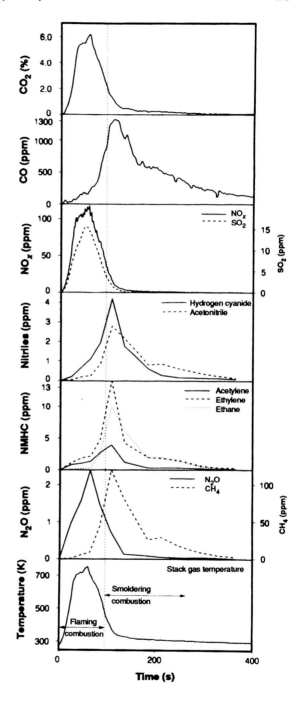

Fig. 7.1 Concentrations of CO_2, CO, and various other gases in the smoke from an experimental fire of Trachypogon grass from Venezuela as a function of time and the stack gas temperature. The *dotted line* separates the flaming phase from the smoldering phase. The flaming stage in this fire lasted for about 96 s. Concentrations are in percent by volume for CO_2, in volume mixing ratios (ppm) for the other species (1 % = 10,000 ppm). Note that CO_2, NO_x, SO_2, and N_2O are mainly emitted in the flaming phase and the other gases in the smoldering phase; NMHC, nonmethane hydrocarbons

Our estimates show very large emissions of CO, between 120 and 510 Tg of C per year. The estimated global source of CO is close to 1000 Tg of C per year [32]; biomass burning is thus one of the main sources of atmospheric CO. Because about 70 % of the OH radicals in background air react with CO, biomass burning can substantially lower the oxidative efficiency of the atmosphere (which is mostly determined by the concentrations of OH), and thus can cause the concentrations of many trace gases to increase.

Methane contributes strongly to the atmospheric greenhouse effect; in this respect it has 20–30 times the efficiency per mole in the atmosphere of CO_2. It resides in the atmosphere long enough to enter the stratosphere. There, the oxidation of each molecule of CH_4 leads to the production of two molecules of H_2O; this process adds substantially to the stratospheric water vapor content. Because reaction with CH_4 also converts active Cl and ClO catalysts (which break down O_3) into inactive HCl, CH_4 plays a substantial role in stratospheric O_3 photochemistry. The pyrogenic emissions of CH_4, 11–53 Tg of C per year, may be about 10 % of the global CH_4 source. On the basis of $^{13}CH_4$ isotope studies, the source of CH_4 from biomass burning was even estimated to be as large as 50–90 Tg per year [33], exceeding our estimated range of pyrogenic CH_4 emissions (Table 7.2). Stevens et al. [34] indicate that the biomass burning source of CH_4 may have been increasing by 2.5–3 Tg per year during the past decade; this rate suggests that global biomass burning may have been increasing by as much as 5 % per year, thus contributing strongly to the observed increases of CH_4 by almost 1 % per year [32].

For H_2, biomass combustion may contribute 5–16 Tg per year. Its global sources and sinks have been estimated to be about 36 Tg per year, mostly due to fossil fuel burning [35]. Biomass burning may thus make a significant contribution to the global source of H_2, which plays a role in stratospheric photochemistry.

Our estimated emission range of CH_3Cl, from 0.5–2 Tg of Cl per year by biomass burning, is large in comparison with its destruction rate of about 2 Tg of Cl per year by reaction with OH radicals [32]. The photochemical breakdown of CH_3Cl is a significant source for active Cl in the stratosphere, so that it plays a role in O_3 depletion. CH_3Cl is often considered to be of natural origin, emanating from the ocean. This view needs to be reconsidered [26, 32].

7.3.3 Nitrogen Gases

Emissions of NO from biomass burning may be in the range of 2–6 Tg of N per year, about 10–30 % of the input from fossil fuel burning and comparable to the natural NO sources: lightning (2–10 Tg of N per year) and soil emissions (5–15 Tg of N per year) [32]. Therefore, biomass burning contributes significantly to total NO emissions. For N_2O, on the other hand, we now estimate that pyrogenic emissions are relatively small (0.1–0.3 Tg of N per year) [36], only a few percent of the global N_2O source of about 14 Tg of N per year [32]. Earlier measurements of N_2O releases by biomass burning [29] have been biased by N_2O production in the collection devices.

The emissions of HCN and CH_3CN (together 0.5–1.7 Tg of N per year, and at a ratio of about 2:1) are significant, if not dominant, contributors to the atmospheric input rates of these compounds. The most important atmospheric sinks of HCN and CH_3CN, their reaction with OH, equals only 0.2 Tg of N per year for HCN and 0.02–0.2 Tg of N per year for CH_3CN [30]. Consequently, other sinks must exist. Hamm and Warneck [37] proposed that these compounds may be taken up by the oceans. Another possibility is that they are consumed by vegetation, in which case they might serve as a minor source of fixed N. The atmospheric budget of NH_3 is not well known. Worldwide emissions are estimated to be in the range of 20–80 Tg of N per year [32, 38, 39], of which microbial release from animal excreta and soils makes up the largest fraction. The pyrogenic source (0.5–2.0 Tg of N per year) is thus only a few percent of the global source.

An important outcome of the burning experiments at our laboratory [36] is that only about 25 % of the plant N is emitted as NO, N_2O, NH_3, HCN, and CH_3CN. At most 20 % of the N may be emitted as high molecular weight compounds, and about 10 % of it is left in the ash. Recent measurements have shown that the remaining fraction, as much as 50 % of fuel N, is emitted as N_2 [36]. Thus biomass burning leads to pyrodenitrification at a global rate of 10–20 Tg of N per year. This rate is 6–20 % of the estimated terrestrial N fixation rate of 100–170 Tg of N per year [39] and therefore of potentially substantial significance. Most of the N loss occurs in the tropics, where it may lead to a substantial nutrient loss, especially from agricultural systems and savannas. Robertson and Rosswall [40] estimated that 8.3 Tg of N are emitted each year from West Africa into the atmosphere by burning, of which about 3.3 Tg could thus be N_2. This is almost three times their estimate of biological denitrification rate from the region.

7.3.4 Sulfur Gases

In contrast to the N species, only relatively small amounts of SO_2 and aerosol sulfate are emitted. Biomass burning contributes only a few percent to the total atmospheric S budget, and only about 5 % of the anthropogenic emissions. Still, because most of the natural emissions are from the oceans and most of the anthropogenic emissions are concentrated in the industrialized regions of the temperate latitudes, biomass burning could make a significant contribution to the S budget over remote continental regions, for example, the Amazon and Congo basins [41]. Here, deposition may be enhanced five times because of tropical biomass burning.

7.3.5 Particles (Smoke)

Even though smoke is the most obvious sign of biomass burning, quantitative estimates on the amounts of particulate matter released are still highly uncertain. On the basis of an emission ratio of 30 g per kg of CO_2–C [27], we estimate that the emission of total particulate matter (TPM) is 36–154 Tg per year (Table 7.2). This amount may appear to be minor compared to the total emission of particulate matter of the order of 1500 Tg per year. However, much of these emissions consists of large dust particles, which only briefly reside in the atmosphere. The smaller smoke particles are much more long-lived and more active in scattering solar radiation. The C content of smoke particles is about 66 % [27], which is consistent with the notion that they consist mostly of partially oxygenated organic matter. This composition leads to an emission of about 30–100 Tg of particulate organic C, which would be about 15–50 % of the organic C aerosol released globally [42]. The content of black elemental C in smoke particles from biomass burning is highly variable. In smoldering fires it is as low as 4 % (weight percent carbon in TPM), whereas in intensively flaming fires it can reach 40 % [43]. We use a value of 18 %, based on our work in Amazonia [27]. From this and the estimate for global TPM emissions of 36–154 Tg per year, we obtain a source estimate for black C aerosol of 6–30 Tg per year. This value already exceeds the earlier estimate of 3–22 Tg per year for the emission of black C from all sources [44].

7.4 Atmospheric Chemical Effects

7.4.1 Long-Range Transport of Smoke Plumes

The hot gases from fires rise in the atmosphere, entraining ambient air. Frequently, clouds form on the smoke plume and usually reevaporate without causing rain. When the plume loses buoyancy, it drifts horizontally with the prevailing winds, often in relatively thin layers, which can extend over a thousand kilometers or more. The height to which the smoke plumes can rise during the dry season is usually limited in the tropics by the trade wind inversion to about 3 km.

The further fate of the smoke-laden air masses depends on the large-scale circulation over the continent in which they originate. In tropical Africa, the plumes will usually travel in a westerly direction and toward the equator. As they approach the Intertropical Convergence Zone (ITCZ), vertical convection intensifies, destroys the layered structure, and causes the pyrogenic emissions to be distributed throughout the lower troposphere. Finally, in the ITCZ region, smoke and gases from biomass burning may reach the middle and upper troposphere, perhaps even the stratosphere. Air masses from the biomass burning regions in South America are usually moving toward the south and southeast, because of the effect of the Andes barrier on the large-scale circulation. Here again, they may become entrained in a convergence zone, the seasonal South Atlantic Convergence Zone (SACZ), which becomes established in austral spring, when biomass burning is abundant. Indeed, the data from the space-borne MAPS (Measurement of Air Pollution from Satellites) instrument typically show high concentrations of CO in the mid- and upper troposphere near the ITCZ and the SACZ [45].

Results of chemical measurements from satellites, space shuttle, aircraft, and research vessels indicate that pyrogenic emissions are transported around the globe. Soot C and other pyrogenic aerosol constituents have been measured during research cruises over the remote Atlantic and Pacific [46]. High levels of O_3 and CO have also been observed from satellites over the tropical regions of Africa and South America, and large areas of the surrounding oceans [45, 47].

7.4.2 Photochemical Smog Chemistry

Biomass fires emit much the same gases as fossil fuel burning in industrial regions: CO, hydrocarbons, and NO_x, the starting ingredients for the formation of O_3 and photochemical smog. Once such a mixture is exposed to sunlight, hydrocarbons, including those naturally emitted by vegetation, are oxidized photochemically first to various peroxides, aldehydes, and so forth, then to CO. This CO is added to the amount directly emitted from the fires and is finally oxidized to CO_2 by reaction with OH. High concentrations of hydrocarbons and CO have been observed during the burning season in the tropics [27, 29, 48]. In the presence of high levels of NO_x, as will be the case in the smoke plumes, the oxidation of CO and hydrocarbons is accompanied by the formation of O_3 [29, 48]. The efficiency of O_3 formation, that is, the amount of O_3 formed per molecule of hydrocarbon oxidized, depends on the spread of the smoke plume and the chemical mix of hydrocarbons, NO_x, and O_3 present in the reaction mixture, and thus on the history of transport and mixing of the air mass [49]. Increased concentrations of O_3 promote high concentrations of OH radicals and thus increase the overall photochemical activity of air masses affected by biomass burning. The effect may be enhanced further by the simultaneous emission of CH_2O ($\approx 2 \times 10^{-3}$ to 3×10^{-3} relative to CO_2 [31]), which is photolyzed in the tropical atmosphere within a few hours; this process leads in part to the production of HO_2 radicals via the formation of H and CHO. A similar effect may be caused by the photolysis of HONO, which may be emitted directly by the fires or formed by reactions on smoke particles [50].

High O_3 concentrations are produced in the plumes that extend over major parts of the tropical and subtropical continents during the dry season [27, 29, 48, 51]. The highest concentrations, typically in the range from 50 to 100 ppb, are usually found in discrete layers at altitudes between 1 and 5 km, in accordance with the transport mechanisms of the burning plumes described above (Fig. 7.2). The concentrations at ground level are substantially lower and show a pronounced daily cycle with minima at night and maxima around midday. This cycle is controlled by the balance of O_3 sources and sinks: at night, O_3 consumption by deposition on the vegetation and reaction with hydrocarbons emitted by the vegetation and with NO emitted from soils reduce the concentration of O_3 near Earth's surface; during the day, these sinks are exceeded by photochemical O_3 formation and downward mixing of O_3-rich air. Also at ground level, O_3 volume mixing ratios in excess of 40 ppb are frequently measured during the dry season [48], similar to average values observed

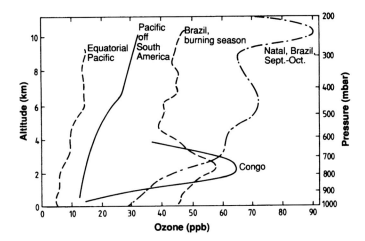

Fig. 7.2 Vertical profiles of O_3 in the tropical troposphere. The profile over the equatorial Pacific shows no influence from biomass burning, whereas the profile over the Pacific off South America suggests O_3 enhancement due to long-range transport from the tropical continents [47]. The O_3 profiles over Brazil [29] and the Congo [51] show high O_3 concentrations at altitudes between 1 and 4 km due to photochemical production in biomass burning plumes. At higher altitudes, O_3 concentrations are also substantially enhanced, possibly also because of O_3 production by reactions in the effluents of biomass burning [48] (Adapted from [47] with permission of the author)

over the polluted industrialized regions of the eastern United States and Europe [52]. Studies in temperate forest regions have linked such levels of O_3 pollution to damage to trees and vegetation, which has become widespread in Europe and North America [53]. In view of the sharp increase of O_3 with altitude frequently observed in the tropics, the risk of vegetation damage by O_3 may be highest in mountainous regions, where O_3 concentrations above 70 ppb could be encountered. Ozone episodes with ground-level concentrations of 80–120 ppb must be expected to occur particularly during the dry season, when photochemically reactive air becomes trapped under the subsiding inversion layer [54]. The regional ecological impact of high concentrations of phytotoxic O_3 on tropical vegetation and food production in the developing world is a matter of concern [55].

7.4.3 Perturbation of Oxidant Cycles in the Troposphere

The global increase of tropospheric O_3, CO, and CH_4 concentrations, which is expected to continue in the future, is an indication of a fundamental change in the chemical behavior of the troposphere. Many gases, particularly hydrocarbons, are continuously emitted into the atmosphere from natural and anthropogenic sources. A buildup of these gases in the atmosphere is prevented by a self-cleaning mechanism, whereby these substances are slowly 'combusted' photochemically to CO_2.

The key molecule responsible for this oxidation process is OH. The reaction chains involved are such that OH is consumed when the concentration of NO_x is low. This is the normal condition of most of the unpolluted troposphere. On the basis of the observed increase of CO and CH_4, it has been suggested that global decreases in OH, the primary sink for CH_4 and CO, could lead through a feedback mechanism to a further increase in CO and CH_4, and that this situation could produce an unstable chemical condition [56]. Injection of large amounts of NO_x from biomass burning and other anthropogenic activities may counteract this feedback, because hydrocarbon oxidation in the presence of elevated amounts of NO_x creates additional O_3 and OH. This counter effect is, however, much more regionally limited because of the much shorter residence time of NO_x compared to that of CO and CH_4.

Model calculations [57] predict that a sixfold increase in regional OH concentrations in the boundary layer could occur as a result of deforestation and biomass burning in the tropics. There are two main reasons for this surprisingly large effect. One is that in the deforested regions NO_x is more easily ventilated to the atmosphere and a smaller portion reabsorbed in the less dense vegetation. The other is that removal of the trees eliminates the large emissions of isoprene (C_5H_8), which would normally react with and strongly deplete OH. We may, therefore, expect a strong enhancement of boundary layer O_3 and OH concentrations over tropical continental areas during the dry season, when vegetation is burned.

However, for the globe as a whole, it is likely that increasing CO and CH_4 emissions, with a large contribution from biomass burning, will lead to decreasing average OH concentrations and thereby to the buildup of the many gases that are removed from the atmosphere by reaction with OH. This change may be an extremely important development in global atmospheric chemistry.

7.5 Climatic and Ecological Effects

7.5.1 Climate Change

With net global CO_2 emissions of 1.1–3.6 Pg of C per year, the clearing of the tropical forests may be responsible for up to 20–60 % of the greenhouse warming caused by the CO_2 emissions from fossil fuel burning. Biomass burning also releases another greenhouse gas, CH_4. In this case, biomass burning accounts for only about 10 % of the global CH_4 sources, but probably for a greater fraction of the increase in global emissions [33]. Estimates of the temporal trends of CH_4 source strengths from 1940 to 1980 [58] suggest that the pyrogenic contribution to the increase in CH_4 emissions over that time period is 10–40 %.

The climatic effect of the smoke aerosols is beyond current understanding because of the complex nature of the interactions involved. Aerosols can influence climate directly by changing Earth's radiation balance. They reflect sunlight back into space. Smoke particles also contain black (elemental) C, which may strongly absorb sunlight and thus cause a heating of the atmosphere and less penetration of solar energy to Earth's surface. Such an effect has an influence on the heat balance of the lower troposphere; it results in less solar heating of the surface, warming of the atmosphere, and more stable meteorological conditions. Robock [59] has shown that large daytime temperature drops can occur below smoke plumes from mid-latitude forest fires. Considering the great extent and expansion of tropical biomass burning, a widespread effect of this kind may well have masked the expected greenhouse temperature rise during the dry season on the tropical continents.

Because the equatorial regions, particularly the Amazon Basin, the Congo Basin, and the area around Borneo, are extremely important in absorbing solar energy and in redistributing this heat through the atmosphere, any change affecting the operation of these "heat engines of the atmosphere" may be highly significant. A matter of considerable interest is the influence of submicrometer-sized pyrogenic particles on the microphysical and optical properties of clouds and climate, an issue that has attracted considerable attention in connection with S emissions into the atmosphere [60, 61]. Cloud droplets form on aerosol particles; these are called cloud condensation nuclei (CCN). The properties of the cloud depend on the number of available CCN: the more CCN, the more droplets that can form and the smaller the droplet size for a given amount of water. Clouds made up of smaller droplets reflect more sunlight back into space, and, because these clouds also are less likely to produce rain, cloud coverage also may increase. Because clouds are one of the most important controls on the heat balance of Earth, any large-scale modification of cloud properties is likely to have a strong impact on climate. Following proposals by Warner and Twomey [62] and by Radke et al. [63], recent studies have shown that many of the submicrometer smoke particles produced by biomass fires can serve as CCN [64]. Aged particles show enhanced CCN activity as their surfaces become coated with water-soluble materials, especially by uptake of HNO_3 and NH_3.

The pyrogenic production of smoke particles (40–150 Tg per year) is of the same magnitude as the input of sulfate particles from the anthropogenic emission of SO_2 from fossil fuel burning ($\approx$70–100 Tg of S per year worldwide) [65]. On a molar basis and because of their larger surface to volume ratio, the emissions of pyrogenic particles may be even larger than those of sulfate aerosol. Consequently smoke particles, in addition to affecting the radiative properties of clouds and Earth's radiation balance, may also disturb the hydrological cycle in the tropics, with potential repercussions for regional and possibly global climate. Altogether, the climatic impact of biomass burning in the tropics may be impressively large. Recent general circulation model calculations by Penner et al. [66] indicate the possibility of a net change in Earth's radiation balance by -1.8 W/m^2, about equal, but opposite to the present greenhouse forcing.

The potential changes in precipitation efficiency add to the perturbation of the hydrological cycle in the tropics caused by deforestation and desertification. Tropical forests are extremely efficient in returning precipitation back to the atmosphere in the form of water vapor. There it can form clouds and rain again, and the cycle can repeat itself many times [67]. A region such as the Amazon Basin can thus retain water (which ultimately comes from the ocean and will return there) for a long time and maintain a large standing stock of water. If the forest is replaced by grassland or, as is often the case, is converted into an essentially unvegetated surface by erosion and loss of topsoil, water runs off more quickly and returns through streams and rivers to the ocean, allowing less recycling. Beyond the unfavorable consequences that such large-scale changes in the availability of water will have on human activities, such a modification of the hydrological cycle may itself perturb tropical weather and maybe even climate [68]. Furthermore, through the introduction of hotter and drier conditions, less evapotranspiration and pre-cipitation, and a lengthening of dry season, there will be a much greater risk of and need for periodic fires [69]. Together with changing biospheric emissions, the decrease in precipitation and cloudiness and changes in other meteorological factors also have the potential to alter the chemistry of the tropical atmosphere in major ways.

7.5.2 Acid Deposition

After acid rain had become a notorious environmental problem in Europe and North America, it came as a surprise to scientists to learn that it was also widespread in the tropics (Table 7.3): acid rain has been reported from Venezuela [70], Brazil [71], Africa [72], and Australia [70, 73]. In all instances, organic acids (especially formic and acetic acids) and nitric acid were shown to account for a large part of the acidity, in contrast to the situation in the industrialized temperate regions, where sulfuric acid and nitric acid predominate. It was originally thought, that the organic acids were largely derived from natural, biogenic emissions, probably from plants [74]. However, more recent evidence shows that acetic acid is produced directly by biomass burning and that both formic and acetic acid are chemically produced in the plumes [75]. Nitric acid is formed photochemically from the NO_x emitted in the fires [57]. Results from modeling the effects of biomass burning and a moderate amount of additional pollution, mostly connected with the activities related to logging and so forth, suggest that during the dry season pH values of ≈ 4.2 can be expected in the tropics as a consequence of the formation of nitric acid alone [57]. For comparison, the mean pH in rain sampled throughout the eastern United States in 1980 was 4.3 [76] (Table 7.3).

Acidic substances in the atmosphere can be deposited onto plants and soils either by rain and fog (wet deposition) or by the direct removal of aerosols and gases onto surfaces (dry deposition). In the humid tropics, wet deposition accounts for most of the deposition flux, whereas in the savanna regions, especially during the dry season, dry deposition dominates. Acid deposition has been linked to forest damage in Europe and the eastern United States [77]. Acid deposition can act on an ecosystem through two major pathways: directly through the deposition of acidic aerosols and gases on leaves, or soil acidification. The danger of leaf injury is serious only at pH levels below 3.5, which is rarely encountered in the tropics [78],

Table 7.3 Rainwater pH and acid deposition at some continental tropical sites and in the eastern United States

Site	pH		Rain-fall (cm)	Deposition (kg H$^+$ ha/per year)	References
	Mean[a]	Range			
Venezuela					
San Eusebio	4.6	3.8–6.2	158	0.39	[70]
San Carlos	4.8	4.4–5.2	–	–	[70]
La Paragua	4.7	4.0–5.0	–	–	[70]
Brazil Manaus, dry season	4.6	3.8–5.0	240[b]	0.29	[71]
Manaus, wet season	5.2	4.3–6.1	–	–	–
Australia					
Grootc Eylandt	4.3	–	–	–	[73]
Katherine	4.8	4.2–5.4	–	–	[70]
Jabiru	4.3	–	–	–	[73]
Ivory coast					
Ayame	4.6	4.0–6.5	179	0.41	[72]
Congo					
Boyele	4.4	–	185	0.74	[72]
Eastern United States	4.3	3.0–5.9	130	0.67	[76]

[a]Volume weighted. [b]Annual average

except perhaps in fog and dew. Nevertheless, the issue deserves some attention, as tropical forests may be inherently more sensitive to foliar damage than temperate forests because of the longer average leaf life of 1–2 years, which promotes cumulative damage.

7.5.3 Alterations of Nutrient Cycles and Effects on Soil Degradation

Savanna and agricultural ecosystems are frequently deficient in N, P, or S [79]. When an area is burned, a substantial part of the N present in the ecosystem is volatilized. If this N were deposited again relatively nearby, this would cause no net gain or loss on a regional basis. If, however, as a result of fires, some 50 % of the fuel N is emitted as N$_2$ [30, 36], a significant loss of nutrient N may result. In addition, long-range transfer of NO$_x$, NH$_3$, and nitriles to other ecosystems (savannas to tropical forests) depletes the fixed N reservoir of frequently burned ecosystems and thus provides a potential for long-term ecological effects. The budget of Robertson and Rosswall [40] for West African savannas implies that this loss of fixed N could deplete the fixed N load of these ecosystems in a few thousand years, a short time in comparison with the period during which humans have been present in these ecosystems. It may therefore be asked, to what extent enhanced N fixation can compensate for the loss of fixed N. This may indeed occur: laboratory research on tallgrass prairie soils has shown an enhancement of nonsymbiotic N_2 fixation after additions of available P in the ash from fires [80]. Although the effects of biomass burning on the N cycle of the fire-affected ecosystems are most obvious, other nutrient elements, especially K, P, Mg, and S are also lost via smoke particles [81] in amounts that may have long-term, ecological consequences.

Regarding the C cycle, two issues appear to be of particular interest:

(1) The burial of pyrogenic charcoal residues that are not subject to microbial oxidation even over geological time scales, and that thus constitute a significant sink for atmospheric CO_2 and consequently a source for O_2 [10]. Because the risk of fires increases with the growing atmospheric O_2 content [82], on geological time scales this may establish a positive feedback loop, which favors O_2 buildup in the atmosphere.

(2) The enhancement of biomass productivity of 30–60 % or more after burning, despite the loss of nutrients, observed in some studies in humid savanna ecosystems [83]. The results depend largely on burning practices, especially timing. Whether enhanced productivity may also increase the pool of organic matter in the soil is unknown. Too little is yet known about the biogeochemical cycling of savanna ecosystems and changes thereof. It was discovered only recently that natural tropical grasslands may be much more productive than hitherto assumed, with productivity comparable to that of tropical forests [84]. With a strong growth of the populations living in savanna regions, there will most likely be more frequent burning in these ecosystems. Significant effects on the global C cycle are possible, either through enhanced sequestering of C as charcoal and root-produced soil organic matter, if optimum burning practices are adopted (our speculation), or through loss of soil C by overly frequent fires and practices that lead to land degradation.

Ecosystems that are not burned, for example, remaining areas of intact rain forest, will receive an increased nutrient input. Studies of rainwater chemistry in the central Amazon Basin suggest that as much as 90 % of the S and N deposited there is from external sources, and that long-range transport of emissions from biomass burning plays a major role [41]. The long-term effects of such increasing inputs of nutrients to the rain forests, in combination with growing acid deposition and O_3 concentrations, are not known.

In addition to the immediate volatilization of N during the burns, enhanced microbial cycling of N in the soils occurs after fires. Emissions of NO and N_2O from soils at experimental sites in the temperate zone after burning were observed to be substantially higher than from soils at unburned sites [85]. This effect persisted for at least 6 months after the fires. Following burning on a Venezuelan savanna site, enhancements in NO emissions by a factor of 10 were found for the 4 days during which the measurements were made [86]. Other studies have also shown that the fluxes of NO_x from soils are enhanced after conversion from forests to grazing land [87], but in these studies the effect of burning was not isolated explicitly. However, in more extensive studies, Luizão et al. [88] did not observe enhanced N_2O fluxes on sites that were only burned and cleared but found a threefold enhanced emission on 3- to 4-year-old pasture sites. According to these researchers, the enhanced emissions may be caused by increased input of oxidizable C from grass roots or rhizomes, or compaction of the soil surface by the cattle.

Although the above studies indicate that the emissions of trace gases increase after land disturbances, the total effect is complex and unclear. According to Robertson and Tiedje [89], denitrification ($N_2 + N_2O$ production) is high in primary forests and at early successional sites but much lower at mid-successional sites. Studies by Sanhueza et al. [90] in a Venezuelan savanna and by Goreau and de Mello [91] on a cleared forest site during the dry season showed that forested areas may emit more N_2O than secondary grassland ecosystems derived by deforestation. Thus, although disturbed tropical forest ecosystems may initially emit more N_2O, this may only be temporary and in the long run less N_2O may be emitted. The issue is, therefore, unclear. Much long-term research is needed to elucidate the effects of biomass burning on nutrient cycling and especially on N volatilization in the tropics. This research is particularly important as there are indications that the main contributions to the total atmospheric N_2O source come from the tropics [92].

7.6   Conclusions

Our, still very uncertain, analysis of tropical biomass burning indicates emissions from about 2–5 Pg of C per year. In comparison, the present net release of CO_2 due to tropical land use change is estimated to range between 1.1 and 3.6 Pg of C per year [14]. Significant amounts of C may be sequestered as charcoal, which may reduce the net release by 0.2–0.6 Pg of C per year. Because of the great importance of biomass burning and deforestation activities for climate, atmospheric chemistry, and ecology, it is clearly of the utmost importance to improve considerably our quantitative knowledge of these processes.

Biomass burning is a major source of many trace gases; especially the emissions of CO, CH_4 and other hydrocarbons, NO, HCN, CH_3CN, and CH_3Cl are of the greatest importance. In the tropical regions during the dry season, these emissions lead to the regional production of O_3 and photochemical smog, as well as increased acid deposition with potential ecological consequences. On a global scale, however, the large and increasing emissions of CO and CH_4, the main species with which OH reacts in the background atmosphere, will probably lead to a decrease in the overall concentration of OH radicals and, therefore, to a decrease in the oxidation efficiency of the atmosphere. As the atmospheric lifetime of NO is only a few days, most of the atmosphere remains in an 'NO-poor' state, where photochemical oxidation of CO and CH_4 leads to further consumption of OH. This in turn will enhance the atmospheric concentrations of CH_4 and CO, leading to a strong photochemical feedback.

Biomass burning is also an important source of smoke particles, a large amount (maybe all) of which act as CCN or can be converted to CCN by atmospheric deposition of hygroscopic substances. The amount of aerosols produced from biomass burning is comparable to that of anthropogenic sulfate aerosol. Through this process, the cloud microphysical and radiative processes in tropical rain and cloud systems can be affected with potential climatic and hydrological consequences.

An important recent finding is the substantial loss of fixed N that may be occurring because of biomass burning (pyrodenitrification). This loss appears to be of the greatest significance for savanna and agricultural ecosystems in the tropics and subtropics. The potential role of the savanna ecosystems in Earth's biogeochemical cycles deserves much more attention than it has been given so far. The savanna regions may play an important role in the global C cycle because of their large productivity, the potential interference of biomass burning with this productivity, and the formation of long-lived elemental C. The geological importance of this C as a sink for atmospheric CO_2 (and source for O_2) should be explored.

References

1. Schüle, W., 1990: in: Goldammer, J.G. (Ed.): *Fire in the Tropical Biota, Ecological Studies,* Vol. 84 (Berlin: Springer): 273–318; Brain, C.K.; Sillen, A., 1988: in: *Nature,* 336: 464.
2. Goldammer, J.G.; Seibert, B., 1989: in: *Naturwissenschaften,* 76: 51; Brünig, E.F., 1969: in: *Erdkunde,* 23: 127; Malingreau, J.P.; Stephens, G.; Fellows, L., 1985: in: *Ambio,* 14: 314.
3. Sanford, R.L., Jr.; Saldarriaga, J.; Clark, K.E.; Uhl, C.; Herrera, R., 1985: in: *Science,* 227: 53.
4. Jones, R., 1979: in: *Annual Review* of *Anthropology,* 8: 445.
5. Griffin, J.J.; Goldberg, E.D., 1979: in: *Science,* 206: 563; in: *Environmental Science and Technology,* 17: 244; Herring, J.R., 1977: (Thesis, University of California, San Diego).
6. Wolbach, W.S.; Lewis, R.S.; Anders, E., 1985: in: *Science,* 230: 167; Wolbach, W.S.; Gilmour, I.; Anders, E.; Orth, C.J.; Brooks, R.R., 1988: in: *Nature,* 334: 665.
7. Levine, J.S. (Ed.), in press: in: *Proceedings of the Chapman Conference on Global Biomass Burning* (Cambridge, MA: MIT Press).
8. International Geosphere-Biosphere Programme, 1990: *A Study of Global Change of the International Council of Scientific Unions* (Stockholm, Sweden: IGBP Secretariat, Royal Swedish Academy of Sciences).
9. Feamside, P.M., in (7).
10. Seiler, W.; Crutzen, P.J., 1980: in: *Climatic Change,* 2: 207.
11. Myers, N., 1989: *Deforestation Rates in Tropical Forests and Their Climatic Implications* (London: Friends of the Earth).
12. Lanly, J.P., 1982: *Tropical Forest Resources* (Rome: FAO, Forestry Pap. 30).
13. Hao, W.M.; Liu, M.H.; Crutzen, P.J., 1990: in: Goldammer, J.G. (Ed.): *Fire in the Tropical Biota, Ecological Studies,* Vol. 84 (Berlin: Springer): 440–462.
14. Houghton, R.A., in press: in: *Climatic Change.*
15. Houghton, J.T.; Jenkins, G.J.; Ephraums, J.J., (Eds.), 1990: *Climate Change, the Intergovernmental Panel on Climate Change Scientific Assessment* (New York: Cambridge University Press).
16. Brown, S.; Gillespie, A.J.R.; Lugo, A., 1989: in: *Forest Science,* 35: 881.
17. Bolin, B.; Degens, E.T.; Duvigneaud, P., Kempe, S., 1979: in: Bolin, B.; Degens, E.T.; Kempe, S.; Ketner, P. (Eds.): *The Global Carbon Cycle, SCOPE 13* (Chichester, England: Wiley): 1–56.

18. Menaut, J.C., in (7).

19. 1989: *Year Book of Forest Products 1987* (1976–1987) (Rome: FAO).

20. Scurlock, J.M.O.; Hall, D.O., 1990: in: *Biomass,* 21: 75.

21. Joshi, V., in (7).

22. Barnard, G.W., in press: in: Pasztor, J.; Kristoferson, L. (Eds.): *Biomass and the Environment* (Boulder, CO: Westview).

23. Ponnamperuma, F.N., 1984: *Organic Matter and Rice* (Los Banos, Philippines: International Rice Research Institute); Strehler, A.; Stützle, W., 1987: in: Hall, D.O.; Overend, R.P., (Eds.): *Biomass* (Wiley, Chichester, England): 75–102.

24. Stocks, B.J., Personal Communication.

25. Bowen, H.J.M., 1979: *Environmental Chemistry of the Elements* (London: Academic Press).

26. Crutzen, P.J.; et al., 1979: in: *Nature,* 282: 253.

27. Andreae, M.O.; et al., 1988: in: *Journal of Geophysical Research,* 93: 1509.

28. Comery, J.A., 1981: (Thesis, University of Washington, Department of Forestry, Seattle).

29. Crutzen, P.J.; et al., 1985: in: *Journal of Atmospheric Chemistry,* 2: 233.

30. Lobert, J.M., 1990: (Thesis, Johannes Gutenberg Universität, Mainz, Germany); Lobert, J.M.; et al. in (7).

31. Griffith, D.W.; Mankin, W.G.; Coffey, M.T.; Ward, D.E.; Ribeau, A., in (7); Cofer III, W.R.; Levine, J.S.; Winstead, E.L.; in (7); Coffer, W.R.; et al., 1988: in: *Journal of Geophysical Research,* 93: 1653.

32. World Meteorological Organization, 1985: *Atmospheric Ozone* (Global Ozone Research and Monitoring Project Report 16, Geneva, Switzerland).

33. Craig, H.; Chou, C.C.; Welhan, J.A.; Stevens, C. M.; Engelkemeir, A., 1988: in: *Science,* 242: 1535; Quay, P.D.; et al., in press: in: *Global Biogeochemical Cycles.*

34. Stevens, C.M.A.; Engelkemeir, E.R.; Rasmussen, A., in (7).

35. Schmidt, U., 1974: in: *Tellus,* 26: 78.

36. Lobert, J.M.; Scharffe, D.H.; Hao, W.M.; Crutzen, P.J., 1990: in: *Nature,* 346: 552; Kuhlbusch, Th., 1990: (Thesis, Johannes Gutenberg Universität, Mainz, Germany).

37. Hamm, S.; Wameck, P., in press: in: *Journal of Geophysical Research.*

38. Galbally, I.E., 1985: in: Galloway, J.N.; Charlson, R.J.; Andreae, M.O.; Rodhe, H., (Eds.): *The Biogeochemical Cycling of Sulfur and Nitrogen in the Remote Atmosphere* (Hingham, MA: Reidel): 27–53.

39. Söderlund, R.; Svensson, B.H., 1976: in: Svensson, B.H.; Söderlund, R. (Eds.): *Nitrogen, Phosphorus, and Sulphur* (*Ecological Bulletins Stockholm* 22: 23).

40. Robertson, G.P.; Rosswall, T., 1986: in: *Ecological Monographs,* 56: 43.

41. Andreae, M.O.; Andreae, T.W., 1988: in: *Journal of Geophysical Research,* 93: 1487; Andreae, M.O.; Talbot, R.W.; Berresheim, H.; Beecher, K.M., 1990: *ibid.* 95: 16987; Andreae, M.O.; et al. *ibid.,* p. 16813; Andreae, M.O.; et al., in press: *ibid.*

42. Duce, R.A., 1978: *Pure Application Geophysics,* 116: 244.

43. Patterson, E.M.; McMahon, C.K., 1984: in: *Atmospheric Environment,* 18: 2541.

44. Turco, R.P.; Toon, O.B.; Whitten, R.C.; Pollack, J.B.; Hamill, P., 1983: in: Pruppacher, H.R.; Semonin, R.G.; Slinn, W.G.N. (Eds.): *Precipitation Scavenging, Dry Deposition, and Resuspension* (Elsevier, Amsterdam): 1337–1351.

45. Reichle, H.G. Jr.; et al., 1986: in: *Journal of Geophysical Research,* 91: 10865; Reichle, H.; et al., 1990: *ibid.* 95: 9845.

46. Andreae, M.O., 1983: in: *Science,* 220: 1148; Andreae, T.W.; Ferek, R.J.; Raemdonck, H., 1984: in: *Science of the Total Environment,* 36: 73.

47. Fishman, J.; Larsen, J.C., 1987: *Journal of Geophysical Research,* 92: 6627.

48. Kirchhoff, V.W.J.H.; Marinho, E.V.A., 1989: in: *Atmospheric Environment,* 23: 461; Kirchhoff, V.W.J.H.; Browell, E.V.; Browell, G.L., 1988: in: *Journal of Geophysical Research,* 93: 15850; Kirchhoff, V.W.J.H.; Setzer, A.W.; Pereira, M.C., 1989: in: *Geophysical Research Letter,* 16: 469; Delany, A.C.; Haagensen, P.; Walters, S.; Wartburg, A.F.; Crutzen, P.J., 1985: in: *Journal of Geophysical Research,* 90: 2425; Kirchhoff, V.W.J.H.; et al., 1983: in: *Geophysical Research Letter,* 8: 1171; Logan, J.A.; Kirchhoff, V.W.J.H., 1986: in: *Journal*

of Geophysical Research, 91: 7875; Kirchhoff, V.W.J.H.; Rasmussen, R.A., 1999: *ibid.* 95, 7521.

49. Chatfield, R.B.; Delany, A.C., 1990: in: *Journal of Geophysical Research,* 95: 18473.
50. Rondon, A.; Sanhueza, E., 1989: in: *Tellus Series B,* 41: 474.
51. Kirchhoff, V.W.J.H.; Nobre, C.A., 1986: in: *Reviews of Geophysics* 24: 95; Cros, B.; Delmas, R.; Clairac, B.; Loemba-Ndembi, J.; Fontan, J., 1987: in: *Journal of Geophysical Research,* 92: 9772; Andreae, M.O.; et al., in press: *ibid.*
52. J. A. Logan, *Journal of Geophysical Research,* 90, 10463 (1985).
53. Prinz, B., 1988: in: Isaksen, I.S.A., (Ed.): *Tropospheric Ozone* (Dordrecht, Reidel): 687–689.
54. Cros, B.; Delmas, R.; Nganga, D.; Clairac, B., (1988): in: *Journal of Geophysical Research,* 93: 8355.
55. Vitousek, P.M.; Matson, P.A., in press: in: *Biotropica.*
56. Crutzen, P.J., 1987: in: Dickinson, R.E. (Ed.): *The Geophysiology of Amazonia* (New York: Wiley): 107–130; Sze, N.D., 1977: in: *Science,* 195: 673.
57. Keller, M.; Jacob, D.J.; Wofsy, S.C.; Harriss, R.C., in press: in: *Climatic Change.*
58. Bolle, H.J.; Seiler, W.; Bolin, B., 1986: in: Bolin, B.; Döös, B.R.; Jäger, J.; Warrick, R.A. (Eds.): *The Greenhouse Effect, Climatic Change, and Ecosystems, SCOPE 29* (Chichester, England: Wiley): 157–203.
59. Robock, A., 1988: in: *Science,* 242: 911.
60. Eagan, R.C.; Hobbs, P.V.; Radke, L.F., 1974: in: *Journal of Applied Meteorology,* 13: 553; Twomey, S., 1977: in: *Journal of the Atmospheric Sciences,* 34: 1149; Piepgrass, M.; Wolfe, T.L., 1984: in: *Tellus Series B* 36: 356.
61. Charlson, R.J.; Lovelock, J.E.; Andreae, M.O.; Warren, S.G., 1989: in: *Nature,* 340: 437; Wigley, T.M.L., 1989: *ibid.* 339: 365 .
62. Warner, J.; Twomey, S., 1967: in: *Journal of the Atmospheric Sciences,* 24: 704.
63. Radke, L.F.; Stith, J.L.; Hegg, D.A.; Hobbs, P.V., 1978: in: *Journal of the Air Pollution Control Association,* 28: 30.
64. Desalmand, F.; Serpolay, R.; Podzimek, J., 1985: in: *Atmospheric Environment,* 19: 1535; Desalmand, F.; Podzimek, J.; Serpolay, R., 1985: in: *Journal of Aerosol Science,* 16: 19 (1985); Rogers, C.F.; Zielinska, B.; Tanner, R.; Hudson, J.; Watson, J., in (7); Radke, L.F.; Hegg, D.A.; Lyons, J.H.; Hobbs, P.V.; Weiss, R.E., in (7).
65. Cullis, C.F.; Hirschler, M.M., 1980: in: *Atmospheric Environment,* 14: 1263; Moller, D., 1984: *ibid.* 18: 19.
66. Penner, J.E.; Ghan, S.J.; Walton, J.J., in (7).
67. Salad, E.; Vose, P.B., 1984: in: *Science,* 225: 129.
68. Shukla, J.; Nobre, C.; Sellers, P., 1990: *ibid.* 247: 1322; Sellers, P.; Mintz, Y.; Sud, Y.C.; Dalcher, A., 1986: in: *Journal of Atmospheric Science,* 43: 505; Dickinson, R.E.; Henderson-Sellers, A., 1988: in: *Quarterly Journal of the Royal Meteorological Society,* 114, 439 (1988); Lean, J.; Warrilow, D.A., 1989: in: *Nature,* 342: 411.
69. Fosberg, M.; Goldammer, J.G.; Price, C.; Rind, D., 1990: in: Goldammer, J.G. (Ed.): *Fire in the Tropical Biota, Ecological Studies,* Vol. 84 (Berlin: Springer): 463–486.
70. Steinhardt, U.; Fassbender, H.W., 1979: in: *Turrialba,* 29: 175; Galloway, J.N.; Likens, G.E.; Keene, W.C.; Miller, J.M., 1982: in: *Journal of Geophysical Research,* 87: 8771; Sanhueza, E.; Elbert, W.; Rondon, A.; Corina Arias, M.; Hermoso, M., 1989: in: *Tellus Series B,* 41: 170.
71. Andreae, M.O.; Talbot, R.W.; Andreae, T.W.; Harriss, R.C., 1988: in: *Journal of Geophysical Research,* 93: 1616.
72. Lacaux, J.P.; Servant, J.; Baudet, J.G.R., 1988: in: *Atmospheric Environment,* 21: 2643; Lacaux, J.P.; et al., 1988: in: *Eos,* 69: 1069.
73. Langkamp, P.J.; Dalling, M.J., 1983: in: *Australian Journal of Botany,* 31: 141; Ayers, G.P.; Gillet, R.W., 1988: in: Rodhe, H.; Herrera, R. (Eds.): *Acidification in Tropical Countries, SCOPE 36* (Chichester, England: Wiley): 347–400.
74. Talbot, R.W.; Beecher, K.M.; Harriss, R.C.; Cofer, W.R., 1988: in: *Journal of Geophysical Research,* 93: 1638; Keene, W.C.; Galloway, J.N., 1986: *ibid.* 91: 14466.
75. Helas, G.; Bingemer, H.; Andreae, M.O., in preparation.

76. Barrie, L.A.; Hales, J.M., 1984: in: *Tellus Series B*, 36: 333.
77. Bormann, F.H., (1985): in: *BioScience*, 35: 434.
78. McDowell, W.H., 1988: in: Rodhe, H.; Herrera, R. (Eds.): *Acidification in Tropical Countries* (Chichester, England: Wiley): 117–139.
79. Sanchez, P.A., 1976: *Properties and Management of Soils in the Tropics* (New York: Wiley).
80. Eisele, K.A.; et al., (1990): in: *Oecologia*, 79: 471.
81. Sanhueza, E.; Rondon, A., (1988): in: *Journal of Atmospheric Chemistry*, 7: 369.
82. Watson, A.J., 1978: (Thesis, Reading University, Reading, England).
83. San Jose, J.J., Medina, E., 1975: in: Golley, F.B.; Medina, E. (Eds.): *Tropical Ecological Systems, Ecological Studies*, Vol. 11 (Berlin: Springer): 251–264; Gillon, D., 1983: in: Bourliere, F. (Ed.): *Tropical Savannas, Ecosystems of the World 13* (Amsterdam: Elsevier): 617–641.
84. Hall, D.O.; Scurlock, J.M.O., in press: in: *Annals of Botany* (London).
85. Anderson, I.C.; Levine, J.S.; Poth, M.A.; Riggan, P.J., 1988: in: *Journal of Geophysical Research*, 93: 3893; Anderson, I.C.; Poth, M.A., 1989: in: *Global Biogeochemical Cycles*, 3: 121; Levine, J.S.; et al., 1988: in: *ibid.* 2: 445.
86. Johansson, C.; Rodhe, H.; Sanhueza, E., 1988: in: *Journal of Geophysical Research*, 93, 7180.
87. Matson, P.A.; Vitousek, P.M.; Ewel, J.J.; Mazzarino, M.J.; Robertson, G.P., 1987: in: *Ecology*, 68: 491.
88. Luizão, F.; Matson, P.; Livingston, G.; Luizão, R.; Vitousek, P., 1989: in: *Global Biogeochemical Cycles*, 3: 281.
89. Robertson, G.P.; Tiedje, J.M., 1988: in: *Nature*, 336: 756.
90. Sanhueza, E.; Hao, W.M.; Scharffe, D.; Donoso, L.; Crutzen, P.J., in press: in: *Journal of Geophysical Research*.
91. Goreau, T.J.; de Mello, W.Z., 1988: in: *Ambio* 17: 275.
92. Prinn, R.G.; et al., in press: in: *Journal of Geophysical Research*.

Chapter 8
A Mechanism for Halogen Release from Sea-Salt Aerosol in the Remote Marine Boundary Layer

Rainer Vogt, Paul J. Crutzen and Rolf Sander

Recent measurements of inorganic chlorine gases [1] and hydrocarbons [2] indicate the presence of reactive chlorine in the remote marine boundary layer; reactions involving chlorine and bromine can affect the concentrations of ozone, hydrocarbons and cloud condensation nuclei. The known formation mechanisms of reactive halogens require significant concentrations of nitrogen oxides [3–5], which are not present in the unpolluted air of the remote marine boundary layer [6]. Here we propose an autocatalytic mechanism for halogen release from sea-salt aerosol: gaseous HOBr is scavenged by the aerosol and converted to only slightly soluble BrCl and Br_2, which are released into the gas phase. Depending on the sea-salt concentration and given a boundary layer that is stable for a few days, gaseous HOCl and HOBr may reach molar mixing ratios of up to 35 pmol mol^{-1}. We calculate that HOBr and HOCl are responsible for 20 and 40 %, respectively, of the sulphur (IV) oxidation [7, 8] that occurs in the aerosol phase. The additional S (IV) oxidation reduces the formation of cloud-condensation nuclei, and hence the feedback between greenhouse warming, oceanic DMS emission and cloud albedo. We also calculate significant bromine-catalysed ozone loss.

Recently, considerable attention [9–13] has been given to the role in the chemistry of the marine boundary layer (MBL) of chlorine atoms, which react with alkanes up to two orders of magnitude faster than do hydroxyl radicals. Chlorine atoms may accordingly serve as an additional oxidant, at concentrations larger than 1×10^3 atoms cm^{-3}. From diurnal measurements of non-methane hydrocarbons [2, 14], or the observation of inorganic chlorine gases, Cl-atom concentrations of the order of 10^4–10^5 atoms cm^{-3} were inferred. In addition, Barrie et al. [15] have suggested—and there is now evidence for this from field measurements [16]—that ozone is destroyed in the MBL during polar sunrise by a mechanism involving Br and BrO.

So far no satisfactory mechanism has been proposed for reactive bromine and chlorine production in the pristine MBL or the Arctic. The gas-phase reactions of

---

This text was first published as "A mechanism for halogen release from sea-salt aerosol in the remote marine boundary layer", as: "Letter to Nature", in: *Nature*, vol. 383, 26 September 1996: 327–330. The permission to republish this article was granted on 19 August 2015 by Ms. Claire Smith, Nature Publishing Group & Palgrave Macmillan, London, UK. The authors thank C. Brühl for calculating photolysis rate constants and B. Luo for calculating activity coefficients. This work was supported by the Deutsche Forschungsgemeinschaft.

© The Author(s) 2016
P.J. Crutzen and H.G. Brauch (eds.), *Paul J. Crutzen: A Pioneer on Atmospheric Chemistry and Climate Change in the Anthropocene*, Nobel Laureates 50, DOI 10.1007/978-3-319-27460-7_8

OH radicals with HCl or HBr are only a minor source of halogen atoms. Singh and Kasting [17] calculated Cl-atom concentrations of 10^3 atoms cm^{-3}, assuming an HCl volume mixing ratio of 1 nmol mol^{-1}, which is much larger than measured in the remote MBL [1, 2, 18]. Very recently, Mozurkewich [19] suggested the reaction of peroxomonosulphuric (Caro's) acid (HSO_5^-) with sea-salt bromide as a source of elemental bromine in the Arctic. However, the mechanism is favoured by low temperatures and high SO_2 concentrations and by itself should not oxidize significant amounts of halides. Mozurkewich also discussed direct bromide oxidation through free radicals, such as OH or HO_2 in the sea-salt aerosol. However, none of these mechanisms is capable of significant chlorine atom production in the unpolluted MBL.

Our proposed mechanism for autocatalytic bromine and chlorine chemistry in the liquid and gas phase is shown in Fig. 8.1. Hypobromous acid, HOBr, which is formed through an initial bromide oxidation (see below), is scavenged by sea-salt

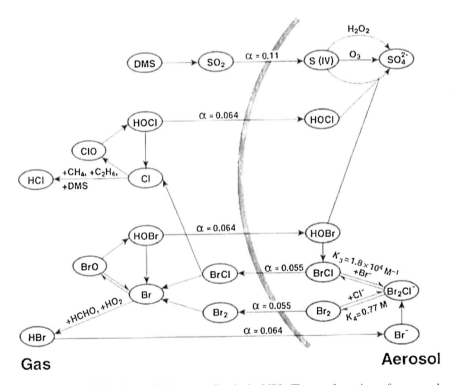

Fig. 8.1 A simplified scheme of halogen cycling in the MBL. The transformations of compounds in the shaded ovals occur in the deliquesced sea-salt aerosol. Also given are the accommodation coefficients, α, and equilibrium constants, K, used in the model calculations

aerosol. Because in sea water (and therefore also in the nascent sea-salt aerosol) the bromide to chloride ratio is $\sim 1/700$, HOBr reacts with Cl^-:

$$HOBr + Cl^- + H^+ \leftrightarrow BrCl + H_2O \qquad (1,-1)$$

Recently, Wang et al. [20] determined a lower limit of the BrCl hydrolysis rate constant, $k_{-1} \geq 1 \times 10^5$ s^{-1}. From the equilibrium constant $K_{1eq} = 5.6 \times 10^4$ M^{-2} Ref. [20], we calculate the rate constant of the forward reaction $k_1 \geq 5.6 \times 10^9$ M^{-2} s^{-1}, which is of the same order of magnitude as that of the comparable reaction of HOBr with Br$^-$ Ref. [21].

$$HOBr + Br^- + H^+ \leftrightarrow Br_2 + H_2O \qquad (2,-2)$$

$$(k_2 = 1.6 \times 10^{10} M^{-2} s^{-1}, \ k_{-2} = 110 \ s^{-1})$$

This reaction has been discussed as potentially important for bromine cycling on sulphate aerosol [22] and for autocatalytic bromine release from sea-salt aerosol [19, 23]. Because of the large $[Cl^-]/[Br^-]$ ratio in sea water, the forward reaction (1) is much more important than forward reaction (2). Although the hydrolysis reaction (-1) is faster than reaction (-2), this is of lesser significance because a substantial fraction of the BrCl will react with Br$^-$ leading to autocatalytic Br activation:

$$HOBr + Cl^- + H^+ \leftrightarrow BrCl + H_2O \qquad (1,-1)$$

$$BrCl + Br^- \leftrightarrow Br_2Cl^- \qquad (3,-3)$$

$$Br_2Cl^- \leftrightarrow Br_2 + Cl^- \qquad (4,-4)$$

$$Br_2 + h\nu \rightarrow 2Br \qquad (5)$$

$$2(Br + O_3) \rightarrow 2(BrO + O_2) \qquad (6)$$

$$2(BrO + HO_2) \rightarrow 2(HOBr + O_2) \qquad (7)$$

$$(net)2HO_2 + H^+ + 2O_3 + Br^- + h\nu \xrightarrow{HOBr,Cl^-} HOBr + 4O_2 + H_2O \qquad (8)$$

Equilibria (3) and (4) were found to be established very rapidly [20], probably at a diffusion-controlled rate. After escape of Br$_2$ to the gas phase, photolysis (5) and reactions (6), (7) and (1) an auto-catalytic cycle of bromine activation is closed. Thus, bromide oxidation is driven by HO$_2$, O$_3$ and aerosol acidity in the presence of sunlight and is catalysed by HOBr and Cl$^-$.

We have incorporated reactions (1)–(7) into a photochemical box model of the MBL [23]. The model treats chemical reactions in the gas phase and deliquesced sea-salt particles, as well as exchange between the two phases; standard O$_3$–NO$_x$– HO$_x$–S chemistry [23] is used. The sea-salt aerosol concentration was 3×10^{-11} m^3

of liquid per m^3 of air (that is, 9.4 μg NaCl per m^3 of air) with a size distribution from Ref. [24], yielding a volume geometric median particle radius of 3 μm, a number density of 0.7 cm^{-3} and a surface area of 30 $μm^2$ cm^{-3}, assuming that sea salt represents the coarse fraction of the aerosol. The liquid-phase concentrations are averaged over the entire particle ensemble which is assigned a lifetime of 2 days. Fresh and slightly alkaline aerosol is added continuously while all liquid-phase compounds are removed with the life-time of the aerosol. We recognize that our treatment of the sea salt is a major simplification, which neglects effects of lower pH that is established in smaller particles.

For some runs we also included a similar scheme of reactions on the sub-micrometre mode of marine aerosol. Typical background aerosol concentration, 1.1×10^{-12} m^3 (liquid) per m^3 (air), size distribution (volume geometric median radius of 0.1 μm), and chemical composition (mainly NH_4HSO_4, $(NH_4)_2SO_4$ and water) are taken from MBL measurements [25]. Scavenging of HOBr, HCl and HBr on the additional surface of the submicrometre sulphate aerosol in combination with reactions (1) and (2) also leads to efficient conversion of Cl^- and Br^- into BrCl and Br_2.

Figure 8.2a–c show the gas-phase halogen, HO_x and NO_x concentrations for the first two days of a model run (no sulphate aerosol) which was started without gas-phase halogens present. Owing to rapid uptake of SO_2 from the gas phase and subsequent oxidation to SO_4^{2-}, the acidity of the originally alkaline sea-salt aerosol particles drops to about pH 5.5. This causes some HCl to evaporate, yielding a gas-phase concentration of HCl of approximately 20 pmol mol^{-1}. During the first 6 h of the model run (night) the only radical species present is NO_3. A fraction of the NO_3 is scavenged by the sea-salt aerosol, where a free-radical chain oxidation of S (IV) Refs. [19, 26] is started, followed by Caro's acid (HSO_5^-) production, oxidation of Br^- to HOBr

$$HSO_5^- + Br^- \rightarrow HOBr + SO_4^{2-} (k_9 = 1 \text{ M}^{-1} \text{ s}^{-1}) \tag{9}$$

and Br_2 formation via reactions (1), (3) and (4). At dawn, Br_2 at a concentration of 0.06 pmol mol^{-1} has formed and after its rapid photolysis the autocatalytic bromide oxidation described above takes over. After the second day of cycling, BrCl and Br_2 reach early-morning maxima of 3.8 and 2.2 pmol mol^{-1}, respectively, and a daytime HOBr concentration of 7 pmol mol^{-1} is calculated. Bromine in particular is very efficiently transferred to the gas phase, causing a bromide deficit of about 90 %. The calculated chloride deficit is about 1 % after two days of cycling.

Steady-state concentrations are reached after ~ 14 model days. The daytime maxima (pmol mol^{-1}) are: [HCl] = 28, [HOCl] = 2.7, [HOBr] = 13 and [HBr] = 1.9. Night-time maxima (pmol mol^{-1}) are [BrCl] = 7.5, [Br_2] = 3, [$ClONO_2$] = 1.3, [$BrONO_2$] = 1.2 and [Cl_2] = 0.4. Although it may be unrealistic to assume that an air mass will stay in the MBL for such a long period, this assumption gives some insight in how rapidly the autocatalytic cycle might work. After only two days of cycling, ~ 50 % of the steady-state halogen concentrations are reached.

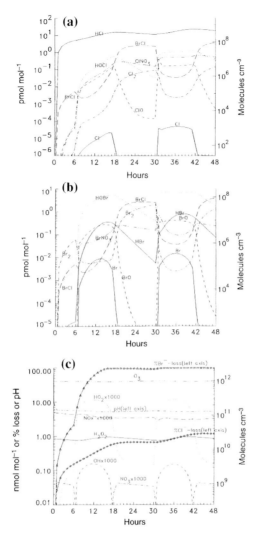

Fig. 8.2 Molar mixing ratios (*left-hand axis*) and number densities (*right-hand axis*) of gaseous chlorine (**a**) and gaseous bromine (**b**) compounds calculated with a photochemical box model during the first 48 h of processing in the MBL. The model was initialized with background values for an unpolluted MBL: $[O_3] = 40$ nmol mol^{-1}, $[CO] = 70$ nmol mol^{-1}, $[CH_4] = 1.800$ nmol mol^{-1}, $[C_2H_6] = 0.5$ nmol mol^{-1}, $[NO_x] = $ pmol mol^{-1}, [DMS] 110 pmol mol^{-1}, $[SO_2] = 70$ pmol mol^{-1}; $T = 293$ K; relative humidity was 76 %. Dry deposition rates are assigned to gas-phase compounds, such as 2 cm s^{-1} for the strong acids HNO$_3$, HCl, HBr and H$_2$SO$_4$, and values between 0.2 cm s^{-1} for the weaker acids, HOCl and HOBr, and 0.5 cm s^{-1} for the moderately soluble compounds, H$_2$O$_2$, HCHO and SO$_2$. Photolysis rates were calculated for 1 April, 45° N Ref. [23]. In **c**, the O$_3$, HO$_x$, and NO$_x$ mixing ratios (nmol mol^{-1}, *left axis* or molecules cm^{-3}, *right axis*), aerosol pH (*left axis*) and chloride and bromide loss for the aerosol (in %, *left axis*) are shown. HO$_2$, OH, NO$_3$ and NO$_x$ mixing ratios are multiplied by 1,000 for clarity

The bromine atoms formed upon photolysis of Br_2 or $BrCl$ react mainly with ozone. Significant amounts of ozone can be destroyed in catalytic cycles, involving reactions (6) and (7), and HOBr photolysis:

$$HOBr + hv \rightarrow OH + Br \qquad (10)$$

The calculated amount of ozone destroyed during the second day of sea-salt processing is ~ 0.14 nmol mol^{-1}. This compares to about 2.5 and 1.2 nmol mol^{-1} d^{-1} of ozone destruction through photolysis ($O_3 \rightarrow O(^1D)$; $O(^1D) + H_2O \rightarrow 2OH$) and chemical reactions ($OH + O_3 \rightarrow HO_2 + O_2$, $HO_2 + O_3 \rightarrow OH + 2O_2$) at O_3 concentrations of 40 or 20 nmol^{-1}, respectively. Although the Br-catalysed ozone destruction during the second model day is only 5–10 % of the total ozone loss, we calculate at steady state a destruction of 20–40 %, if recycling of HBr and HCl on sulphate aerosol is also considered.

A crucial parameter is the sea-salt aerosol content, which is largely determined by the wind speed. If we assume a 5 times higher aerosol content (15×10^{-11} m^3 (liquid), per m^3 (air), that is, 47 μg NaCl per m^3 (air)), which is at the upper range of values typically found in the MBL, the reactive halogen concentrations increase dramatically. We estimate maximum concentrations (pmol mol^-) after two days of $[HCl] = 70$, $[HOCl] = 33$, $[HOBr] = 35$, $[HBr] = 3$, $[BrCl] = 38$, $[Br_2] = 10$ and $[Cl_2] = 1.5$. Catalytic ozone loss through bromine and chlorine during the second day increases to 1.4 or 0.6 nmol mol^{-1}, respectively, corresponding to 58 % or 25 % of the ozone lost by photolysis and HO_x chemistry.

The proposed autocatalytic mechanism also represents a source of chlorine atoms via the photolysis of BrCl. The product of reaction (1), BrCl, like Br_2 or Cl_2 is only slightly soluble. In the gas phase BrCl will undergo fast photolysis:

$$BrCl + hv \rightarrow Br + Cl \qquad (11)$$

Chlorine-atom concentrations reach 1×10^3 atoms cm^{-3} during the second day of simulation, and 2.5×10^3 atoms cm^{-3} at steady state. Including recycling reactions on sulphate aerosol particles has a strong effect: Cl-atom concentrations are increased by a factor of 3.5 and reach 10^4 atoms cm^{-3} at steady state. With a five times higher sea-salt aerosol content, we estimate for the second day of processing $[Cl] = 3 \times 10^4$ atoms cm^{-3}, and $[Cl] = 5 \times 10^4$ atoms cm^{-3}, if we also include sulphate aerosol particles. The strong nonlinear response of the Cl concentration to the sea-salt aerosol content is caused by the multiplication effect of higher HOBr concentration and the reaction chain (1), (11), (6–7).

Our mechanism of Br activation can thus also account for the presence of Cl atoms in MBL [1, 2]. The simulated overnight accumulation of a Cl– and Br-atom precursor in conjunction with the daytime formation of HCl agrees with diel patterns observed by Pzenny et al. [2]. The main fate of Cl atoms is reaction with ozone, and to a smaller extent with methane. The additional methane loss on the second model day is small (~ 2 %, considering recycling on sulphate aerosol), but may reach 18 %, if the sea-salt aerosol content is increased by a factor of five. If recycling on sulphate aerosol is also taken into account, the additional methane loss is 26 %. For ethane, we calculate an additional loss of the order of 10 % or 30 % considering recycling on sulphate particles. If the sea-salt aerosol content is increased by a factor of five, ethane loss is dominated by Cl and the total loss is 3.9 times larger than by OH reaction alone.

HOBr and HOCl are scavenged by the aerosol (Fig. 8.1). There they are capable of very rapid S (IV) oxidation [7, 8]:

$$SO_3^{2-} + HOCl \rightarrow HSO_4^- + Cl^- \ (k_{12} = 7.6 \times 10^8 \, M^{-1} \, s^{-1}) \qquad (12)$$

$$SO_3^{2-} + HOBr \rightarrow HSO_4^- + Br^- \ (k_{13} = 5 \times 10^9 \, M^{-1} \, s^{-1}) \qquad (13)$$

The rate coefficients of reactions (12) and (13) have been measured at high pH values of 10–9. The rate constants were found to increase at lower pH. If we adopt for the analogous reactions of HSO_3^- (which is the predominant S(IV) species at pH range 3–6) the same rate constants as for SO_3^{2-}, we calculate on the second model day that ~ 40 % of the SO_2 scavenged by the aerosol is oxidized by HOCl and ~ 20 % by HOBr. The remaining oxidation occurs by ozone, H_2O_2 Ref. [26] and free radicals, such as Cl_2^- and Br_2^-. But the total amount of S(IV) oxidized in sea-salt aerosol does not only depend on the availability of oxidants, but also on the transport of SO_2 into the aerosol particle and the solubility of SO_2. We estimate that the total amount of S (IV) oxidized in sea-salt aerosol is increased by 0.4–0.8. If we include recycling reactions on sulphate aerosol, these values increase to 2.4–3.6. The additional SO_2 oxidation on pre-existing particles reduces the formation of new cloud-condensation nuclei and therefore the feedback between greenhouse warming, oceanic dimethyl sulphide emissions and increased cloud albedo [27, 28].

The proposed aulocalalytic mechanism critically depends on the rale of reaction (1). Our calculations stress the need for improved kinetic data of hypohalogenic acids in concentrated halogenide solutions. Because the H^+ concentration determines the rate of reaction (1), better knowledge of aerosol pH is also required. Furthermore the Henry's law constant *(H)* of HOBr needs to be determined. We have assumed $H_{HOBr} = 92.6 \, M \, atm^{-1}$, which is 1/10 of the solubility constant of HOCl at 293 K Ref. [29]. In sensitivity studies, we find that the value of H_{HOBr} can be lowered to $\sim 0.1 \, M \, atm^{-1}$ until HOBr solubility limits the mechanism suggested.

The calculated reactive halogen concentrations of 10–40 pmol mol^{-1} should be detectable in the field. The main daytime species are HOBr and HOCl, whereas at night BrCl and Br_2 are most abundant. Our calculated daytime BrO concentrations reach values of $\sim 10^7$–10^8 molecules cm^{-3}, which is in the detection range of differential optical absorption spectroscopy (DOAS). Some experimental evidence already exists [30] for the proposed Br^- catalysed production of Cl from sea-salt aerosol, which was irradiated in a large reaction chamber in the presence of O_3.

References

1. Pszenny, A.A.P.; et al., 1993: in: *Geophysical Research Letters*, 20: 699–702.
2. Singh, H.B.; et al., 1996: in: *Journal of Geophysical Research*, 101: 1907–1917.
3. Finlayson-Pitts, B., 1993: in: *Journal Research on Chemical Intermediates*, 19: 235–249.
4. Zetzsch, C.; Behnke, W., 1992: in: *Berichte der Bunsengesellschaft für physikalische Chemie*, 96: 488–493.
5. Finlayson-Pitts, B.J.; Livingston, F.E.; Berko, H.N., 1989: in: *The Journal of Physical Chemistry*, 93: 4397–4400.
6. Rohrer, F.; Bruening, D., 1992: in: *Journal of Atmospheric Chemistry*, 15: 253–267.
7. Fogelman, K.D.; Walker, D.M.; Margerum, D.W., 1989: in: *Inorganic Chemistry*, 28: 986–993.
8. Troy, R.C.; Margerum, D.W., 1993: in: *Inorganic Chemistry*, 30: 3538–3543.
9. Finlayson-Pitts, B.J., 1993: in: *Journal of Geophysical Research*, 98D: 14991–14993.
10. Parrish, D.D.; et al., 1993: in: *Journal of Geophysical Research*, 98D: 14995–14997.
11. McKeen, S.A.; Liu, S.C., 1993: in: *Geophysical Research Letters*, 20: 2363–2366.
12. Wingenter, O.W.; et al., 1996: in: *Journal of Geophysical Research*, 101: 4331–4340.
13. Graedel, T.E.; Keene, W.C., 1995: in: *Global Biogeochemical Cycles*, 9,1: 47–77.
14. Jobson, B.T.; et al., 1994: in: *Journal of Geophysical Research*, 99D: 25355–25368.
15. Barrie, L.A.; Bottenheim, J.W.; Schnell, R.C.; Crutzen, P.J.; Rasmussen, R.A., 1988: in: *Nature*, 334: 138–141.
16. Hausmann, M.; Platt, U., 1994: in: *Journal of Geophysical Research*, 99D: 25399–25413.
17. Singh, H.B.; Kasting, J.F., 1988: in: *Journal of Atmospheric Chemistry*, 7: 261–285.
18. Harris, G.W.; Klemp, D.; Zenker, T., 1992: in: *Journal of Atmospheric Chemistry*, 15: 327–332.
19. Mozurkewich, M., 1995: in: *Journal of Geophysical Research*, 100(D): 14199–14207.
20. Wang, T.X.; Kelley, M.D.; Cooper, J.N.; Beckwith, R.C.; Margerum, D.W., 1994: in: *Inorganic Chemistry*, 33: 5872–5878.
21. Eigen, M.; Kustin, K., 1962: in: *Journal of the American Chemical Society*, 84: 1355–1361.
22. Fan, S.-M.; Jacob, D.J., 1992: in: *Nature*, 359: 522–524.
23. Sander, R.; Crutzen, P.J., 1996: in: *Journal of Geophysical Research*, 101: 9121–9138.
24. Kim, Y.; Sievering, H.; Boatman, J., 1990: in: *Global Biogeochemical Cycles*, 4: 165–177.
25. Kim, Y.; Sievering, H.; Boatman, J.; Wellman, D.; Pszenny, A., 1995: in: *Journal of Geophysical Research*, 100: 23027–23038.
26. Chameides, W.L.; Stelson, A.W., 1992: in: *Journal of Geophysical Research*, 97D: 20565–20580.
27. Charlson, R.J.; Lovelock, J.E.; Andreac, M.O.; Warren, S.G., 1987: in: *Nature*, 326: 655–661.
28. Sievering, H.; et al., 1992: in: *Nature* 360: 571–573.
29. Huthwelker, T.; et al., 1995: in: *Journal of Atmospheric Chemistry*, 21: 81–95.
30. Behnke, W.; Elend, M.; Krüger, H.U.; Scheer, V.; Zetzsch, C., in Press: in: Borrell, P.M.; et al. (Eds.): *Proceedings of Eurotrac Symposium* (Southampton: Computational Mechanics).

Chapter 9
The Indian Ocean Experiment: Widespread Air Pollution from South and Southeast Asia

Jos Lelieveld, Paul J. Crutzen, V. Ramanathan, M.O. Andreae,
C.A.M. Brenninkmeijer, T. Campos, G.R. Cass, R.R. Dickerson,
H. Fischer, J.A. de Gouw, A. Hansel, A. Jefferson, D. Kley,
A.T.J. de Laat, S. Lal, M.G. Lawrence, J.M. Lobert,
O. Mayol-Bracero, A.P. Mitra, T. Novakov, S.J. Oltmans,
K.A. Prather, T. Reiner, H. Rodhe, H.A. Scheeren, D. Sikka
and J. Williams

The Indian Ocean Experiment (INDOEX) was an international, multiplatform field campaign to measure long-range transport of air pollution from South and Southeast Asia toward the Indian Ocean during the dry monsoon season in January to March 1999. Surprisingly high pollution levels were observed over the entire northern Indian Ocean toward the Intertropical Convergence Zone at about 6 °S. We show that agricultural burning and especially biofuel use enhance carbon monoxide concentrations. Fossil fuel combustion and biomass burning cause a high aerosol loading. The growing pollution in this region gives rise to extensive air quality degradation with local, regional, and global implications, including a reduction of the oxidizing power of the atmosphere.

9.1 Introduction

Until recently, North America and Europe dominated the use of fossil fuels, resulting in strong carbon dioxide emissions and global warming [1]. The fossil energy-related CO_2 release per capita in Asia is nearly an order of magnitude

This text was first published as: Lelieveld, J., P.J. Crutzen, M.O. Andreae, C.A.M. Brenninkmeijer, T. Campos, G.R. Cass, R.R. Dickerson, H. Fischer, J.A. de Gouw, A. Hansel, A. Jefferson, D. Kley, A.T.J. de Laat, S. Lal, M.G. Lawrence, J.M. Lobert, O. Mayol-Bracero, A.P. Mitra, T.Novakov, S.J. Oltmans, K.A. Prather, V. Ramanathan, T. Reiner, H. Rodhe, H.A. Scheeren, D. Sikka and J. Williams (2001) The Indian Ocean Experiment: Widespread air pollution from South and South-East Asia. *Science 291*, 1031–1036. The copyright belongs to the authors. We are grateful for the support by many funding agencies, notably the NSF, the Department of Energy, NASA, the National Oceanic and Atmospheric Administration, the Indian Space Research Organization, the European Union, the Max-Planck-Gesellschaft, the French Centre National d'Etudes Spatiales, the Netherlands Supercomputing Facility, and the Netherlands Organization for Scientific Research. We thank J. Olivier for his help with emission estimates (16, 39).

P.J. Crutzen and H.G. Brauch (eds.), *Paul J. Crutzen: A Pioneer on Atmospheric Chemistry and Climate Change in the Anthropocene*, Nobel Laureates 50, DOI 10.1007/978-3-319-27460-7_9

smaller than in North America and Europe [2]. However, Asia is catching up. About half of the world's population lives in South and East Asia, and hence the potential for growing pollutant emissions is large. In China, many pollution sources reduce air quality [3–5]. In rural residential areas, notably in India, the burning of biofuels, such as wood, dung, and agricultural waste, is a major source of pollutants [6]. In urban areas, the increasing energy demand for industry and transport propels fossil fuel utilization [7].

Here we evaluate measurements of the Indian Ocean Experiment (INDOEX) to characterize the atmospheric chemical composition of the outflow from South and Southeast Asia, from January to March 1999 during the dry winter monsoon [8]. During this season, the northeasterly winds are persistent, and convection over the continental source regions is suppressed by large-scale subsidence, thus limiting upward dispersion of pollution [9]. Our analysis is based on measurements from a C-130 and a Citation aircraft operated from the Maldives near 5 °N, 73 °E, the research vessels *Ronald H. Brown* and *Sagar Kanya,* and the Kaashidhoo Climate Observatory (KCO) on the Maldives (Fig. 9.1). During the campaign, the location of the Intertropical Convergence Zone (ITCZ) varied between the equator and 12 ° S. Hence, transport of primary pollutants and reaction products toward the ITCZ could be studied over an extended ocean area where pollutant emissions are otherwise minor. By performing measurements across the ITCZ, the polluted air masses could be contrasted against comparatively clean air over the southern Indian Ocean. Furthermore, we used the measurements to evaluate the numerical representation of these processes in a chemistry general circulation model (GCM) [10]. The model was subsequently applied to calculate the large-scale atmospheric chemical effects of the measured pollution.

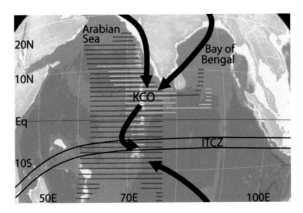

Fig. 9.1 Schematic over-view of the INDOEX measurement domain, traversed by two ships (*red hatching*) and two aircraft (43 flights; *yellow hatching*), the mean location of the ITCZ, and 1- to 2-week boundary layer air mass trajectories during January to March 1999 (*arrows*). KCO is the Kaashidhoo Climate Observatory at 5 °N, 73.5 °E

Aerosol chemical and optical measurements were performed from both aircraft, the R/V *Brown,* and KCO. The latter is located on a small island about 500 km southwest of India and more than 1000 km from the main pollution centers. At KCO, we measured the size distribution and chemical composition of fine particles, collected on filters and cascade impactors [11]. The filter analysis shows an average dry mass concentration of ~ 17 µg/m^3 (Fig. 9.2). The aerosol contained substantial amounts of both inorganic and organic pollutants, including black carbon (BC). Mass spectrometric particle analysis shows that the BC particles were always mixed with organics and sulfate, indicating substantial chemical processing. Very similar results were obtained from KCO, the boundary layer flights by the C-130 aircraft, and the R/V *Brown,* which shows that the aerosol composition was remarkably uniform over the northern Indian Ocean.

The aerosol mass loading observed over the Indian Ocean is quite comparable to suburban air pollution in North America and Europe [12]. However, the BC content was relatively high (Table 9.1), which gives the aerosol a strong sunlight-absorbing character, yielding a single scattering albedo at ambient relative humidity between 0.8 and 0.9. This aerosol, with a mean optical depth of 0.2–0.4 (at 0.63 µm wavelength), reduces solar heating of the northern Indian Ocean by about 15 % (~ 25 W/m^2) and enhances the heating of the boundary layer by about 0.4 K/day (~ 12 W/m^2), which substantially perturbs the regional hydrological cycle and climate [13, 14].

The BC aerosol and fly ash are unquestionably human produced because natural sources are negligible. Likewise, non-sea-salt sulfate can be largely attributed to anthropogenic sources. Filter samples collected on board the R/V *Brown* in the clean marine boundary layer south of the ITCZ reveal a fine aerosol sulfate concentration of about 0.5 µg/m^3, probably from the oxidation of naturally emitted

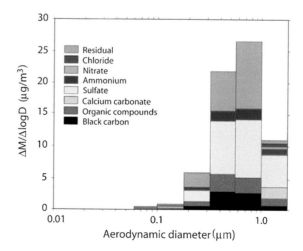

Fig. 9.2 Average mass (M) composition of fine aerosol on KCO (Maldives) as a function of the logarithm of the particle diameter (D) in February 1999. The residual includes mineral dust, fly ash, and unknown compounds (11)

Table 9.1 Mean fine and coarse mass fractions of aerosols collected on filters on board the C-130 aircraft in the boundary layer (34 samples) and at KCO (24 samples)

Compound	$D < 1$ μm (%)	$D > 1$ μm (%)
Sulfate	32	25
Organics	26	19
Black carbon	14	10
Mineral dust	10	11
Ammonium	8	11
Fly ash	5	6
Potassium	2	1
Nitrate	<1	4
Sea salt, MSA	<1	12
Rest	2	1
Total mass (μg/m^3)	22	17

D is diameter. MSA is methane sulfonic acid. 'Rest' includes magnesium, calcium, oxalate, formate, and unidentified material

dimethyl sulfide. The sulfate concentration over the northern Indian Ocean was close to 7 μg/m^3, and we thus infer an anthropogenic fraction of more than 90 %. Similarly, the ammonium concentration south of the ITCZ, from natural ocean emissions, was 0.05 μg/m^3, indicating an anthropogenic contribution of more than 95 % to the nearly 2 μg/m^3 of ammonium observed north of the ITCZ.

It is more difficult to attribute the organic aerosol fraction to a particular source category. Secondary organic particles from natural hydrocarbon sources are probably of minor importance because India is scarcely forested. Moreover, the BC/total carbon ratio of 0.5, as derived from the filter samples, is typical for aerosols from fossil fuel combustion [15]. In the aerosol south of the ITCZ, organic compounds were negligible, whereas over the northern Indian Ocean, it was almost 6 μg/m^3. We thus infer that most of the particulate organics north of the ITCZ were of anthropogenic origin. INDOEX aerosol components of natural origin included a total mass fraction of 1 % sea salt and 10 % mineral dust. Nevertheless, some of the mineral aerosol likely originated from road dust and agricultural emissions. Taken together, the human-produced contribution to the aerosol was at least 85 %. Because precipitation is scarce during the winter monsoon, the aerosol can spread over the entire northern Indian Ocean before entering the ITCZ, where it is largely removed in deep convective clouds.

To evaluate gaseous pollution sources with our model, we adopted the Emission Database for Global Atmospheric Research (EDGAR) [16]. Table 9.2 indicates that the South and East Asian region is a substantial source of global pollution. For example, the total carbon monoxide (CO) release is estimated to be 50 % larger than the combined emissions from Europe and North America. Table 9.2 also indicates that the nature of the pollution is different from that in Europe and North America. Particularly in India, the use of biofuels and agricultural burning causes substantial CO emissions.

Table 9.2 Global anthropogenic CO_2, CO, NO_x, SO_2, and NMHC emissions (India region includes Bangladesh, Maldives, Sri Lanka, Myanmar, Nepal, and Pakistan. China region includes Cambodia, Vietnam, Laos, Mongolia, and North Korea. East Asia includes Japan, South Korea, Indonesia, Malaysia, Philippines, and Thailand) [16]

Source category	Global	North America	Europe	India	China	East Asia
Carbon dioxide (Pg of CO_2 per year)						
Total	29.8	6.2 (21%)	4.9 (16%)	2.2 (7%)	4.0(13%)	2.5 (8%)
Fossil fuel use	21.9	5.6	4.5	0.7	2.6	1.7
Industrial processes	0.6	0.1	0.2	–	0.1	0.1
Biofuel use	5.5	0.5	0.2	1.4	1.2	0.5
Agriculture	1.8	–	–	0.1	0.1	0.2
Carbon monoxide (Tg of CO per year)						
Total	975	107 (11%)	85 (9%)	110 (11%)	111 (11%)	69 (7%)
Fossil fuel use	263	74	53	4	34	16
Industrial processes	35	2	8	1	5	6
Biofuel use	181	9	2	47	40	19
Agriculture	496	22	22	58	32	28
Nitrogen oxides (Tg of NO_2 per year)						
Total	102	26 (25%)	16 (16%)	6 (6%)	11 (10%)	6 (6%)
Fossil fuel use	72	24.3	13.6	2.6	7.2	4.3
Industrial processes	5	0.4	1.1	0.2	0.9	0.7
Biofuel use	5	0.5	0.2	1.1	1.5	0.4
Agriculture	20	0.8	0.7	2.0	1.1	1.0
Sulfur dioxide (Tg of SO_2 per year)						
Total	148	24.5 (17%)	33.3 (23%)	5 (3%)	28 (19%)	7 (5%)
Fossil fuel use	120	22.8	26.4	4.0	25.0	5.0
Industrial processes	23	1.2	6.4	0.3	2.8	1.7
Biofuel use	2	0.4	0.4	0.2	0.3	0.1
Agriculture	4	0.1	0.1	0.4	0.2	0.2
Nonmethane hydrocarbons (Tg of NMHC per year)						
Total	178	22 (12%)	21 (12%)	19 (11%)	17 (10%)	16 (9%)
Fossil fuel use	69	12	12	1.5	3	6
Industrial processes	34	7	7	3	4	4
Biofuel use	31	1	0.2	8.5	6	3
Agriculture	44	2	2	6	4	3

Emissions from biomass burning are difficult to estimate because they usually occur scattered over large rural areas. Moreover, the burning process is not well defined because the fuel type and the combustion phase (flaming, smoldering) strongly affect the smoke composition [17]. Many people in the Indian region still live in rural areas where domestic energy consumption largely depends on biofuels, whereas in urban areas, soft coke, kerosene, and other liquid fuels are also used. In Asia, about one-quarter of the energy use depends on biofuels, whereas in India, this fraction is larger, close to 50 % [18, 19]. It has been estimated that in India, firewood contributes about two-thirds to biofuel consumption, whereas the burning of dung and agricultural wastes contribute roughly equally to the remaining one-third [20–22]. A particularly useful indicator of biomass burning is the relative abundance of methyl cyanide (CH_3CN) to that of CO [23]. The biomass burning emission of both gases mostly takes place from smoldering. The $\Delta CH_3CN/\Delta CO$ ratio measured on the C-130 aircraft and the R/V *Brown* was about 0.2 % (Fig. 9.3) [24]. This is close to the values obtained from controlled biomass fires in the laboratory [23]. Without other substantial sources of CH_3CN, it follows that biomass burning was a major source of CO over the northern Indian Ocean. Measurements in air masses transported from southwestern Asia, mostly west of India (in blue), show a much lower $\Delta CH_3CN/\Delta CO$ ratio (Fig. 9.3), illustrating the importance of fossil fuel combustion as a pollution source to these air masses in addition to biomass burning [25]. From our model simulations, which are in good agreement with the measurements, we infer that 60–90 % of the CO originated from biomass burning (Fig. 9.4).

This model estimate is supported by a comparison of radiocarbon monoxide (^{14}CO) in low-latitude clean Southern Hemispheric air with that over the northern Indian Ocean, as measured from samples taken from the R/V *Brown*. The clean air samples south of the ITCZ contained on average 55 parts per billion by volume (ppbv) of CO and 6.2 molecules of $^{14}CO/cm^3$ whereas north of the ITCZ, this was 155 ppbv and 9.7 molecules/cm^3 [26]. The ^{14}CO difference between these air masses must be of biogenic origin, i.e., mainly biomass burning, because fossil fuels are radiocarbon-depleted. Previous analysis has shown that biomass burning adds 0.038 molecules of $^{14}CO/cm^3$ per ppbv of CO [26]. If we assume further that

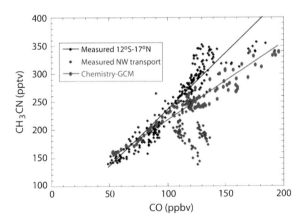

Fig. 9.3 Methyl cyanide (CH_3CN) versus carbon monoxide (CO) mixing ratios measured from the R/V *Brown* and calculated with a chemistry GCM. Average values are shown by the *straight lines*. The measurements (*black*) were performed between 12 °S, 73 °E and 17 °N, 69 °E. The measurements in *blue* represent air masses transported from the northwest, as determined by *back-*trajectory calculations [25]. Because our chemistry GCM is unable to distinguish the air mass history, because it mixes the air masses at 1.8° resolution, the slope of the *red line* is less steep than of the *black line*

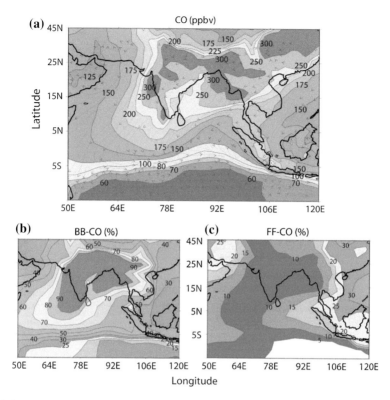

Fig. 9.4 **a** Mean CO (ppbv) near the surface over the Indian Ocean during February 1999, as calculated with our chemistry GCM [10]. Average winds are shown by streamlines. Marked tracers indicate the percentage of CO from **b** biomass burning (BB)—mostly biofuel use and agricultural waste burning—and **c** fossil fuel (FF) combustion. The remainder largely originates from hydrocarbon oxidation

about a third of the 55 ppbv of background CO is also related to biomass burning, as calculated with our model (Fig. 9.4b), it follows that the average contribution of biomass burning to CO over the northern Indian Ocean was 70–75 %.

The highest pollution levels originated from the area around the Bay of Bengal (Table 9.3). The impact of these air masses over the Indian Ocean was largest in February. In March, the region was more strongly influenced by air that originated north of the Arabian Sea (Fig. 9.1). Although this air was generally cleaner, it also carried desert dust, which contributed to the aerosol load. The aircraft measurements also show substantially enhanced methyl cyanide and methyl chloride (CH_3Cl) concentrations, particularly in air from the Bay of Bengal region. The latter points to the extensive use of chlorine-rich fuels such as agricultural waste and dung [27]. Levels of NO only rarely exceeded the instrument detection limit of 40 parts per trillion by volume (pptv) (only in fresh pollution plumes and downwind of ITCZ lightning), hence these are not shown.

Table 9.3 Mean results from boundary layer Citation aircraft measurements (25 flights) between the Maldives and the ITCZ during February to March 1999. The two main source regions of the measured air pollution have been determined by back-trajectory calculations [25] (standard deviations in parentheses)

	Source	Region
	Bay of Bengal	Arabian Sea
CO (ppbv)	208 (42)	135 (16)
O_3 (ppbv)	15 (5)	13 (4)
$CH_3C(O)CH_3$ (ppbv)	2.2 (0.4)	1.6 (0.2)
CH_3CN (pptv)	288 (72)	266 (39)
C_2H_6 (pptv)	817 (251)	465 (134)
C_2H_2 (pptv)	291 (179)	81 (34)
C_3H_8 (pptv)	50 (36)	36 (41)
C_6H_6 (pptv)	99 (42)	40 (18)
CH_3Cl (pptv)	757 (64)	650 (30)

We observed strongly enhanced CO levels over the northern Indian Ocean [28]. Average CO mixing ratios at KCO in February were close to 200 ppbv. Such high CO concentrations are comparable to polluted air down-wind of North America and Europe. The KCO measurements show that aerosol absorption and scattering were highly correlated with CO, which indicates that the trace species of various origins were well mixed in the marine boundary layer (BL). Especially in February and early March, pollution levels at KCO varied strongly on a 3- to 7-day time scale. CO typically ranged from 120–250 ppbv. These changes were associated with tropical cyclones that transported cleaner air from the south [9]. Later in March, the pollution levels near the surface were lower, largely associated with the air mass trajectory change from the northeast to the northwest. The aerosol optical thickness, however, was higher than in February. This indicates that particularly in March, substantial pollution transport took place above the BL.

Pollution variations over the northern Indian Ocean are also influenced by tropical waves that alter the intensity of ITCZ convection, acting on a 1- to 2-month time scale [known as the Madden Julian Oscillation (MJO)]. Strong convection ventilates the BL and increases the monsoonal flow [9]. Furthermore, variations on an interannual time scale are affected by the El Niño-Southern Oscillation. During the recent El Niño in February 1998, for example, pollution transport from India was reduced, so that CO concentrations at KCO were only 110–140 ppbv. In February 1999, on the other hand, the monsoonal flow was strong, and hence pollution transport was efficient. In March 1999, the ITCZ convection intensified during an active phase of the MJO, which ventilated BL pollution from the Indian Ocean.

Considering that the pollution occurs at low latitudes, one expects strong photochemical activity, possibly giving rise to ozone (O_3) buildup. Because of its important role in atmospheric chemistry, O_3 was measured from all platforms and ground stations, as well as through balloon soundings from KCO and the R/V *Brown* [29]. In several O_3 profiles over KCO (Fig. 10.5a), sharp peaks can be discerned, with a particularly pronounced O_3 maximum above the BL. The O_3 minimum within the BL, which extended to an altitude of 0.5–1 km, and the maximum directly above are not well reproduced by the model. This is related to a sea breeze circulation at the Indian coast that is not resolved. During daytime, the convective BL over land extends to about 2–3 km, whereas further down-wind, the marine BL only reaches about 1-km altitude or less [30, 31]. The sea breeze causes upward transport over land that adds pollution to a stable layer that develops over the Indian Ocean between about 1 and 3 km in the monsoonal outflow from India. Because cumulus convection is weak in the Indian outflow, the layer can remain intact, which constitutes a 'residual' pollution layer.

Typical altitude profiles of pollutants downwind of India, measured from the C-130 aircraft, also show the residual layer (Fig. 9.5b). In general, this layer was more pronounced in March than in February, related to the growing convection over land as surface heating increases toward the end of winter. Some of the profiles also show a secondary maximum between 3- and 4-km altitude. Meteorological analysis indicates that these air masses were transported from the east, carrying pollution from Southeast Asia. On several occasions, it was observed that the vertical layering, shown in Fig. 9.5, can be maintained as far south as the Maldives, whereas further toward the ITCZ, trade wind cumulus convection causes breakup, vertical mixing, and partial dispersion into the free troposphere.

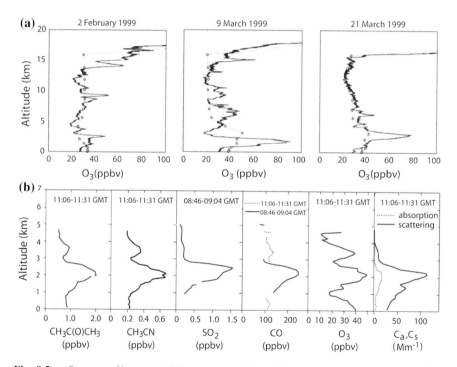

Fig. 9.5 a Ozone profiles over KCO as measured from balloon sondes and calculated with a chemistry GCM (dashed lines). The soundings show instances where the pronounced layering of the lower troposphere has remained intact as far south as 5 °N. **b** Pollutant profiles downwind of India (7.5 °N, 72 °E), including aerosol absorption and scattering, observed from the C-130 aircraft on 13 March 1999

Although O_3 concentrations near the Indian coast were about 50 ppbv and peak values in the residual layer even reached 80–100 ppbv, photochemical destruction of O_3 prevents its accumulation over the Indian Ocean. Typically, O_3 decreased from ~ 50 ppbv at 15 °N to ~ 10 ppbv near the ITCZ, which implies an O_3 loss rate in the BL of 1.5–2 ppbv per degree of latitude, or about 10 %/day. Much pollution originates from biomass burning. In particular, smoldering fires produce relatively little NO_x, a necessary ingredient for photochemical O_3 formation ($NO_x = NO + NO_2$). Nevertheless, several hundred pptv equivalent nitrate was measured in the coarse aerosols, which indicates that NO_x emissions are not negligible. However, NO_x is converted into nitrate by nighttime heterogeneous reactions on aerosols and daytime reaction with hydroxyl (OH) radicals, followed by uptake of HNO_3 by sea salt and dust particles. As a result, the NO_x lifetime is half a day or less, and its mixing ratio was generally quite low in the marine BL (NO < 10 pptv) [32], favoring chemical O_3 destruction rather than O_3 formation [33–36].

The combined anthropogenic NO_x source (S_N) from South and Southeast Asia is proportionally much smaller than the total CO and hydrocarbon source (S_C) as compared with Europe and North America. Thus, the ratio S_N/S_C (mol/mol) is comparatively low in Asia. The North American and European emissions, largely associated with high-temperature fossil fuel combustion, contain much more NO_x. This implies not only that O_3 photochemistry in the south-southeast Asian plume is strongly NO_x limited but also that OH regeneration by NO is inefficient [37, 38]. On a global scale, OH regeneration by NO_x is about equally as important as the primary OH production by O_3 photodissociation [38]. From our chemistry GCM, using the EDGAR emission database, we infer that the S_N/S_C ratio is more than four times lower in South and East Asia than in North America and Europe. Our model calculations indeed indicate that human-produced emissions from South and East Asia reduce OH concentrations, whereas European and North American pollution has the opposite effect. Because OH is the foremost oxidant that removes natural and human-produced gases, the Asian pollution reduces the oxidizing power of the atmosphere. For example, it increases the lifetime of methane (CH_4), an important greenhouse gas.

Our results show that during the winter monsoon, South and Southeast Asian emissions cause considerable air quality degradation over an area in excess of 10 million km^2. The nature of the pollution deviates from that in Europe and North America, a consequence of widespread biofuel use and agricultural burning, in support of the emission estimates in Table 9.2. In the next decades, emission trends in the region will likely reflect the additional use of fossil fuels, more strongly associated with NO_x emissions, boosting photochemical O_3 formation and the production of BC and sulfate, comparable to Europe and North America during the 1970s [39]. However, considering the population size, the situation in Asia may become more serious. In southern Asia, the pollution buildup will be strongest in the winter monsoon under large-scale subsidence and cloud-free conditions. Unless international control measures are taken, air pollution in the Northern Hemisphere will continue to grow into a global plume across the developed and the developing world.

References

1. Houghton, J.T.; et al. (Ed.), 1996: *Climate Change 1995: The Science of Climate Change* (Cambridge: Cambridge University Press).
2. Marland, G.; et al., 1999: *Global, Regional and National CO_2 Emissions* (Oak Ridge, TN: Carbon Dioxide Information Analysis Center, Oak Ridge National Laboratory); at: http://cdiac.esd.ornl.gov.
3. Elliot, S.; et al., 1997: in: *Geophysical Research Letters,* 24: 2671.
4. Arndt, R.L.; Carmichael, G.R.; Streets, D.G.; Bhatti, N., 1997: in: *Atmospheric Environment,* 31: 1553.
5. Chameides, W.L.; et al., 1999: in: *Geophysical Research Letters,* 26: 867.
6. Hall, D.O.; Rosillo-Calle, F.; Woods, J., 1994: in: *Chemosphere,* 29: 711.
7. Sathaye, J.; Tyler, S.; Goldman, N., 1994: in: *Energy,* 19: 573.
8. Ramanathan, V.; et al., 1996: *Indian Ocean Experiment (INDOEX)* (C[4] publication #162, University of Cal-ifornia: San Diego, CA).
9. Krishnamurti, T.N.; Jha, B.; Rasch, P.J.; Ramanathan, V., 1997: in: *Meteorology and Atmospheric Physics,* 64: 123 (1997). For a review of the monsoon meteorology and the role of El Niño, see Webster, P.J.; et al., 1996: in: *Journal of Geophysical Research,* 103: 14451.
10. Computer simulations with an interactive chemistry GCM have been performed at T63 resolution (1.8° latitude and longitude). The chemistry GCM has been described by G.J. Roelofs and J. Lelieveld [*Tellus* 49B, 38 (1997)] and E. Roeckner et al. [*Journal of Climate,* 12: 3004 (1997)]. To represent actual meteorology, we assim-ilated analyses from the European Centre for Medium-Range Weather Forecasts (ECMWF), as described by A.T.J. de Laat et al. (35). The T63 resolution model version has been evaluated by A.S. Kentarchos, G.J. Roelofs, and J. Lelieveld [*Journal of the Atmospheric Sciences,* 57: 2824 (2000)]. The chemistry scheme includes a representation of higher hydrocarbons as described by G.J. Roelofs and J. Lelieveld [*Journal of Geophysical Research,* 105: 22697 (2000)].
11. Aerosol sampling, as used at KCO, the R/V *Brown,* and the C-130 aircraft, and associated chemical anal-ysis techniques are described by C. Leck and C. Persson [*Tellus,* 48B: 272 (1996)]; T. Novakov, D.A. Hegg, and P.V. Hobbs [*Journal of Geophysical Research,* 102: 30023 (1997)]; L.S. Hughes et al. [*Environmental Science and Technology,* 32: 1153 (1998)]; M.O. Andreae et al. [*Tellus,* 52B: 1066 (2000)]; and H. Maring et al. [*Journal of Geophysical Research* 105: 14677 (2000)]. Single-particle analysis was per-formed by aerosol time-of-flight spectrometry, as described by P.J. Silva and K.A. Prather [*Environmental Science and Technology,* 31: 3074 (1997)] and E.E. Gard et al. [*Science,* 279: 1184 (1998)]. Aerosol measurements on the Citation aircraft were performed according to F. Schröder and J. Ström [*Atmospheric Research,* 44: 333 (1997)]. On KCO and the C-130 aircraft, aerosol optical properties were measured according to T.L. Anderson et al. [*Journal of Geophysical Research,* 104: 26793 (1999)] and P.J. Sheridan and J.A. Ogren [*Journal of Geophysical Research,* 104:16793 (1999)]. At KCO, four cascade impactors were operated simultaneously, each with six stages (size ranges). The number of cascade impactor substrates analyzed, resulting in Figure 10.2, was 192.
12. According to a review by J. Heintzenberg [*Tellus,* 41B: 149 (1989)], in North America and Europe, urban fine aerosols typically contain 28 % sulfate, 31 % organics, 9 % BC, 8 % ammonium, 6 % nitrate, and 18 % other material (mean mass = 32 $\mu g/m^3$); suburban aerosols contain 37 % sulfate, 24 % organics, 5 % BC, 11 % ammonium, 4 % nitrate, and 19 % other material (mean mass = 15 $\mu g/m^3$); and remote continental aerosols contain 22 % sulfate, 11 % organics, 3 % BC, 7 % ammonium, 3 % nitrate, and 56 % other material (mean mass = 4.8 $\mu g/m^3$). Additional data on different aero-sol types, consistent with this review, are presented by J.H. Seinfeld and S.N. Pandis [*Atmospheric Chemistry and Physics* (Wiley: New York, 1998)].
13. Ackerman, A.S.; et al. 2000: in: *Science* 288: 1042.

14. Satheesh, S.K.; Ramanathan, V., in press: in: *Nature* 405: 60; V. Ramanathan, unpublished data.

15. Cooke, W.F.; et al., 1999: in: *Journal of Geophysical Research,* 104: 22137. The BC/total carbon and organic carbon/BC ratios observed during INDOEX much resemble those in urban Japan and Korea [T. Novakov et al., *Journal of Geophysical Research,* 27: 4061 (2000)]. These ratios were nearly constant over the Indian Ocean, whereas total carbon and BC were highly correlated ($r^2 = 0.81$), providing strong indications of the primary nature of the INDOEX aerosol carbon.

16. J.G.J. Olivier et al., *Description of EDGAR Version 2* (RIVM Report 771060002, National Institute for Public Health and Environment, Bilthoven, Netherlands, 1996) at: http://www.rivm.nl. An update is presented by J.A. van Aardenne et al.*(Global Biogeochem. Cycles,* in press. The EDGAR estimates of Asian CO emissions from biofuel combustion are about 20 % lower compared with the inventory by D.G. Streets and S.T. Waldhoff [*Energy,* 24: 841 (1999)].

17. Crutzen, P.J.; Andreae, M.O., 1990: in: *Science,* 250: 1669.

18. Srivastava, L., 1997: in: *Energy Policy,* 25: 941.

19. Streets, D.G.; Waldhoff, S.T. 1998: in: *Energy,* 23: 1029.

20. Ravindranath, N.H.; Ramakrishna, J., 1997: in: *Energy Policy,* 25:63.

21. Sinha, C.S.; Sinha, S.; Joshi, V., 1998: in: *Biomass Bioenergy,* 14: 489.

22. Mahapartra, A.K.; Mitchell, C.P., 1999: in: *Biomass Bioenergy,* 17: 291.

23. The $\Delta CH3CN/\Delta CO$ molar ratio refers to the enhancement of these gases compared with background air. The lowest concentrations of CH_3CN (140 pptv) and CO (50 ppbv), observed south of the ITCZ, have been used as background values (pptv is parts per trillion by volume or pmol/mol; ppbv is parts per billion by volume or nmol/mol). CH3CN/CO from biomass burning varies according to the fuel type and the burning temperature. The measurements by J.M. Lobert et al. [in *Global Biomass Burning: Atmospheric, Climatic and Biospheric Implications,* J.S. Levine, Ed. (MIT Press, Cambridge, MA, 1991), pp. 289–304] and by R. Holzinger et al. [*Geophysical* Research *Letters,* 26:1161 (1999)] indicate a mean range of 0.13–0.25 %.

24. The instrumentation on board the Citation aircraft has been described by J. Lelieveld et al. [*Geophysical* Research *Letters,* 104: 8201 (1999)]. On the Citation, C-130 air-craft, and the R/V *Brown,* CH_3CN and $CH_3C(O)CH_3$ have been measured by proton-transfer-reaction mass spectrometry, as described by W. Lindinger, A. Hansel, and A. Jordan [*International Journal of Mass Spectrometry* and *Ion* Processes, 173: 191 (1998)]. On the C-130 aircraft, these gases and SO_2 were also measured by chemical ionization mass spectrometry. On the Citation aircraft, CO was measured by tunable diode laser spectrometry (TDLAS), as described by F.G. Wienhold et al. [Journal of *Applied Physics,* B67: 411 (1998)]; on the R/V *Brown,* CO was measured by TDLAS, as described by H. Fischer et al. [*Journal of Geophysical Research,* 102: 23559 (1997)].

25. The source analysis of measured pollution has been supported by back-trajectory calculations. The trajec-tory model uses high-resolution ECMWF-analyzed meteorological data, as described by M.P. Scheele, P.C. Siegmund, and P.F.J. van Velthoven [*Meteorological Applications,* 3: 267 (1996)].

26. The measurements refer to nine 600-liter canister samples, three collected south and six north of the ITCZ, analyzed according to the method of C.A.M. Brenninkmeijer [C.A.M. Brenninkmeijer, *Journal of Geophysical Research*, 98: 10595 (1993); C.A.M. Brenninkmeijer et al., *Chemosphere Global Change Science,* 1: 33 (1999)]. ^{14}CO can result from biomass burning and the oxidation of natural hydrocarbons. Forest emissions of hydrocar-bons, however, are small, whereas hydrocarbon oxidation is minimal during winter.

27. We observed a mean molar enhancement of CH_3Cl relative to CO of 1.98×10^{-3} in the marine BL, a strong indication of biomass burning emissions; see W.C. Keene et al., *Journal of Geophysical Research,* 104: 8429 (1999); J. M. Lobert et al., *Journal of Geophysical Research,* 104: 8373 (1999). We found high correlations ($r^2 = 0.9$) of CH_3Cl with CO, CH_3CN, C_2H_2, and C_6H_6, all products from in-complete (biomass) combustion.

28. On the R/V *Brown* and on the C-130 aircraft, CO was measured according to R.R. Dickerson and A.C. Delany [*Journal* of Atmospheric and *Oceanic Technology,* 5, 424 (1988)]; on KCO, CO was measured according to M. Cogan and J.M. Lobert [*Proceedings of the 43rd Annual ISA Analysis Division Symposium*, vol. 31 (Instrument So-ciety of America, Research Triangle Park, NC, 1998), pp. 229–234. See also (24).

29. Ozone measurements were performed from balloons with Electrochemical Concentration Cell sondes cou-pled to Väisälä radiosondes, on KCO as described by W.D. Komhyr et al. [*Journal of Geophysical Research*, 100: 9231 (1995)], and on the R/V *Brown* as described by D. Kley et al. [*Quarterly Journal of the Royal Meteorological Society,* 123: 2009 (1997)].

30. Venkata Ramana, M.; et al., 1999: in: *Current Science*, 76: 931.

31. Madan, O.P.; et al., 1999: *Meteorological Analysis During INDOEX Intensive Field Phase—1999* (New Delhi, India: Centre for Atmospheric Sciences).

32. NO, as measured from the Citation aircraft, was generally below the detection limit of 40 pptv. On the R/V *Brown*, NO was generally below 10 pptv (T.P. Carsey, R.R. Dickerson, M.L. Farmer, unpublished data).

33. Rhoads, K.P.; et al., 1997: in: *Journal of Geophysical Research*, 102: 18981.

34. Lal, S.; Naja, M.; Jayaraman, A., 1998: in: *Journal of Geophysical Research*, 103: 18907.

35. de Laat, A.T.J.; et al., 1999: in: Journal of Geophysical Research, 104: 13881.

36. Naja, M.; et al.,1999: in: *Current Science,* 76: 931.

37. Wang, Y.; Jacob, D.J., 1998: in: *Journal of Geophysical Research*, 103: 31123.

38. Crutzen, P.J.; Lawrence, M.G.; Pöschl, U., 1999: in: *Tellus,* 51AB: 123. Primary OH production occurs through O_3 photodissociation by solar shortwave radiation in the presence of water vapor [the reaction $O(^1D) + H_2O \rightarrow 2OH$]. After the initial loss of OH through its reaction with a pollutant gas, e.g., CO, a peroxy radical (e.g., HO_2) is formed, which can react with NO, yielding NO_2 and regenerating OH. NO_2 subsequently photodissociates and forms O_3.

39. The Asian emissions of CO, nonmethane hydrocarbons (NMHC), NO_x, and SO_2 will strongly depend on the fuel mix used (coal, oil, and biofuels) and the efficiency of industrial and traffic emissions. Two-stroke engines, for example, which are widely used in India, burn at relatively low temperatures so that NO_x emissions are limited and CO and NMHC emissions are large. The Intergovernmental Panel on Climate Change [*Emission Scenarios* (Cambridge University Press, Cambridge, 2000)] estimates that CO_2, CO, NO_X, SO_2, and NMHC emissions in Organization for Economic Cooperation and Development countries will change from 2000 to 2020 by 2–24 %, −14 to 27 %, −13 to 30 %, −60 to −49 %, and −8 to 5 %, respectively (hence partly reductions), whereas in Asia, these emissions will grow by 41 to 104 %, 7 to 34 %, 50 to 81 %, 15 to 114 %, and 9 to 89 %, respectively.

Chapter 10
Geology of Mankind

Paul J. Crutzen

> The Anthropocene could be said to have started in the late eighteenth century, when analyses of air trapped in polar ice showed the beginning of growing global concentrations of carbon dioxide and methane.

For the past three centuries, the effects of humans on the global environment have escalated. Because of these anthro-pogenic emissions of carbon dioxide, global climate may depart significantly from natural behaviour for many millennia to come. It seems appropriate to assign the term 'Anthropocene' to the present, in many ways human-dominated, geological epoch, supplementing the Holocene—the warm period of the past 10–12 millennia. The Anthropocene could be said to have started in the latter part of the eighteenth century, when analyses of air trapped in polar ice showed the beginning of growing global concentrations of carbon dioxide and methane. This date also happens to coincide with James Watt's design of the steam engine in 1784.

Mankind's growing influence on the environment was recognized as long ago as 1873, when the Italian geologist Antonio Stoppani spoke about a "new telluric force which in power and universality may be compared to the greater forces of earth," referring to the "anthropozoic era". And in 1926, V.I. Vernadsky acknowledged the increasing impact of mankind: "The direction in which the processes of evolution must proceed, namely towards increasing consciousness and thought, and forms having greater and greater influence on their surroundings." Teilhard de Chardin and Vernadsky used the term 'noösphere'—the 'world of thought'—to mark the growing role of human brain-power in shaping its own future and environment.

The rapid expansion of mankind in numbers and per capita exploitation of Earth's resources has continued apace. During the past three centuries, the human population has increased tenfold to more than 6 billion and is expected to reach 10 billion in this century. The methane-produc-ing cattle population has risen to 1.4 billion. About 30–50 % of the planet's land surface is exploited by humans. Tropical rainforests disappear at a fast pace, releasing carbon dioxide and strongly

The text was first published as: Paul J. Crutzen: Geology of mankind—The Anthropocene. *Nature*, **415**, 2002: 23. The permission to republish this artiucle was granted on 19 August 2015 by Ms. Claire Smith, Nature Publishing Group & Palgrave Macmillan, London, UK.

© The Author(s) 2016
P.J. Crutzen and H.G. Brauch (eds.), *Paul J. Crutzen: A Pioneer on Atmospheric Chemistry and Climate Change in the Anthropocene*, Nobel Laureates 50, DOI 10.1007/978-3-319-27460-7_10

increasing species extinction. Dam building and river diver-sion have become commonplace. More than half of all accessible fresh water is used by mankind. Fisheries remove more than 25 % of the primary production in upwelling ocean regions and 35 % in the temperate continental shelf. Energy use has grown 16-fold during the twentieth century, causing 160 million tonnes of atmospheric sulphur dioxide emissions per year, more than twice the sum of its natural emissions. More nitrogen fertilizer is applied in agriculture than is fixed naturally in all terrestrial ecosystems; nitric oxide prod-uction by the burning of fossil fuel and biomass also overrides natural emissions. Fossil-fuel burning and agriculture have caused sub-stantial increases in the concen-trations of 'greenhouse' gases—carbon dioxide by 30 % and methane by more than 100 %—reaching their highest levels over the past 400 millennia, with more to follow.

So far, these effects have largely been caused by only 25 % of the world population. The consequences are, among others, acid precipitation, photochemical 'smog' and climate warming. Hence, according to the latest estimates by the Intergovernmen-tal Panel on Climate Change (IPCC), the Earth will warm by 1.4–5.8 °C during this century.

Many toxic substances are released into the environment, even some that are not toxic at all but nevertheless have severely damaging effects, for example the chloro-fluorocarbons that caused the Antarctic 'ozone hole' (and which are now regulated). Things could have become much worse: the ozone-destroying properties of the halo-gens have been studied since the mid-1970s. If it had turned out that chlorine behaved chemically like bromine, the ozone hole would by then have been a global, year-round phenomenon, not just an event of the Antarctic spring. More by luck than by wisdom, this catastrophic situation did not develop.

Unless there is a global catastrophe—a meteorite impact, a world war or a pan-demic—mankind will remain a major environmental force for many millennia. A daunting task lies ahead for scientists and engineers to guide society towards environ-mentally sustainable management during the era of the Anthropocene. This will require appropriate human behaviour at all scales, and may well involve internationally accepted, large-scale geo-engineering pro-jects, for instance to 'optimize' climate. At this stage, however, we are still largely treading on *terra incognita*.

Box 10.1: Texts by Paul J. Crutzen and Colleagues on the Anthropocene: Setting the Scientific and Policy Agenda and Initiating a Global Debate on a New Era of Earth and Human History

1. Crutzen, P.J. and E.F. Stoermer, 2000: The "Anthropocene". IGBP Newsletter, 41, 17–18.
2. Crutzen, P.J., 2002: Geology of mankind—The Anthropocene. Nature, 415, 23.

3. Crutzen, P.J., 2002: The effects of industrial and agricultural practices on atmospheric chemistry and climate during the anthropocene. J. Environ. Sci. Health, 37, 423–424.
4. Crutzen, P.J., 2002: Atmospheric Chemistry in the "Anthropocene". In: Challenges of a Changing Earth. Proceedings of the Global Change Open Science Conference, Amsterdam, The Netherlands, 10–13 July 2001. W. Steffen, J. Jäger, D.J. Carson and C. Bradshaw (Eds.), Springer Verlag, 2002, 45–48.
5. Crutzen, P.J., 2002: The "anthropocene". In: ERCA, Vol. 5, From the Impacts of Human Activities on our Climate and Environment to the Mysteries of Titan. C Boutron (Ed.), EDP Sciences, 2002, 1–5
6. Crutzen, P.J., 2006: The "Anthropocene". In: Earth System Science in the Anthropocene—Emerging Issues and Problems. E. Ehlers, T. Krafft (Eds.), Springer, Heidelberg, pp. 13–18.
7. Crutzen, P.J.: Atmospheric chemistry and climate in the Anthropocene. In: Making Peace with the Earth—What Future for the Human Species and the Planet? J. Bindé (Ed.), Berghahn Books, UNESCO Publishing, pp. 113–120. 2007.
8. Crutzen, P.J., 2010: Anthropocene man. Nature, 467, S10.
9. Ajai, L. Bengtsson, D. Breashears, P.J. Crutzen, S. Fuzzi, W. Haeberli, W.W. Immerzeel, G. Kaser, C. Kennel, A. Kulkarni, R. Pachauri, T.H. Painter, J. Rabassa, V. Ramanathan, A. Robock, C. Rubbia, L. Russell, M. Sánchez Sorondo, H.J. Schellnhuber, S. Sorooshian, T.F. Stocker, L. G. Thompson, O.B. Toon, D. Zaelke and J. Mittelstraß, 2011: Fate of Mountain Glaciers in the Anthropocene. A Report by the Working Group Commissioned by the Pontifical Academy of Sciences.
10. Crutzen, P.J. and C. Schwägerl, 2011: Living in the Anthropocene: Toward a New Global Ethos. Yale Environment 360.
11. Crutzen, P.J., 2012: Climate, Atmospheric Chemistry and Biogenic Processes in the Anthropocene. In: 100 Jahre Kaiser-Wilhelm-/ Max-Planck-Institut für Chemie (Otto-Hahn-Institut). Horst Kant und Carsten Reinhardt (Herausgeber), Archiv der Max-Planck-Gesellschaft, Band 22.

Selected Peer-reviewed Texts by Paul J. Crutzen and Colleagues on the Anthropocene: Providing the Scientific Evidence

12. Crutzen, P.J. and W. Steffen, 2003: How long have you been in the Anthropocene Era? An Editorial Comment, Climatic Change, 61, 251–257
13. Crutzen, P.J. and V. Ramanathan, 2004: Atmospheric Chemistry and Climate in the Anthropocene. Where are we Heading? In: Earth System Analysis for Sustainablility. Dahlem Workshop Report. H.J. Schellnhuber, P.J. Crutzen, W.C. Clark, M. Claussen and H. Held (Eds.), pp. 265–292, MIT Press, Cambridge, USA

14. W. Steffen, W., M.O. Andreae, P.M. Cox, P.J. Crutzen, U. Cubasch, H. Held, N. Nakicenovic, L. Talaue-McManus and B.L. Turner II, 2004: Group Report: Earth System Dynamics in the Anthropocene. In: Earth System Analysis for Sustainablility. Dahlem Workshop Report. H.J. Schellnhuber, P.J. Crutzen, W.C. Clark, M. Claussen and H. Held (Eds.), pp. 313–340, MIT Press, Cambridge, USA.Steffen, W., P.J. Crutzen and J.R. McNeill, 2007: The Anthropocene: Are Humans Now Overwhelming the Great Forces of Nature? Ambio, 36, 614–621

15. Zalasiewicz, J., M. Williams, W. Steffen and P. Crutzen, 2010: The New World of the Anthropocene. Environ. Sci. Technol., 44, 2228–2231.

16. Zalasiewicz J., M. Williams, W. Steffen and P. Crutzen, 2010: Response to "The Anthropocene forces us to reconsider adaptationist models of human-environment interactions". Environ. Sci. Technol., 44 (16), 6008, doi:10.1021/es102062w.

17. Zalasiewicz, Jan, Colin N. Waters, Mark Williams, Anthony D. Barnosky, Alejandro Cearreta, Paul Crutzen, Erle Ellis, Michael A. Ellis, Ian J. Fairchild, Jacques Grinevald, Peter K. Haff, Irka Hajdas, Reinhold Leinfelder, John McNeill, Eric O. Odada, Clement Poirier, Daniel Richter, Will Steffen, Colin Summerhayes, James P.M. Syvitski, Davor Vidas, Michael Wagreich, Scott L. Wing, Alexander P. Wolfe, An Zhisheng, Naomi Oreskes, 2015: When did the Anthropocene begin? A mid-twentieth century boundary level is stratigraphically optimal. Quaternary International, xxx.

Texts from the Anthropocene Symposium: Honouring Paul J. Crutzen on his 80th Birthday (2013)[1]

1. Hartmut Graßl: Shaping Germany's Role in Ozone and Climate Policy. The Push by Paul Crutzen

2. Ralph J. Cicerone: Stratospheric Ozone and Climate Change: different Human Causes and Responses

3. Mario J. Molina: Climate Change Science and Policy

4. Susan Solomon: Ozone Depletion: An Enduring Challenge

THE ANTHROPOCENE SYMPOSIUM

ZUSAMMENFASSUNG DER WISSENSCHAFTLICHEN BEITRÄGE

Konferenz zu Ehren von Atmosphärenchemiker und Nobelpreisträger Paul J. Crutzen am 2. Dezember 2013 in Mainz.

---

[1]These presentations can be listened to on the website of the Max-Planck Institute on Chemistry (MPIC) at: http://www.mpic.de/aktuelles/pressemeldungen/news/konferenz-zu-ehren-von-atmosphaerenchemiker-paul-crutzen.html. The English programme is at: http://www.mpic.de/fileadmin/user_upload/images_presse/Images_PIs/Crutzen_Symposium/Program_Anthropocene_Symposium_SB11_small.pdf. The presentations can also be approached on YouTube at: https://www.youtube.com/watch?v=g0HuKpbMREU.

5. Veerabhadran Ramanathan: The Two worlds in the Anthropocene: A new Approach for Climate Change Mitigation
6. Henning Rodhe: The Anthropocene Sulfur Cycle
7. Jack Fishman: Tropospheric Ozone in the Anthropocene: Are We Creating a Toxic Atmosphere?
8. John P. Burrows: Living in and Observing the Anthropocene from Space
9. Klaus Töpfer: The Anthropocene—Sustainability in a World of 9 Billion People
10. Meinrat O. Andreae: 400,000.036 years of Biomass Burning

References

Berger, A.; Loutre, M.-F., 1996: in: *Comptes Rendus de l'Académie des Sciences,* 323,IIA: 1–16.

Clark, W.C.; Munn, R.E. (Eds.), 1986: *Sustainable Development of the Biosphere Chap. 1* (Cambridge: Cambridge University Press).

Crutzen, P.J.; Stoermer, E.F., 2000: *IGBP Newsletter* 41 (Stockholm: Royal Swedish Academy of Sciences).

Houghton, J.T.; et al. (Eds.), 2001: *Climate Change 2001: The Scientific Basis* (Cambridge: Cambridge University Press).

Marsh, G.P., 1965: *Man and Nature* (1864) (Reprinted as the earth as modified by human action) (Cambridge, Massachusetts: Belknap Press).

McNeill, J.R., 2000: *Something New Under the Sun: An Environmental History of the Twentieth-Century World* (New York: W. W. Norton).

Schellnhuber, H.J., 1999: in: *Nature,* 402: C19–C23.

Turner, B.L.; et al. 1990 *The Earth as Transformed by Human Action* (Cambridge: Cambridge University Press).

Vernadski, V.I., 1998: *The Biosphere* (Translated and Annotated Version from the Original of 1926) (New York: Springer).

Chapter 11
Albedo Enhancement by Stratospheric Sulfur Injections: A Contribution to Resolve a Policy Dilemma? An Editorial Essay

Paul J. Crutzen

Fossil fuel burning releases about 25 Pg of CO_2 per year into the atmosphere, which leads to global warming (Prentice et al. 2001). However, it also emits 55 Tg S as SO_2 per year (Stern 2005), about half of which is converted to sub-micrometer size sulfate particles, the remainder being dry deposited. Recent research has shown that the warming of earth by the increasing concentrations of CO_2 and other greenhouse gases is partially countered by some backscattering to space of solar radiation by the sulfate particles, which act as cloud condensation nuclei and thereby influence the micro-physical and optical properties of clouds, affecting regional precipitation patterns, and increasing cloud albedo (e.g., Rosenfeld 2000; Ramanathan et al. 2001; Ramaswamy et al. 2001). Anthropogenically enhanced sulfate particle concentrations thus cool the planet, offsetting an uncertain fraction of the anthropogenic increase in greenhouse gas warming. However, this fortunate coincidence is 'bought' at a substantial price. According to the World Health Organization, the pollution particles affect health and lead to more than 500,000 premature deaths per year worldwide (Nel 2005). Through acid precipitation and deposition, SO_2 and sulfates also cause various kinds of ecological damage. This creates a dilemma for environmental policy makers, because the required emission reductions of SO_2, and also anthropogenic organics (except black carbon), as dictated by health and ecological considerations, add to global warming and associated negative consequences, such as sea level rise, caused by the greenhouse gases. In fact, after earlier rises, global SO_2 emissions and thus sulfate loading have been declining at the rate of 2.7 % per year, potentially explaining the observed reverse from dimming to brightening in surface solar radiation at many stations worldwide (Wild et al. 2005).

This text was first published as: Crutzen, Paul J. "Albedo enhancement by stratospheric sulfur injections: a contribution to resolve a policy dilemma?". *Climatic Change* (Springer) 77 (3–4): 211–219. The permission to include this text here was granted on 28 August 2015 by Rights and Permissions, Springer Science + Business Media in Heidelberg. Thanks go to many colleagues, in particular Ron Nielsen for advice on cost estimates, and to him, and colleagues V. Ramanathan, Jos Lelieveld, Carl Brenninkmeijer, Mark Lawrence, Yoya Joseph, and Henning Rodhe for advice and criticism on this paper. Part of this study was conducted during a stay at the International Institute of Advanced Systems Analysis (IIASA) in Laxenburg, Austria and discussed with Bob Ayres and Arnulf Grübler.

P.J. Crutzen and H.G. Brauch (eds.), *Paul J. Crutzen: A Pioneer on Atmospheric Chemistry and Climate Change in the Anthropocene*, Nobel Laureates 50, DOI 10.1007/978-3-319-27460-7_11

The corresponding increase in solar radiation by 0.10 % per year from 1983 to 2001 (Pinker et al. 2005) contributed to the observed climate warming during the past decade. According to model calculations by Brasseur/Roeckner (2005), complete improvement in air quality could lead to a decadal global average surface air temperature increase by 0.8 K on most continents and 4 K in the Arctic. Further studies by Andreae et al. (2005) and Stainforth et al. (2005) indicate that global average climate warming during this century may even surpass the highest values in the projected IPCC global warming range of 1.4–5.8 °C (Cubasch et al. 2001).

By far the preferred way to resolve the policy makers' dilemma is to lower the emissions of the greenhouse gases. However, so far, attempts in that direction have been grossly unsuccessful. While stabilization of CO_2 would require a 60–80 % reduction in current anthropogenic CO_2 emissions, worldwide they actually increased by 2 % from 2001 to 2002 (Marland et al. 2005), a trend, which probably will not change at least for the remaining 6-year term of the Kyoto protocol, further increasing the required emission restrictions.

Therefore, although by far not the best solution, the usefulness of artificially enhancing earth's albedo and thereby cooling climate by adding sunlight reflecting aerosol in the stratosphere (Budyko 1977; NAS 1992) might again be explored and debated as a way to defuse the Catch-22 situation just presented and additionally counteract the climate forcing of growing CO_2 emissions. This can be achieved by burning S_2 or H_2S, carried into the stratosphere on balloons and by artillery guns to produce SO_2. To enhance the residence time of the material in the stratosphere and minimize the required mass, the reactants might be released, distributed over time, near the tropical upward branch of the stratospheric circulation system. In the stratosphere, chemical and micro-physical processes convert SO_2 into sub-micrometer sulfate particles. This has been observed in volcanic eruptions e.g., Mount Pinatubo in June, 1991, which injected some 10 Tg S, initially as SO_2, into the tropical stratosphere (Wilson et al. 1993; Bluth et al. 1992). In this case enhanced reflection of solar radiation to space by the particles cooled the earth's surface on average by 0.5 °C in the year following the eruption (Lacis/Mishchenko 1995). Although climate cooling by sulfate aerosols also occurs in the troposphere (e.g., Ramaswamy et al. 2001), the great advantage of placing reflective particles in the stratosphere is their long residence time of about 1–2 years, compared to a week in the troposphere. Thus, much less sulfur, only a few percent, would be required in the stratosphere to achieve similar cooling as the tropospheric sulfate aerosol (e.g., Dickinson 1996; Schneider 1996; NAS 1992; Stern 2005). This would make it possible to reduce air pollution near the ground, improve ecological conditions and reduce the concomitant climate warming. The main issue with the albedo modification method is whether it is environmentally safe, without significant side effects.

We will next derive some useful metrics. First, a loading of 1 Tg S in the stratosphere yields a global average vertical optical depth of about 0.007 in the visible and corresponds to a global average sulfur mixing ratio of ~ 1 nmol/mole, about six times more than the natural background (Albritton et al. 2001). Second, to derive the radiative forcing caused by the presence of 1 Tg S in the stratosphere, we adopt a simple approach based on the experience gained from the Mount Pinatubo volcanic eruption. For the Mount Pinatubo eruption, Hansen et al. (1992) calculated a radiative cooling of 4.5 W/m^2 caused by 6 Tg S, the amount of S that remained in the stratosphere as sulfate six months after the eruption from initially 10 Tg S (Bluth et al. 1992). Linear downscaling results in a sulfate climate cooling efficiency of 0.75 W/m^2 per Tg S in the stratosphere. The estimated annual cost to put 1 Tg S in the stratosphere, based on information by the NAS (1992), at that time would have been US $25 billion (NAS 1992; Ron Nielsen, personal communication). Thus, in order to compensate for enhanced climate warming by the removal of anthropogenic aerosol [an uncertain mean value of 1.4 W/m^2, according to Crutzen/Ramanathan (2003)], a stratospheric sulfate loading of 1.9 Tg S would be required, producing an optical depth of 1.3 %. This can be achieved by a continuous deployment of about 1–2 Tg S per year for a total price of US $25–50 billion, or about $25–50 per capita in the affluent world, for stratospheric residence times of 2–1 year, respectively. The cost should be compared with resulting environmental and societal benefits, such as reduced rates of sea level rise. Also, in comparison, current annual global military expenditures approach US $1000 billion, almost half in the U.S.A. The amount of sulfur that is needed is only 2–4 % of the current input of 55 Tg S/year (Stern 2005). Although the particle sizes of the artificial aerosols are smaller than those of the volcanic aerosol, because of greater continuity of injections in the former, the radiative forcings are rather similar for effective particle radii ranging between 0.1 and 1 μm (see Table 2.4, page 27, Lacis/Mishchenko 1995). However the smaller particles have a longer stratospheric residence time, so that less material needs to be injected to cool climate, compared to the volcanic emission case. It should be mentioned that Anderson et al. (2003a, b) state that the radiative cooling by the aerosol could be much larger than the figure of 1.4 W/m^2, derived by Crutzen/Ramanathan (2003), which is based on the assumption of constant relative humidity in the troposphere. If Anderson et al. (2003a, b) are indeed correct, the result might be a stronger climate heating from air pollution cleanup than derived above (see also Andreae et al. 2005).

To compensate for a doubling of CO_2, which causes a greenhouse warming of 4 W/m^2, the required continuous stratospheric sulfate loading would be a sizeable 5.3 Tg S, producing an optical depth of about 0.04. The Rayleigh scattering optical depth at 0.5 μm is about 0.13, so that some whitening on the sky, but also colorful sunsets and sunrises would occur. It should be noted, however, that considerable whitening of the sky is already occurring as a result of current air pollution in the continental boundary layer.

Locally, the stratospheric albedo modification scheme, even when conducted at remote tropical island sites or from ships, would be a messy operation. An alternative may be to release a S-containing gas at the earth's surface, or better from balloons, in the tropical stratosphere. A gas one might think of is COS, which may be the main source of the stratospheric sulfate layer during low activity volcanic periods (Crutzen 1976), although this is debated (Chin/Davis 1993). However, about 75 % of the COS emitted will be taken up by plants, with unknown long-term ecological consequences, 22 % is removed by reaction with OH, mostly in the troposphere, and only 5 % reaches the stratosphere to produce SO_2 and sulfate particles (Chin/Davis 1993). Consequently, releasing COS at the ground is not recommended. However, it may be possible to manufacture a special gas that is only processed photochemically in the stratosphere to yield sulfate. The compound should be non-toxic, insoluble in water, non-reactive with OH, it should have a relatively short lifetime of less than about 10 years, and should not significantly contribute to greenhouse warming, which for instance disqualifies SF_6.

The albedo modification scheme presented here has been discussed before, however, without linking opposite climate warming and improved air quality considerations. Instead of sulfur, it has also been proposed to launch reflecting small balloons or mirrors, or to add highly reflective nano-particles of other material than sulfur (Teller et al. 1997; Keith 2000). An interesting alternative could be to release soot particles to create minor "nuclear winter" conditions. In this case earth's albedo would actually decrease, but surface temperatures would, nevertheless, decline. Only 1.7 % of the mass of sulfur would be needed to effect similar cooling at the earth's surface, making the operations much cheaper and less messy. However, because soot particles absorb solar radiation very efficiently, differential solar heating of the stratosphere could change its dynamics. It would, however, also counteract stratospheric cooling by increasing CO_2 and may even prevent the formation of polar stratospheric cloud particles, a necessary condition for ozone hole formation.

Since it is likely that the greenhouse warming is substantially negated by the cooling effect of anthropogenic aerosol in the troposphere, by 25–65 % according to an estimate by Crutzen/Ramanathan (2003), but possibly greater (Anderson et al. 2003a, b), air pollution regulations, in combination with continued growing emissions of CO_2, may bring the world closer than is realized to the danger described by Schneider (1996): "Supposing, a currently envisioned low probability but high consequence outcome really started to unfold in the decades ahead (for example, 5 °C warming in this century) which I would characterize as having potential catastrophic implications for ecosystems... Under such a scenario, we would simply have to practice geo-engineering..."

There are some worrying indications of potentially large climate changes: for instance the locally drastic atmospheric warming by up to 3 W/m^2 per decade in Alaska due to surface albedo decreases through tree and shrub expansion (Chapin III et al. 2005), the projected increase in surface temperatures by 2–3 K by the middle of this century in Africa even with the Kyoto protocol in force (B. Hewitson, University of Cape Town, quoted by Cherry 2005) with great impacts on biodiversity, and potentially also the 30 % slowdown in the north Atlantic over-turning circulation during the past half century (Bryden et al. 2005). Given the grossly disappointing international political response to the required greenhouse gas emissions, and further considering some drastic results of recent studies (Andreae et al. 2005; Stainforth et al. 2005), research on the feasibility and environmental consequences of climate engineering of the kind presented in this paper, which might need to be deployed in future, should not be tabooed. Actually, considering the great importance of the lower stratosphere/upper troposphere (LS/UT) for the radiation balance, chemistry, and dynamics of the atmosphere, its research should anyhow be intensified. For instance, it is not well known how much of the large quantities of anthropogenic SO_2 emitted at ground level reaches the LS/UT to produce sulfate particles, what regulates temperatures, water vapour concentrations and cirrus cloud formation in the LS/UT region, and how these factors may change in response to growing CO_2 concentrations, which are already 30–40 % higher than ever experienced during the past 650,000 years (Siegenthaler et al. 2005). Progress in the understanding of the complicated earth climate system is generally slow. Therefore it is recommended to intensify research in order to challenge the climate modification idea here presented, starting with model investigations and, dependent on their outcome, followed step by step by small scale atmospheric tests. Also, as natural sulfur injection experiments occur intermittently in the form of explosive volcanic eruptions, often at low latitudes, they provide excellent opportunities for model development and testing (e.g., Robock 2000).

Researchers at the Lawrence Livermore Laboratory are so far the only ones who have modelled the stratospheric albedo modification scheme. In a first study, Govindasamy/Caldeira (2000) simulated this by reducing the solar luminosity by 1.8 %, to balance future climate warming by a doubling of CO_2. Although solar radiative forcing has a different physics and spatial distribution than the infrared effects caused by CO_2, the model results indicated that the global temperature response by both perturbations at the Earth' surface and atmosphere largely cancelled out. Although these preliminary model results would be in favor a stratospheric sulfur injection operation, the required annual S inputs are large, so that the possibility of adverse environmental side effects needs to be fully researched before the countermeasure to greenhouse warming is attempted. What has to be done first, is to explore whether using a sulfur injection scheme with advanced micro-physical and radiation process descriptions will show similar model results as the simple solar luminosity adjustment scheme of Govindasamy/Caldeira (2000). Further studies, following those conducted by Govindasamy (2003), should address the biological effects of the albedo modification scheme. As already mentioned, injection of soot may be an alternative, but in need of critical analysis. Such studies by themselves, even when the experiment is never done, will be very informative.

Among possible negative side effects, those on stratospheric ozone first spring to mind. Fortunately, in this case one can build on the experience with past volcanic eruptions, such as El Chichón in 1982 and Mount Pinatubo in 1991, which injected 3–5 Tg S (Hofmann/Solomon 1989) and 10 Tg S (Bluth et al. 1992), respectively, in the stratosphere. Local ozone destruction in the El Chichon case was about 16 % at 20 km altitude at mid-latitudes (Hofmann/Solomon 1989). For Mount Pinatubo, global column ozone loss was about 2.5 % (Kinnison et al. 1994). For the climate engineering experiment, in which the cooling effect of all tropospheric anthropogenic aerosol is removed, yielding a radiative heating of 1.4 W/m^2 (Crutzen/Ramanathan 2003), a stratospheric loading of almost 2 Tg S, and an input of 1–2 Tg S/year is required, depending on stratospheric residence times. In this case, stratospheric sulfate injections would be 5 times less than after the Mount Pinatubo eruption, leading to much smaller production of ozone-destroying Cl and ClO radicals, whose formation depends on particle surface-catalyzed heterogeneous reactions (Wilson et al. 1993). Compensating for a CO_2 doubling would lead to larger ozone loss but not as large as after Mount Pinatubo. Furthermore, the amounts of stratospheric chlorine radicals, coming from past production of the chloro-fluoro-carbon gases, are now declining by international regulation, so that ozone will significantly recover by the middle of this century. If instead of SO_2, elemental carbon would be injected in the stratosphere, higher temperatures might prevent the formation of polar stratospheric ice particles and thereby hinder the formation of ozone holes. This and the consequences of soot deposition on polar glaciers should be checked by model calculations.

In contrast to the slowly developing effects of greenhouse warming associated with anthropogenic CO_2 emissions, the climatic response of the albedo enhancement experiment would start taking effect within about half a year, as demonstrated by the Mount Pinatubo eruption (Hansen et al. 1992). Thus, provided the technology to carry out the stratospheric injection experiment is in place, as an escape route against strongly increasing temperatures, the albedo adjustment scheme can become effective at rather short notice, for instance if climate heats up by more than 2 °C globally or when the rates of temperatures increase by more than 0.2 °C/decade), i.e. outside the so-called "tolerable window" for climate warming (e.g., Bruckner/Schellnhuber 1999). Taking into account the warming of climate by up to 1 °C by air pollution reduction (Brasseur/Roeckner 2005), the tolerable window for greenhouse gas emissions might be as low as 1 °C, not even counting positive biological feedbacks. As mentioned before, regionally more rapid climate changes are already happening in the Arctic (Chapin et al. 2005) or are in petto for Africa (Cherry 2005). Already major species extinctions by current climate warming have been reported by Pounds et al. (2005) and Root et al. (2003). If sizeable reductions in greenhouse gas emissions will not happen and temperatures rise rapidly, then climatic engineering, such as presented here, is the only option available to rapidly reduce temperature rises and counteract other climatic effects. Such a modification could also be stopped on short notice, if undesirable and unforeseen side effects become apparent, which would allow the atmosphere to return to its prior state within a few years. There is, therefore, a strong need to estimate negative, as well as positive, side effects of the proposed stratospheric modification schemes. If positive effects are greater than the negative effects, serious consideration should be given to the albedo modification scheme.

Nevertheless, again I must stress here that the albedo enhancement scheme should only be deployed when there are proven net advantages and in particular when rapid climate warming is developing, paradoxically, in part due to improvements in worldwide air quality. Importantly, its possibility should not be used to justify inadequate climate policies, but merely to create a possibility to combat potentially drastic climate heating (e.g. Andreae et al. 2005; Stainforth et al. 2005; Crutzen/Ramanathan 2003; Anderson et al. 2003a, b). The chances of unexpected climate effects should not be underrated, as clearly shown by the sudden and unpredicted development of the antarctic ozone hole. Current CO_2 concentrations are already 30–40 % larger than at any time during the past 650,000 years (Siegenthaler et al. 2005). Climate heating is known to be particularly strong in arctic regions (Chapin et al. 2005), which may trigger accelerated CO_2 and CH_4 emissions in a positive feedback mode. Earth system is increasingly in the non-analogue condition of the Anthropocene.

Reductions in CO_2 and other greenhouse gas emissions are clearly the main priorities (Socolow et al. 2004; Lovins 2005). However, this is a decades-long process and so far there is little reason to be optimistic. There is in fact a serious additional issue. Should the proposed solutions to limit CO_2 emissions prove unsuccessful and should CO_2 concentrations rise to high levels with risk of acidification of the upper ocean waters, leading to dissolution of calcifying organisms (Royal Society 2005; Orr et al. 2005), underground CO_2 sequestration (Lackner 2003), if proven globally significant, will be needed to bring down atmospheric CO_2 concentrations. However, that kind of sequestration does not allow for rapid remedial response. Reforestation could do so, but has its own problems. A combination of efforts may thus be called for, including the stratospheric albedo enhancement scheme.

In conclusion: The first modelling results and the arguments presented in this paper call for active scientific research of the kind of geo-engineering, discussed in this paper. The issue has come to the forefront, because of the dilemma facing international policy makers, who are confronted with the task to clean up air pollution, while simultaneously keeping global climate warming under control. Scientific, legal, ethical, and societal issues, regarding the climate modification scheme are many (Jamieson 1996; Bodansky 1996). Building trust between scientists and the general public would be needed to make such a large-scale climate modification acceptable, even if it would be judged to be advantageous. Finally, I repeat: the very best would be if emissions of the greenhouse gases could be reduced so much that the stratospheric sulfur release experiment would not need to take place. Currently, this looks like a pious wish.

References

Albritton, D.L.; et al., 2001: "Technical Summary, In Climatic Change 2001, The Scientific Basis, Intergovernmental Panel for Climate Change", in Houghton, J.T. et al. (Ed.): (United Kingdom and New York, NY, USA: Cambridge University Press).

Anderson, T.L.; et al., 2003a: "Climate Forcing by Aerosols—A Hazy Picture", in: *Science,* 300: 1103–1104.

Anderson, T.L.; et al., 2003b: "Response to P. J. Crutzen and V. Ramanathan, op cited", in: *Science,* 302: 1680–1681.

Andreae, M.O.; Jones, C.D.; Cox, P.M., 2005: "Strong Present-Day Aerosol Cooling Implies a Hot Future", in: *Nature,* 435: 1187–1190.

Bluth, G.J.S.; Doiron, S.D.; Schnetzler, C.C.; Krueger, A.J.; Walter, L.S., 1992: "Global Tracking of the SO_2 Clouds from the June 1991 Mount Pinatubo Eruptions", in: *Geophysical Research Letters,* 19: 151–154.

Bodansky, D., 1996: "May we Engineer the Climate?", in: *Climatic Change,* 33: 309–321.

Brasseur, G.P.; Roeckner, E., 2005: "Impact of Improved Air Quality on the Future Evolution of Climate", in: *Geophysical Research Letters,* 32: L23704. doi:10.1029/2005GL023902.

Bruckner, T.; Schellnhuber, H.J., 1999: "Climate Change Protection: The Tolerable Windows Approach", *IPTS Report,* 34: 6.

Bryden, H.L.; Longworth, H.R.; Cunningham, S.A., 2005: "Slowing of the Atlantic Meridional Overturning Circulation at 25 N", in: *Nature,* 438: 655–657.

Budyko, M.I., 1977: *Climatic Changes, American Geophysical Society* (Washington, D.C.): 244 pp.

Chapin III, F.S.; et al., 2005: "Role of Land-Surface Changes in Arctic Summer Warming", in: *Science,* 310: 657–660.

Cherry, M., 2005: "Ministers Agree to Act on Warnings of Soaring Temperatures in Africa", in: *Nature,* 437: 1217.

Chin, M.; Davis, D.D., 1993: "Global Sources and Sinks of OCS and CS_2 and their Distributions", in: *Global Biogeochemical Cycles,* 7: 321–337.

Climate Change 2001: "The Scientific Basis, Third Assessment Report of the Intergovernmental Panel on Climate Change". in: Houghton, J.T. et al. (eds.): (Cambridge, U.K. and New York, N.Y., USA: Cambridge University Press).

Climate Change 2001: The Scientific Basis, Third Assessment Report of the Intergovernmental Panel on Climate Change. In: Houghton, J.T. et al. (Eds.), Cambridge University Press, Cambridge, U.K. and New York, N.Y., USA.

Crutzen, P.J., 1976: "The Possible Importance of COS for the Sulfate Layer of the Stratosphere", in: *Geophysical Research Letters,* 3: 73–76.

Crutzen, P.J.; Ramanathan, V., 2003: "The Parasol Effect on Climate", in: *Science,* 302: 1679–1681.

Cubasch, U.; et al., 2001: "Projections of Future Climate Change", Chapter 9, pp. 525–582.

Dickinson, R.E., 1996: "Climate Engineering. A Review of Aerosol Approaches to Changing the Global Energy Balance", in: *Climatic Change,* 33, 279–290.

Govindasamy, B.; Caldeira, K., 2000: "Geoengineering Earth's Radiative Balance to Mitigate CO_2-Induced Climatic Change", in: *Geophysical Research Letters,* 27: 2141–2144.

Govindasamy, B.; et al., 2002: "Impact of Geoengineering Schemes on the Terrestrial Biosphere", in: *Geophysical Research Letters,* 29,22: 2061.

Hansen, J.; Lacis, A.; Ruedy, R.; Sato, M., 1992: "Potential Climate Impact of Mount Pinatubo eruption", in: *Geophysical Research Letters,* 19: 215–218.

Hofmann, D.J.; Solomon, S., 1989: "Ozone Destruction Through Heterogeneous Chemistry Following the Eruption of El Chichón". in: *Journal of Geophysical Research* 94,D4: 5029–5041.

Jamieson, D., 1996: "Ethics and Intentional Climate Change", in: *Climatic Change,* 33: 323–336.

Keith, D.W., 2000: "Geoengineering the Climate: History and Prospect", in: *Annual Review of Energy and the Environment,* 25: 245–284.

Kinnison, D.E., et al., 1994: "The Chemical and Radiative Effects of the Mount Pinatubo Eruption", in: *Journal of Geophysical Research,* 99: 25705–25731.

Lacis, A.A.; Mishchenko, M.I., 1995: "Climate Forcing, Climate Sensitivity, and Climate Response: A Radiative Modelling Perspective on Atmospheric Aerosols", in: Charlson R.J.; Heinztenberg, J. (Eds.): *Aerosol Forcing of Climate* (Chichester: Wiley): 416 pp, 11–42.

Lackner, K.S., 2003: "A Guide to CO_2 Sequestration", in: *Science,* 300: 1677–1678.

Lovins, A.B., 2005: "More Profit with Less Carbon", in: *Scientific American,* 293: 52–61.

Marland, G.; Boden, T.A.; Andres, R.J., 2005: "Global, Regional, and National CO_2 Emissions", in: Trends: A Compendium of Data on Global Change. Carbon Diozide Information Analysis Center, Oak Ridge National Laboratory, US. Department of Energy, Oak Ridge, Tenn.

National Academy of Sciences (NAS): 1992, Policy Implications of Greenhouse Warming: Mitigation, Adaptation, and the Science Base, Panel on Policy Implications of Greenhouse Warming, Committee on Science, Engineering, and Public Policy, National Academy Press, Washington DC, 918 pp.

Nel, A., 2005: "Air Pollution-Related Illness: Effects of Particles", in: *Science,* 308: 804.

Orr, J.C.; et al., 2005: "Anthropogenic Ocean Acidification Over the Twenty-First Century and its Impact on Calcifying Organisms", in: *Science,* 437, 681–686.

Pinker, R.T.; Zhang, B.; Dutton, E.G., 2005: "Do Satellites Detect Trends in Surface Solar Radiation?", in: *Science* 308: 850–854.

Pounds, J.A.; et al., 2005: "Widespread Amphibian Extinctions from Epidemic Disease Driven by Global Warming", in: *Nature* 439: 161–165.

Prentice, I.C.; et al., 2001: "The Carbon Cycle and Atmospheric Carbon Dioxide", Chapter 3, in: Houghton, J.T. et al. (Ed.): *Third Assessment Report of the Intergovernmental Panel on Climate Change* (U.K. and New York, USA: Cambridge University Press): 183–238.

Ramanathan, V.; Crutzen, P.J.; Kiehl, J.T.; Rosenfeld, D., 2001: "Aerosols, Climate and the Hydrological Cycle", in: *Science,* 294: 2119–2124.

Ramaswamy, V.; et al., 2001: "Radiative Forcing of Climate Change', Chap. 6, pp. 349–416.

Robock, A., 2000: "Volcanic Eruptions and Climate", in: *Reviews of Geophysics,* 38: 191–219.

Root, T.L.; et al., 2003: "Fingerprints of Global Warming on Wild Animals and Plants", in: *Nature,* 421: 57–60.

Rosenfeld, D., 2000: "Suppression of Rain and Snow by Urban and Industrial Air Pollution", in: *Science,* 287: 1793–1796.

Royal Society: June 2005, Ocean Acidification Due to Increasing Atmospheric Carbon Dioxide, 57 pp.

Schneider, S.H., 1996: "Geoengineering: Could-or-Should-we do it", in: *Climatic Change,* 33: 291–302.

Siegenthaler, U.; et al., 2005: "Stable Carbon Cycle-Climate Relationship During the Late Pleistocene", in: *Science,* 310: 1313–1317.

Socolow, R.; et al., 2004: "Solving the Climate Problem", in: *Environment,* 46: 8–19.

Stainforth, D.A.; et al., 2005: "Uncertainty in Predictions of the Climate Response to Rising Levels of Greenhouse Gases", in: *Nature,* 433: 403–406.

Stern, D.I., 2005: "Global Sulfur Emissions from 1850 to 2000", in: *Chemosphere,* 58: 163–175.

Teller, E.; Wood, L.; Hyde, R., 1997: "Global Warming and Ice Ages: 1. Prospects for Physics Based Modulation of Global Change", UCRL-JC-128157, Livermore National Laboratory, Livermore, CA.

Wild, M.; et al., 2005: "From Dimming to Brightening: Decadal Changes in Solar Radiation and Earth's Surface", 308: 847–850.

Wilson, J.C.; et al., 1993: "*In-situ* Observations of Aerosol and Chlorine Monoxide after the 1991 Eruption of Mount Pinatubo: Effect of Reactions on Sulfate Aerosol". in: *Science,* 261: 1140–1143.

Chapter 12
N₂O Release from Agro-biofuel Production Negates Global Warming Reduction by Replacing Fossil Fuels

Paul J. Crutzen, A.R. Mosier, K.A. Smith and W. Winiwarter

Abstract The relationship, on a global basis, between the amount of N fixed by chemical, biological or atmospheric processes entering the terrestrial biosphere, and the total emission of nitrous oxide (N_2O), has been re-examined, using known global atmospheric removal rates and concentration growth of N_2O as a proxy for overall emissions. For both the pre-industrial period and in recent times, after taking into account the large-scale changes in synthetic N fertiliser production, we find an overall conversion factor of 3–5 % from newly fixed N to N_2O–N. We assume the same factor to be valid for biofuel production systems. It is covered only in part by the default conversion factor for 'direct' emissions from agricultural crop lands (1 %) estimated by IPCC (2006), and the default factors for the 'indirect' emissions (following volalilization/deposition and leaching/runoff of N: 0.35–0.45 %) cited therein. However, as we show in the paper, when additional emissions included in the IPCC methodology, e.g. those from livestock production, are included, the total may not be inconsistent with that given by our "top-down" method. When the extra N_2O emission from biofuel production is calculated in "CO_2-equivalent" global warming terms, and compared with the quasi-cooling effect of 'saving' emissions of fossil fuel derived CO_2, the outcome is that the production of commonly used biofuels, such as biodiesel from rapeseed and bioethanol from corn (maize), depending on N fertilizer uptake efficiency by the plants, can contribute as much or more to global warming by N_2O emissions than cooling by fossil fuel savings. Crops with less N demand, such as grasses and woody coppice species, have more favourable climate impacts. This analysis only considers the conversion of biomass to biofuel. It does not take into account the use of fossil fuel on the farms and for

The text was originally published as: W. Asman, C. Brenninkmeijer, R. Cicerone, L. Ganzeveld, B. Hermann, M. Lawrence, A. Leip, O. Oenema, G. Pearman, V. Ramanathan, H. Rodhe and E. Smeets for discussions and comments. Edited by R. Cohen. The text was first published as: "N₂O release from agro-biofuel production negates global warming reduction by replacing fossil fuels", in: *Atmospheric Chemistry and Physics*, 8 (2008): 389–395. The copyright rests with the authors under the Creative Commons Attribution 3.0 License who granted permission for its inclusion in this volume.

P.J. Crutzen and H.G. Brauch (eds.), *Paul J. Crutzen: A Pioneer on Atmospheric Chemistry and Climate Change in the Anthropocene*, Nobel Laureates 50, DOI 10.1007/978-3-319-27460-7_12

227

fertilizer and pesticide production, but it also neglects the production of useful co-products. Both factors partially compensate each other. This needs to be analyzed in a full life cycle assessment.

12.1   Introduction

N_2O, a by-product of fixed nitrogen application in agriculture, is a "greenhouse gas" with a 100-year average global warming potential (GWP) 296 times larger than an equal mass of CO_2 (Prather et al. 2001). As a source for NO_x, i.e. NO plus NO_2, N_2O also plays a major role in stratospheric ozone chemistry (Crutzen 1970). The increasing use of biofuels to reduce dependence on imported fossil fuels and to achieve "carbon neutrality" will further cause atmospheric N_2O concentrations to increase, because of N_2O emissions associated with N-fertilization. Here we propose a global average criterion for the ratio of N to dry matter in the plant material, which indicates to what degree the reduced global warming ("saved CO_2") achieved by using biofuels instead of fossil fuel as energy sources is counteracted by release of N_2O. This study shows that those agricultural crops most commonly used at present for biofuel production and climate protection can readily lead to enhanced greenhouse warming by N_2O emissions.

12.2   A Global Factor to Describe N_2O Yield from N Fertilization

We start this study by deriving the yield of N_2O from fresh N input, based on data compiled by Prather et al. (2001), Galloway et al. (2004) with some analysis of our own. Fresh fixed N input includes N, which is produced by chemical, biological and atmospheric processes. The pre-industrial, natural N_2O sink and source at an atmospheric mixing ratio of 270 nmol/mol is calculated to be equal to 10.2 Tg N_2O–N/year (Prather et al. 2001), which includes marine emissions. By the start of the present century, at an atmospheric volume mixing ratio of 315 nmol/mol, the stratospheric photochemical sink of N_2O was about 11.9 Tg N_2O–N/year. The total N_2O source at that time was equal to the photochemical sink (11.9 Tg N_2O–N/year) plus the atmospheric growth rate (3.9 Tg N_2O–N/year), together totalling 15.8 Tg N_2O–N/year (Prather et al. 2001). The anthropogenic N_2O source is the difference between the total source strength, 15.8 Tg N_2O–N/year, and the current natural source, which is equal to the preindustrial source of 10.2 Tg N_2O–N/year minus an uncertain 0–0.9 Tg N_2O–N, with the latter number taking into account a decreased natural N_2O source due to 30 % global deforestation (Klein Goldewijk 2001).

Thus we derive an anthropogenic N_2O source of 5.6–6.5 Tg N_2O–N/year. To obtain the agricultural contribution, we subtract the estimated industrial source of 0.7–1.3 Tg N2O–N/year (Prather et al. 2001), giving a range of 4.3–5.8 Tg N_2O–N/year. This is 3.8–5.1 % of the anthropogenic 'new' fixed nitrogen input of 114 Tg N/year for the early 1990s; the input value is derived from the 100 Tg of N fixed by the Haber-Bosch process, plus 24.2 Tg of N fixed due to fossil fuel combustion and 3.5 Tg difference from biological N fixation, BNF, between current and pre-industrial times (Galloway et al. 2004), reduced by the 14 Tg of Haber-Bosch N not used as fertilizer (Smeets et al. 2007). (This total of 114 TgN is very similar to the sum of the different values for N from fertilizer and BNF given by Smeets et al.: 81 + 38 = 119 Tg.) In an earlier study (Mosier et al. 1998) the source of N_2O from agriculture was estimated to be even larger, 6.3 Tg N_2O–N, giving an N_2O yield of 5.5 %. In comparison, the N_2O–N emission estimated by Prather et al. (2001) is 2.9–6.3 Tg N_2O–N/year, or 3.4–6.8 Tg N_2O–N/year if we also include biomass and biofuel burning (which we consider an agricultural source), leading to N_2O–N yields of 2.6–5.5 % or 3.0–6.0 %, respectively.

Because of good knowledge of the chemical processing of N_2O in the atmosphere and its tropospheric concentrations, obtained from air enclosure in ice cores, its natural sources and sinks are well known and can be calculated with models. Thus, pre-industrial, natural conditions provide additional information on the yield of N_2O from fixed N input. For that period, the global source and sink of N_2O was 10.2 Tg N_2O–N/year with 6.2–7.2 Tg N_2O–N/year coming from the land and coastal zones (Prather et al. 2001), derived from a fresh fixed N input of 141 Tg N/year (Galloway et al. 2004), giving an N_2O–N yield of 4.4–5.1 %. Both for the pre-Haber-Bosch natural terrestrial emissions and the agricultural emissions in the Haber-Bosch era, we find that the ratio y = N_2O out- put/fresh fixed N input is 3–5 %. This is a parametric relationship, based on the global budgets of N_2O and fixed N input, and atmospheric concentrations and known lifetime of N_2O, and thus is not dependent on detailed knowledge of the terrestrial N cycle. We assume that this global ratio will be the same in agro/biofuel production systems. This is a reasonable assumption, as similar agricultural plants are currently used as feedstocks for biofuel production as those grown in regular agriculture. Some correction is needed for the use of animal manure in biofuel crop production, but this is quite small: Cassman et al. (2002) noted that approximately 11 % of total N input to world's cropland came from animal manures.

A comparison of our "top-down" estimates of N_2O emissions from inputs of newly fixed N with the "bottom-up" estimates that are made with the IPCC inventory methodology (Mosier et al. 1998; IPCC 2006) is presented in Appendix 1. A key feature of our methodology is that the 114 Tg of newly fixed N entering agricultural systems (synthetic fertilizer N and N from biological nitrogen fixation (BNF)) is regarded as the source of all agriculture related N_2O emissions.

12.3 N$_2$O Release Versus CO$_2$ Saved in Biofuels

As a quick indicator to describe the consequence of this 'background' N$_2$O production we compare its global warming with the cooling due to replacement of
fossil fuels by biofuels. Here we will only consider the climatic effects of conversion of biomass to biofuel and not a full life cycle, leaving out for instance the
input of fossil fuels for biomass production, on the one hand, and the use of
co-products on the other hand.

  We assume that the fixed nitrogen which is used to grow the biofuels is used
with an average efficiency of 40 % (see below) and that this factor determines how
much newly fixed N must be supplied to replenish the fields over time. We also
obtain the fossil CO$_2$ emissions avoided from the carbon processed in the harvested
biomass to yield the biofuel. With these assumptions, we can compare the climatic
gain of fossil fuel-derived CO$_2$ 'savings', or net avoided fossil CO$_2$ emissions, with
the counteracting effect of enhanced N$_2$O release resulting from fixed N input. Our
assumptions lead to expressions per unit mass of dry matter harvested in biofuel
production to avoid fossil CO$_2$ emissions, "saved CO$_2$", (M), and for "equivalent
CO$_2$", (Meq), the latter term accounting for the global warming potential (GWP) of
the N$_2$O emissions. We derive M from carbon contained in biomass as the lower
heat value per carbon, and consequently the CO$_2$ emissions per energy unit, are
almost identical for the fossil fuels and biofuels discussed here (JRC 2007):

$$M = r_C * \mu_{CO_2}/\mu_c * cv \qquad (1)$$

$$Meq = r_N * y * \mu_{N_2O}/\mu_{N_2} * GWP/e \qquad (2)$$

  In these formulae r_C is in g carbon per g dry matter in the feedstock; r_N is the
mass ratio of N to dry matter in g N/kg; cv is the mass of carbon in the biofuel per
mass of carbon in feedstock biomass (maize, rapeseed, sugar cane); e is a surrogate
for the uptake efficiency of the fertilizer by the plants; $y = 0.03$–0.05, the range of
yields of N$_2$O–N from fixed N application; GWP $= 296$; $\mu_{CO_2}/\mu_c = 44/12$,
$\mu_{N_2O}/\mu_{N_2} = 44/28$, where the μ terms are the molar weights of N$_2$O, N$_2$, CO$_2$, and
C. Inserting these values in Eqs. (1) and (2) we thus obtain, with expressions in
parentheses representing ranges,

$$M = 3.667 \cdot cv \cdot r_C \qquad (3)$$

$$Meq = (14 - 23.2)r_N/e \qquad (4)$$

$$Meq/M = (3.8 - 6.3)r_N/(e \cdot cv \cdot r_C) \qquad (5)$$

  The latter term is the ratio between the climate warming effect of N$_2$O emissions
and the cooling effect due to the displacement of fossil fuels by biofuels.

  These equations are valid for all above-ground harvested plant material, and
separately also for the products and residues which are removed from the

agricultural fields. If Meq > M, there will be net climate warming, the greenhouse warming by increased N$_2$O release to the atmosphere then being larger than the quasi-cooling effect from "saved fossil CO$_2$". There will neither be net climate warming nor cooling by biofuel production when Meq = M, which occurs for

$$r_N = (0.158 - 0.263) \cdot (e \cdot cv \cdot r_C) \tag{6}$$

Under current agricultural practices, worldwide, the average value for e $\approx$ 0.4 (or 40 %) (Cassman et al. 2002; Galloway et al. 2003; Balasubramanian et al. 2004). This value reflects the considerable amounts of N lost to the atmosphere via ammonia volatilization and denitrification (N$_2$) and by leaching and runoff to aquatic systems. Fertilizer N use efficiency much higher than this (e.g. Rauh/Berenz 2007) is certainly possible when fertilizer N is made available according to plant uptake requirements, but this does not reflect the agricultural practice in many countries of the world.

Nonetheless, we recognise the possibility of better efficiencies in future, as has been possible in special circumstances on a research basis. Below we derive values for r_N based on both e = 0.4 and e = 0.6.

The data (and their sources) used to calculate the carbon contents, r_C, and the conversion efficiency factors, cv, and the calculations themselves, are given in Appendix 2. As r_C we use 0.61, 0.44 and 0.43 for rapeseed, maize, and sugar cane, respectively. We derive values of cv = 0.58 for rapeseed bio-diesel, cv = 0.37 for maize bio-ethanol, and cv = 0.30 for sugar cane ethanol production.

Consequently, for e = 0.4,

r_N = 22.3–37.2 g N/kg dry matter for rapeseed bio-diesel,
r_N = 10.3–17.1 g N/kg dry matter for maize bio-ethanol,
r_N = 8.1–13.6 g N/kg dry matter for sugar cane bio-ethanol.

Similarly, for e = 0.6,

r_N = 33.5–55.8 g N/kg dry matter for rapeseed bio-diesel,
r_N = 15.4–25.7 g N/kg dry matter for maize bio-ethanol,
r_N = 12.2–20.4 g N/kg dry matter for sugar cane bioethanol.

For each of these biofuels, a larger value of r_N in the plant matter than this range implies that use of the fuel causes a net positive climate forcing.

Note that our analysis only considers the conversion of biomass to biofuels, emphasizing the role of N$_2$O emissions. It does not take into account the supply of fossil fuel for fertilizer production, farm machinery and biofuel process facility, which require a considerable fraction of the energy gained (Hill et al. 2006). Furthermore, we assume that biofuel production is based on mineral fertilizer only (substitution of manure for synthetic fertilizer would offset our result by the percentage of synthetic fertilizer that is not used). The energy content gained from by-products will largely be offset from additional energy needed to produce it (Hill et al. 2006), here we also neglect its potential to replace other animal feed crops (and the associated N_2O emissions). We are aware that integrated processes exist which better connect biofuel production with animal husbandry, but we believe this cannot be taken for granted on a global scale.

12.4 Results and Discussion

12.4.1 Nitrogen Content in Biofuels

Data on r_N for several agricultural products, in g (N)/kg dry matter (Velthof/Kuikman 2004; Biewinga/van der Bijl 1996), are presented in Table 12.1, together with results on "relative warming". They show net climate warming, or considerably reduced climate cooling, by fossil fuel "CO_2 savings", due to N_2O emissions. The r_N value for maize is equal to 15 g N/kg dry matter, leading to a relative climate warming of 0.9–1.5 compared to fossil fuel CO_2 savings. The effect of the high nitrogen content of rapeseed is particularly striking; it offsets the advantages of a high carbon content and energy density for biodiesel production. World-wide, rapeseed is the source of >80 % of bio-diesel for transportation, and has been particularly promoted for this purpose in Europe. For bio-diesel derived from rapeseed, this analysis indicates that the global warming by N_2O is on average about 1.0–1.7 times larger than the quasi-cooling effect due to "saved fossil CO_2" emissions. For sugar cane/ethanol the relative warming is 0.5–0.9, based on a r_N value of 7.3 g N/kg dry matter (Isa et al. 2005), causing climate cooling with respect to N_2O (not necessarily for the whole process, as fossil energy input is not considered).

Although there are possibilities for improvements by increasing the efficiency, e.g. for the uptake of N fertilizer by plants (Cassman et al. 2002)—which is much needed in regular agriculture as well—on a globally averaged basis the use of agricultural crops for energy production, with the current nitrogen use efficiencies, can readily be detrimental for climate due to the accompanying N_2O emissions, as indicated here for the common biofuels: rapeseed/bio-diesel, and maize/ethanol. However, if nitrogen use efficiency can be increased to $e = 0.6$, then as the calculations above and in Table 12.2 show, maize/ethanol and rapeseed/biodiesel may

Table 12.1 Relative warming derived from N_2O production against cooling by "saved fossil CO_2" by crops as a function of the actual nitrogen content $r_{N(actual)}$

Crop	$r_{N(actual)}$ (gN/kg dry matter)	Relative warming (Meq/M) (N-efficiency $e = 0.4$)	Type of fuel produced
Rapeseed	39	1.0–1.7	Bio-diesel
Maize	15	0.9–1.5	Bio-ethanol
Sugar cane	7.3	0.5–0.9	Bio-ethanol

Uncertainty ranges presented derive from the uncertainty of the yield factor y (see text)

Table 12.2 Sensitivity analysis, showing the impact on relative warming (Meq/M) resulting from changes to parameters used for Table 11.1

Crop	Increased N-efficiency ($e = 0.6$)	High share of manure (20 %) in fertilizer for biofuels	Efficient use of by-products: considerable fraction (50 %) of N harvested for biofuel production replaces crops that would need N fertilizer
Rapeseed	0.7–1.2	0.8–1.4	0.5–0.9
Maize	0.6–1.0	0.7–1.2	0.4–0.7
Sugar cane	0.4–0.6	0.4–0.7	0.3–0.4

The calculations depend on assumptions made about the global agricultural practice of biofuel production. In each column, values differ from those presented in Table 12.1 by one parameter only as indicated in the relevant column heading

be climate-neutral or beneficial. Also the effect of other assumptions on our result (substitute manure; replace other crops) is tested in Table 12.2.

More favourable conditions for bio-energy production, with much lower nitrogen to dry matter ratios (Tillman et al. 2006), resulting in smaller N_2O emissions, exist for special "energy plants", for instance perennial grasses (Christian et al. 2006) such as switch grass *(Panicum virgatum)* and elephant grass *(Miscanthus × giganteus* hybrid), with a r_N of 7.3 g N/kg dry matter. The production of biofuel from palm oil, with a r_N of 6.4 g N/kg dry matter (Wahid et al. 2005), may also have moderately positive effects on climate, viewed solely from the perspective of N_2O emissions. Other favourable examples are ligno-cellulosic plants, e.g. eucalyptus, poplar and willow.

The importance of N_2O emissions for climate also follows from the fact that the agricultural contribution of 4.3–5.8 Tg N2O–N/year gives the same climate radiative forcing as that provided by 0.55–0.74 Pg C/year, that is 8–11 % of the greenhouse warming by fossil fuel derived CO_2. Increased emissions of N_2O will also lead to enhanced NO_x concentrations and ozone loss in the stratosphere (Crutzen 1970). Further, NO is also produced directly in the agricultural N cycle. Adopting the relative yield of NO to N_2O of 0.8 (Mosier et al. 1998), and the agricultural contribution to the N_2O growth rate of 4.3–5.8 Tg N₂O–N/year, the global NO production from agriculture is equal to 3.4–4.6 Tg N/year, about 20 % of that caused by fossil fuel burning (Prather et al. 2001), affecting tropospheric chemistry in significant ways.

12.4.2 Potential Impact on Life Cycle Analysis

An abridged analysis as presented above, yielding N/C ratios to indicate whether biofuels are GHG-positive or GHG-negative, can not replace a full life cycle assessment. In recent years, a number of such assessments have become available

(Adler et al. 2007; Kaltschmitt et al. 2000; von Blottnitz et al. 2006; Farrell et al. 2006; Hill et al. 2006). At this stage, we can not discuss the differences between these respective approaches, which also affect conclusions. But we may look into the release rate of N_2O-N used, presented as a function of applied fertilizer N. In these life cycle studies, release rates typically are based on the default values estimated by IPCC (2006) for 'direct' emissions which were derived from plot-scale measurements (1 % of the fertilizer N applied, or, in a previous version, 1.25 %). Only a few studies (Adler et al. 2007) also incorporate the corresponding default values for 'indirect' emissions also specified by IPCC (totalling less than 0.5 % and which, together with the direct emissions, add up to c. 1.5 % of fertilizer N), whereas our global analysis indicates a value of 3–5 %. Past studies seem to have underestimated the release rates of N_2O to the atmosphere, with great potential impact on climate warming. The effect of applying higher N_2O yields can be assessed using the openly accessible EBAMM model (Farrell et al. 2006).

12.5   Conclusions

As release of N_2O affects climate and stratospheric ozone chemistry by the production of biofuels, much more research on the sources of N_2O and the nitrogen cycle is needed. Here we have shown that the yield of N_2O–N from fixed nitrogen application in agro-biofuel production can be in the range of 3–5 %, 3–5 times larger than assumed in current life cycle analyses, with great importance for climate. We have also shown that the replacement of fossil fuels by biofuels may not bring the intended climate cooling due to the accompanying emissions of N_2O. There are also other factors to consider in connection with the introduction of biofuels. Here we concentrated on the climate effects due only to required N fertilization in biofuel production and we have shown that, depending on N content, the current use of several agricultural crops for energy production, at current total nitrogen use efficiencies, can lead to N_2O emissions large enough to cause climate warming instead of cooling by "saved fossil CO_2". What we have discussed is one important step in a life cycle analysis, i.e. the emissions of N_2O, which must be considered in addition to the fossil fuel input and co-production of useful chemicals in biofuel production. We have not yet considered the extent to which any loss by volatilisation of part of the fertilizer N may stimulate CO_2 uptake from the atmosphere, following deposition on natural ecosystems; estimates for this effect are very uncertain (de Vries et al. 2006; Magnani et al. 2007; Hyvönen et al. 2007). We conclude that the relatively large emission of N_2O exacerbates the already huge challenge of getting global warming under control.

Appendix 1: Comparison Between the Present and the IPCC Method to Estimate the Global N$_2$O Yields

The basis of our methodology is that the newly fixed N entering agricultural systems (synthetic fertilizer N and N from biological nitrogen fixation (BNF)) is regarded as the source of all related N$_2$O emissions, and furthermore these emissions may not all happen in the season of application, but involve longer cycling times (which are nonetheless short compared with the lifetime of N$_2$O in the environment). These emissions can be conveniently considered in three categories:

- direct emissions from N-fertilized soils;
- 'secondary' emissions resulting from the complex transformations of N compounds in the various flows within agricultural systems; and
- indirect emissions (in the IPCC meaning of the phrase) arising from leached N leaving agricultural fields and entering water systems, and from volatilized N deposited onto natural ecosystems.

Examples of the 'secondary' emission sources are:

- crop residues ploughed in as fertilizer for a successor crop;
- dung and urine from livestock (both grazing and housed) fed variously on N-fertilized grain crops, feeds containing BNF-N (e.g. soya bean meal, alfalfa, clover-rich pasture and silage in Europe, and tropical grasses with *Azospirillum* associations in Brazil); and
- N mineralized from soil organic matter and root residues following cultivation or grassland renewal.

In contrast, in the IPCC approach, emissions from crop residues and mineralization are included in the 'direct' emissions and have the same emission factor (*EF*); separate *EF*s are used for emissions from grazing animals, and the N source here is quantified on the basis of the N excreted, and essentially is treated as a 'new' N source, not as fertilizer- or BNF-derived N. The fractions of the N applied to fields that are lost by leaching, runoff and volatilization have additional *EF*s applied to them. The aggregate emissions from agriculture are arrived at by summing all these individual sources. The IPCC's 1 % *EF* for direct N$_2$O emissions contains an uncertainty of one-third to 3 times the default value. The default *EF* for emissions from cattle, poultry and pigs is 2 % of the N excreted, with a range of 0.7–6 %—again, from one-third to 3 times the default value. The *EF*s for N derived from N volatilization and re-deposition and N derived from leaching and runoff are 1 % (uncertainty range 0.2–5 %) and 0.75 % (0.05–2.5 %), respectively. At default volatilization fractions of 10 % (mineral fertilizer) or 20 % (animal manure), and default leaching fraction of 30 %, indirect emissions amount to 0.35–0.45 % of N applied. Each of the source terms in the bottom–up, IPCC method is very uncertain. However, their sum is not inconsistent with the total derived by the top-down methodology.

Appendix 2: Calculation of CV Values

(a) Bio-ethanol Production from Maize

Yield = 2.66 US gallons per US bushel (mean of values for wet and dry milling processes) (USDA 2002, cited in UK Dept for Transport 2006)

$=2.66 \times 3.785 = 10.071$ ethanol/25.4 kg maize

$=7.945$ kg ethanol/25.4 kg maize

$=0.313$ kg ethanol/kg maize.

C content of ethanol (C_2H_5OH, mol. wt. 46) by weight = 24/46 = 522 g/kg.

C content of maize (r_C) $\hat{=}$ 0.44 g/g $\hat{=}$ 440 kg/t.

$Cv = (0.313 \times 522)/440 = 0.37$.

(b) Bio-diesel Production from Rapeseed

- the average oil yield is 45 % (450 kg/t rapeseed) (Elaine Booth, SAC Aberdeen, personal communication)
- the average composition of the oil is adequately represented by the triglyceride of the dominant fatty acid, erucic acid, i.e. $(C_{22}H_{41}O_2)_3(C_3H_5)$, mol. wt. 1052, then

C content of the oil by weight = 828/1052 = 0.787 kg/kg.

Thus the C content of the oil = $(450 \times 0.787) = 354$ kg/t rape-seed.

The conversion to bio-diesel involves conversion to the methyl ester:

$$(C_{22}H_{41}O_2)_3(C_3H_5) \rightarrow 3C_{22}H_{41}O_2CH_3$$

but the C content of the bio-diesel is almost unchanged from that of the natural oil: mol. wt. of methyl ester = 352, and C content = $(276/352) \times 450 = 353$ kg/t rapeseed.

Oil content of original rapeseed = 45 % (450 kg/t), and non-oil components $\hat{=}$ 550 kg/t, of which

- protein is 40 % (=220 kg/t original rapeseed), with a C content of 510 g/kg;
- the remainder (60 %, = 330 kg/t original rapeseed) is dominantly carbohydrate,

(Colin Morgan, SAC Edinburgh, personal communication)

Thus the C content of the protein fraction in the original rape-seed = $220 \times 510/1000 = 112$ kg/t; and the C content of the carbohydrate fraction (for which a C content of 440 g/kg can be adopted, as for grains) = $330 \times 440/1000 = 145$ kg/t.

The overall C content of the original rapeseed ($r_C = C_{oil} + C_{protein} + C_{CHO}$) = 354 + 112 + 145 = 612 kg/t.

$cv = 353/612 = 0.58$.

(c) **Bio-ethanol Production from Sugar Cane**

Yield is 861 dry ethanol (density 0.79 kg/1) per tonne sugar cane harvested at a water content of 72.5 %, or 247 kg ethanol per tonne dry sugar cane (Macedo et al. 2004; as cited by JRC 2007).

C content of ethanol (C_2H_5OH, mol. wt. 46) by weight = 24/46 = 522 g/kg.

C content of dry sugar cane is determined by its structural material, cellulose, and its sugar content (polysaccharides: 440 g/kg; saccharose: 420 g/kg), we use r_C = 430 g/kg

$$cv = (0.247 \times 522)/430 = 0.30.$$

References

Adler, P.R.; Del Grosso, S.J.; Parton, W.J., 2007: "Life-Cycle Assessment of Net Greenhouse-Gas Flux for Bioenergy Cropping Systems", in: *Ecological Applications*, 17: 675–691.

Balasubramanian, V.; Alves, B.; Aulakh, M.; Bekunda, M.; Cai, Z.; Drinkwater, L.; Mugendi, D.; van Kessel, C.; Oenema O., 2004: "Crop, Environmental, and Management Factors Affecting Nitrogen Use Efficiency", in: Mosier, A.R.; Syers, J.K.; Freney, J.; SCOPE (Eds.): *Agriculture and the Nitrogen Cycle* (Washington: Island Press): 65, 19–33.

Biewinga, E.E.; van der Bijl, G., 1996: *Sustainability of Energy Crops in Europe, Centre for Agriculture and Environment (CLM)* (Utrecht, The Netherlands).

Cassman, K.G.; Dobermann, A.; Walters, D.T., 2002: "Agroecosystems, Nitrogen-Use Efficiency, and Nitrogen Management", in: *Ambio*, 31: 132–140.

Christian, D.G.; Poulton, P.R.; Riche, A.B.; Yates, N.E., 2006: "The Recovery over Several Seasons of ¹⁵N-labelled Fertilizer Applied to Miscanteus x Giganteus Ranging from 1 to 3 Years Old", in: *Biomass and Bioenergy*, 30: 125–133.

Crutzen, P.J., 1970: "The Influence of Nitrogen Oxides on the Atmospheric Ozone Content", in: *Quarterly Journal of the Royal Meteorological Society*, 96: 320–325.

de Vries, W.; Reinds, G.J.; Gundersen, P.; Sterba, H., 2006: "The Impact of Nitrogen Deposition on Carbon Sequestration in European Forests and Forest Soils", in: *Global Change Biology*, 12: 1151–1173.

Farrell, A.E.; Plevin, R.J.; Turner, B.T.; Jones, A.D.; O'Hare, M.; Kammen, D.M.: "Ethanol Can Contribute to Energy and Environmental Goals", in: *Science*, 311: 506–508.

Galloway, J.N.; Dentener, F.J.; Capone, D.G.; Boyer, E.W.; Howarth, R.W.; Seitzinger, S.P.; Asner, G.P.; Cleveland, C.C.; Green, P.A.; Holland, E.A.; Karl, D.M.; Michaels, A.F.; Porter, J.H.; Townsend, A.R.; Vörösmarty, C.J., 2004: "Nitrogen Cycles: Past, Present, and Future", in: *Biogeochemistry*, 70: 153–226.

Galloway, J.N.; Aber, J.D.; Erisman, J.W.; Seitzinger, S.P.; Howarth, R.H.; Cowling, E.B.; Cosby B.J., 2003: "The Nitrogen Cascade", in: *Bioscience*, 53: 341–356.

Hill, J.; Nelson, E.; Tilman, D.; Polasky, S.; Tiffany, D., 2006: "Environmental, Economic, and Energetic Costs and Benefits of Biodiesel and Ethanol Biofuels", in: *Proceedings of the National Academy of Sciences*, 103: 11206–11210.

Hyvönen, R.; Persson, T.; Andersson, S.; Olsson, B.; Ågren, G.I.; Linder, S., 2007: "Impact of Long-term Nitrogen Addition on Carbon Stocks in Trees and Soils in Northern Europe", in: *Biogeochemistry*.

IPCC, 2006: "IPCC Guidelines for National Greenhouse Gas Inventories, Prepared by the National Greenhouse Gas Inventories Programme", in: Eggleston, H.S.; Buendia, L.; Miwa, K.; Ngara, T.; Tanabe, K. (Eds.): *Volume 4, Chapter 11, N₂O Emissions from Managed Soils, and C₂O Emissions from Lime and Urea Application* (Hayama, Japan: IGES).

Isa, D.W.; Hofman, G.; van Cleemput, O., 2005: "Uptake and Balance of Fertilizer Nitrogen Applied to Sugarcane", in: *Field Crops Research*, 95: 348–354.

JRC, 2007: *Well-to-Wheels Analysis of Future Automotive Fuels and Powertrains in the European Context* (Well-to-Tank Report, Version 2c, Joint Research Centre, Ispra, Italy).

Kaltschmitt, M.; Krewitt, W.; Heinz, A.; Bachmann, T.; Gruber, S.; Kappelmann, K.-H.; Beerbaum, S.; Isermeyer, F.; Seifert, K., 2000: Gesamtwirtschafiliche Bewertung der Energiegewinnung aus Biomasse unter Berücksichtigung externer und makroökonomischer Effekte (Externe Effekte der Biomasse), Final Report (in German), IER (Germany: University of Stuttgart).

Klein Goldewijk, C.G.M., 2001: "Estimating Global Land Use Change over the Past 300 Years: The HYDE Data Base", in: *Global Biogeochemical Cycles*, 15: 415–434.

Magnani, F.; Mencuccini, M.; Borghetti, M.; et al., 2007: "The Human Footprint in the Carbon Cycle of Temperate and Boreal Forests", in: *Nature*, 447: 848–850.

Mosier, A.; Kroeze, C.; Nevison, C.; Oenema, O.; Seitzinger, S.; van Cleemput, O., 1998: "Closing the Global N_2O Budget: Nitrous Oxide Emissions Through the Agricultural Nitrogen Cycle", in: *Nutrient Cycling in Agroecosystems*, 52: 225–248.

Prather, M.; Ehhalt, D.; et al., 2001: "Atmospheric Chemistry and Greenhouse Gases", in: Houghton, J.T.; Ding, Y.; Griggs, D.J.; et al. (Eds.): *Climate Change 2001: The Scientific Basis* (Cambridge: Cambridge University Press): 239–287.

Rauh, S.; Berenz, S., 2007: "Interactive Comment on N_2O Release from Agro-Biofuel Production Negates Global Warming Reduction by Replacing Fossil Fuels", in: Crutzen, P.J.; et al. (Eds.): *Atmospheric Chemistry and Physics Discussions*, 7: S4616–4619.

Smeets, E.; Bouwman, A.F.; Stehfest, E., 2007: "Interactive Comment on N_2O Release from Agro-biofuel Production Negates Global Warming Reduction by Replacing Fossil Fuels", in: Crutzen, P.J.; et al. (Eds.): *Atmospheric Chemistry and Physics Discussions*, 7: S4937–4941.

Tillman, D.; Hill, J.; Lehman, C., 2006: "Carbon-Negative Biofuels from Low-Input High-Diversity Grassland Biomass", in: *Science*, 314: 1598–1600.

UK Department for Transport International Resource Costs of Biodiesel and Bioethanol; (26 January 2006).

Velthof, G.L.; Kuikman, P.J., 2004: *Beperking van lachgasemissie uit gewasresten, Alterra rapport 114.3* (in Dutch) (Wageningen, The Netherlands).

von Blottnitz, H.; Rabl, A.; Boiadjiev, D.; Taylor, T.; Arnold, S., 2006: "Damage Costs of Nitrogen Fertilizer in Europe and Their Internalization", in: *Journal of Environmental Planning and Management,* 49: 413–433.

Wahid, M.B.; Abdullah, S.N.A.; Henson, I.E., 2005: "Oil Palm–Achievements and Potential", in: *Plant Production Science*, 8: 288–297.

Nobel Prize in Chemistry in 1995

The Royal Swedish Academy of Sciences has decided to award the 1995 Nobel Prize in Chemistry to Professor Paul Crutzen, Max Planck Institute for Chemistry, Mainz, Germany (Dutch citizen), Professor Mario Molina, Department of Earth, Atmospheric and Planetary Sciences and Department of Chemistry, MIT, Cambridge, MA, USA and Professor F. Sherwood Rowland, Department of Chemistry, University of California, Irvine, CA, USA *for their work in atmospheric chemistry, particularly concerning the formation and decomposition of ozone.*

Paul J. Crutzen	**Mario J. Molina**	**F. Sherwood Rowland**
Prize share: 1/3	**Prize share:** 1/3	**Prize share:** 1/3

- **Paul Crutzen** was born in 1933 in Amsterdam. Dutch citizen. Doctor's degree in meteorology, Stockholm University, 1973. Member of the Royal Swedish Academy of Sciences, the Royal Swedish Academy of Engineering Sciences and Academia Europaea. Professor **Paul Crutzen,** Max Planck Institute for Chemistry, P.O. Box 3060 D-55020 Mainz, Germany.
- **Mario Molina** was born in 1943 in Mexico City, Mexico. Ph.D. in physical chemistry, University of California, Berkeley. Member of the US National Academy of Sciences. Professor **Mario Molina,** Department of Earth,

P.J. Crutzen and H.G. Brauch (eds.), *Paul J. Crutzen: A Pioneer on Atmospheric Chemistry and Climate Change in the Anthropocene,* Nobel Laureates 50, DOI 10.1007/978-3-319-27460-7

Atmospheric and Planetary Sciences MIT 54—1312, Cambridge MA 02139, USA.

- **F. Sherwood Rowland** was born in Delaware, Ohio, USA, 1927 and deceased in 2012. Doctor's degree in chemistry, University of Chicago, 1952. Member of the American Academy of Arts and Sciences and of the US National Academy of Sciences, where he was Foreign Secretary.

Source: "Press Release: The 1995 Nobel Prize in Chemistry". *Nobelprize.org*. Nobel Media AB 2014. Web. 3 Jan 2015. http://www.nobelprize.org/nobel_prizes/ chemistry/laureates/1995/press.html and at: http://www.nobelprize.org/nobel_ prizes/chemistry/laureates/1995/illpres/reading.html and for the Award Ceremony Speech by Professor Ingmar Grenthe of the Royal Swedish Academy of Sciences is at: http://www.nobelprize.org/nobel_prizes/chemistry/laureates/1995/presentation-speech.html.

Interview with *Paul Crutzen* by Astrid Gräslund at the meeting of Nobel Laureates in Lindau, Germany, June 2000. Paul Crutzen talks about family background, early education and interest in natural science; his work in the Institute of Meteorology in Stockholm (5:02); his discovery (6:55); the ozone layer (15:42); the Greenhouse Effect (19:10); ozone holes (23:43); and the consequences of a 'Nuclear winter' (27:09). See a Video of the Interview, 34 min., at: http://www. nobelprize.org/mediaplayer/index.php?id=734.

Prof. Paul J. Crutzen on the eve of his 80th birthday on 2 December 2013 at the symposium The Anthropocene organized in his honour by the Max Planck Institute of Chemistry in Mainz. Photo by Carsten Costard

Max Planck Institute for Chemistry

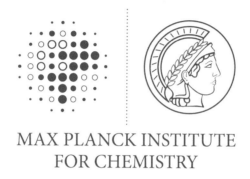

MAX PLANCK INSTITUTE
FOR CHEMISTRY

Aiming at an Integral Scientific Understanding of Chemical Processes in the Earth System from Molecular to Global Scales

Current research at the Max Planck Institute for Chemistry in Mainz aims at an integral understanding of chemical processes in the Earth system, particularly in the atmosphere and biosphere. Investigations address a wide range of interactions between air, water, soil, life and climate in the course of Earth history up to today's human-driven epoch, the *Anthropocene:* see at: http://www.mpic.de/en/employees/honors-and-awards/the-anthropocene.html.

Research at the Max Planck Institute for Chemistry has been at the forefront of science throughout its existence. Since the Institute's foundation in 1912, three of its directors were awarded with the Nobel Prize for Chemistry: Richard Willstätter in 1915 for the revelation of the structure of chlorophyll and other plant pigments, Otto Hahn in 1944 for the discovery of nuclear fission, and Paul Crutzen in 1995 for the elucidation of atmospheric ozone chemistry.

The research departments and focal points of the Institute have gone through a history of change and scientific evolution. What began in 1912 with classical organic, inorganic and physical chemistry at the Kaiser Wilhelm Institute for Chemistry in Berlin evolved into radiochemistry and nuclear physics in the 1930s,

P.J. Crutzen and H.G. Brauch (eds.), *Paul J. Crutzen: A Pioneer on Atmospheric Chemistry and Climate Change in the Anthropocene*, Nobel Laureates 50, DOI 10.1007/978-3-319-27460-7

leading to the discovery of nuclear fission by Otto Hahn, Lise Meitner and Fritz Strassmann.

As the Institute was severely damaged towards the end of World War II it was moved to the Swabian Alps in today's Baden-Wuerttemberg. There the chemists worked provisionally from 1944 to 1949 until the Institute moved a second time to the campus of the newly founded Johannes Gutenberg University in Mainz. At the same time it was integrated into the Max Planck Society, the successor of the Kaiser Wilhelm Society, Institute for Chemistry. Since 1959 the Institute also carries the name "Otto Hahn Institute" in honor of its previous director and the first president of the Max Planck Society.

In the 1960s and 1970s the Institute's research portfolio was extended from Physical Chemistry, Nuclear Physics and Mass Spectrometry to Cosmochemistry, Isotope Cosmology and Air Chemistry. Meteorites and moon dust samples were studied and the interplay of atmospheric gases, particles and meteorology were investigated. In the 1980s new departments for Geochemistry and Biogeochemistry were founded, in 2001 the Particle Chemistry department was established jointly with the Institute for Atmospheric Physics at the Johannes Gutenberg University of Mainz. In 2012 the Multiphase Chemistry Department and in 2015 the Climate Geochemistry Department was founded.

Nowadays, the research focus of the Max Planck Institute for Chemistry is on Earth System science, in particular on the chemical processes occurring in the atmosphere and their interactions with the biosphere and oceans. It also includes the influence of humans, as unprecedented urbanization and industrialization in the past centuries have changed the course of natural processes on our planet, in an epoch now known as the Anthropocene.

Currently, the institute employs some 300 staff in five departments (Atmospheric Chemistry, at: http://www.mpic.de/en/research/atmospheric-chemistry.html, Biogeochemistry, at: http://www.mpic.de/en/research/biogeochemistry.html, Climate Geochemistry at: http://www.mpic.de/en/research/111561.html, Multiphase Chemistry, at: http://www.mpic.de/en/research/multiphase-chemistry.html and Particle Chemistry, at: http://www.mpic.de/en/research/particle-chemistry.html) and additional research groups. Scientists conduct laboratory experiments, collect samples and record measurement data during field campaigns utilizing airplanes, ships, and vehicles. The practical work is complemented with mathematical models that simulate chemical, physical, and biological processes from molecular to global scales. One of the major goals is to find out how air pollution, including reactive trace gases and aerosols, affect the atmosphere, biosphere, climate, and public health.

A description of current research topics is given in the **Institute Reports** http://www.mpic.de/fileadmin/user_upload/images_Institut/Fachbeirat/MPIC_Scientific_Report_2012-2014_screen_sm_rev.pdf.

About the Author

Photo by Carsten Costard

The research of **Paul J. Crutzen** has been mainly concerned with the role of chemistry in climate and biogeochemistry, and in particular the photochemistry of ozone in the stratosphere and troposphere. In 1970 he hypothesized that natural ozone production by the action of solar ultraviolet radiation on molecular oxygen (O_2) is mainly balanced by destruction processes, involving NO and NO_2 as catalysts. These catalysts in turn result from the oxidation of N_2O, a product of microbiological nitrogen conversion in soils and waters. He and Prof. Harold Johnston of the University of California, Berkeley, pointed out that NO emissions from large fleets of supersonic aircraft could cause substantial ozone losses in the stratosphere.

In the years 1972–1974 Crutzen proposed that NO and NO_2 could catalyse ozone production in the background troposphere by reactions occurring in the CO and CH_4 oxidation chains. Additional photochemical reactions leading to ozone loss were likewise identified. These gross ozone production and destruction terms are each substantially larger than the downward flux of ozone from the stratosphere, which until then had been considered the main source of tropospheric ozone.

In 1979–1980 Crutzen and co-workers drew attention to the great importance of the tropics in atmospheric chemistry. In particular, some measurement campaigns in Brazil clearly showed that biomass burning in the tropics was a major source of air pollutants, on a par with, or larger than, industrial pollution in the developed world.

In 1982 Crutzen, together with Prof. John Birks of the University of Colorado, drew attention to the risk of darkness and strong cooling at the earth surface as a consequence of heavy smoke production by extensive fires in a nuclear war ('nuclear winter'). This study and additional studies by R. Turco, B. Toon, T. Ackerman, J. Pollack and C. Sagan and by the Scientific Committee on Problems of the Environment (SCOPE), to which Crutzen contributed, showed that more people could die from the indirect consequences of a nuclear war than by the direct impacts of the nuclear explosions.

© The Author(s) 2016
P.J. Crutzen and H.G. Brauch (eds.), *Paul J. Crutzen: A Pioneer on Atmospheric Chemistry and Climate Change in the Anthropocene*, Nobel Laureates 50, DOI 10.1007/978-3-319-27460-7

In 1986, together with Dr. F. Arnold of the Max Planck Institute of Nuclear Physics in Heidelberg, Crutzen showed that nitric acid and water vapour could co-condense in the stratosphere at higher temperatures than required for water ice formation, providing a significant part of a chain of events leading to rapid ozone depletion at high latitudes during late winter and spring (the so-called Antarctic 'ozone hole').

His most recent research is concerned with the role of clouds in atmospheric chemistry as well as photochemical reactions taking place in the marine boundary layer, involving catalysis by halogen gases produced by marine organisms. Also, his current research deals with the chemical and climatic effects of the heavy air pollution which is found over Asia and other regions in the developing world: the so-called ABC (Atmospheric Brown Clouds) phenomenon.

Address: Prof. Paul J. Crutzen, via Ms. Astrid Kaltenbach, Max Planck Institute for Chemistry, Otto-Hahn-Institut, Hahn-Meitner-Weg 1, 55128 Mainz, Germany.
Email: astrid.kaltenbach@mpic.de
Website: http://www.mpic.de/index.php?id=31&type=0.

The photo of Pope Francis with Prof. Dr. Paul J. Crutzen was taken in the Vatican in October 2014. On the *left* Bishop-Chancellor Marcelo Sánchez Sorondo, Pontifical Academy of Sciences and Pontifical Academy of Social Sciences; in the *back centre* Prof. Dr. Veerabhadran Ramanathan. Image right and copyright reserved to the Photographic Service of L'Osservatore Romano who granted permission to use the photo on 16 November 2015

About this Book

This book contains texts by the Nobel laureate Paul J. Crutzen who is best known for his research on *ozone depletion*. It comprises Crutzen's autobiography, several pictures documenting important stages of his life, and his most important scientific publications. The Dutch atmospheric chemist is one of the world's most cited scientists in geosciences. His political engagement makes him a tireless ambassador for environmental issues such as climate change. He popularized the term 'Anthropocene' for the current geological era acknowledging the enduring influence of humankind on planet Earth. This concept conceives humans to be a geologic factor, influencing the evolution of our globe and the living beings populating it. The selection of texts is representing Paul Crutzen's scientific oeuvre as his research interests span from ozone depletion to the climatic impacts of biomass burning, the consequences of a worldwide atomic war—the Nuclear Winter—to geoengineering and the Anthropocene.

- Comprehensive collection of key scientific papers of Prof. Crutzen. Autobiographic chapter with photos on his life and career
- He made major contributions to atmospheric chemistry on ozone depletion
- Paul J. Crutzen received in 1995 the Nobel Prize in Chemistry with Mario J. Molina and F. Sherwood Rowland
- Paul J. Crutzen triggered global policy debates on nuclear winter, Anthropocene and sustainability.

Part 1: On Paul J. Crutzen—1: The Background of an Ozone Researcher; 2: Complete Bibliography of His Writings (1965–2015)

Part 2: Scientific Texts by Paul J. Crutzen—3: Influence of Nitrogen Oxides on Atmospheric Ozone Content (Crutzen); 4: Biomass Burning as a Source of Atmospheric Gases CO_2, H_2, N_2O, NO, CH_3CL and COS (Crutzen/Heidt/Krasnec/Pollock/Seiler); 5 The Atmosphere After a Nuclear War: Twilight at Noon (Crutzen/Birks); 6: Nitric–Acid Cloud Formation in the Cold Antarctic Stratosphere—A Major Cause for the Springtime Ozone Hole (Crutzen/Arnoldt); 7: Biomass Burning in the Tropics: Impact on Atmospheric Chemistry and Biogeochemical Cycles (Crutzen/Andreae); 8: A Mechanism for Halogen Release from Sea-Salt Aerosol in the Remote Marine Boundary Layer (Vogt/Crutzen/Sander): 9: The Indian Ocean Experiment: Widespread Air Pollution

© The Author(s) 2016

P.J. Crutzen and H.G. Brauch (eds.), *Paul J. Crutzen: A Pioneer on Atmospheric Chemistry and Climate Change in the Anthropocene*, Nobel Laureates 50, DOI 10.1007/978-3-319-27460-7

from South and South-East Asia (Lelieveld/Crutzen/Ramanathan/Andreae et al.);
10: Geology of Mankind (Crutzen): 11: Albedo Enhancement by Stratospheric
Sulfur Injections: A Contribution to Resolve a Policy Dilemma? An editorial essay
(Crutzen); 12: N_2O Release from Agro-biofuel Production Negates Global
Warming Reduction by Replacing Fossil Fuels (Crutzen/Mosier/Smith et al.)

Nobel Prize in Chemistry, Max Planck Institute for Chemistry, About the Author

A book website with additional information on Prof. Dr. Paul J. Crutzen,
including videos is at: http://afes-press-books.de/html/SpringerBriefs_PSP_
Crutzen.htm.

Printed in the United States
By Bookmasters